Studies on Eumalacostraca: a homage to Masatsune Takeda

CRUSTACEANA MONOGRAPHS constitutes a series of books on carcinology in its widest sense. Contributions are handled by the Series Editor and may be submitted through the office of KONINKLIJKE BRILL Academic Publishers N.V., P.O. Box 9000, NL-2300 PA Leiden, The Netherlands.

Published in this series:

CRM 001 - *Stephan G. Bullard*	Larvae of anomuran and brachyuran crabs of North Carolina
CRM 002 - *Spyros Sfenthourakis et al. (eds.)*	The biology of terrestrial isopods, V
CRM 003 - *Tomislav Karanovic*	Subterranean Copepoda from arid Western Australia
CRM 004 - *Katsushi Sakai*	Callianassoidea of the world (Decapoda, Thalassinidea)
CRM 005 - *Kim Larsen*	Deep-sea Tanaidacea from the Gulf of Mexico
CRM 006 - *Katsushi Sakai*	Upogebiidae of the world (Decapoda, Thalassinidea)
CRM 007 - *Ivana Karanovic*	Candoninae (Ostracoda) from the Pilbara region in Western Australia
CRM 008 - *Frank D. Ferrari & Hans-Uwe Dahms*	Post-embryonic development of the Copepoda
CRM 009 - *Tomislav Karanovic*	Marine interstitial Poecilostomatoida and Cyclopoida (Copepoda) of Australia
CRM 010 - *Carrie E. Schweitzer et al.*	Systematic list of fossil decapod crustacean species
CRM 011 - *Peter Castro et al. (eds.)*	Studies on Brachyura: a homage to Danièle Guinot
CRM 012 - *Patricio R. De los Ríos-Escalante*	Crustacean zooplankton communities in Chilean inland waters
CRM 013 - *Katsushi Sakai*	Axioidea of the world and a reconsideration of the Callianassoidea (Decapoda, Thalassinidea, Callianassida)
CRM 014 - *Charles H.J.M. Fransen et al. (eds.)*	Studies on Malacostraca: Lipke Bijdeley Holthuis Memorial Volume
CRM 015 - *Akira Asakura et al. (eds.)*	New Frontiers in Crustacean Biology: Proceedings of the TCS Summer Meeting, Tokyo, 20-24 September 2009
CRM 016 - *Danielle Defaye et al. (eds.)*	Studies on Freshwater Copepoda: a Volume in Honour of Bernard Dussart

In preparation (provisional titles):

CRM01x - *Masahiro Dojiri & Ju-shey Ho*	Systematics of three caligiform families, copepods parasitic on marine fishes
CRM01y - *Patricio R. De los Ríos-Escalante (ed.)*	Freshwater malacostracans in Chilean inland waters
CRM01z - *Darren C. Yeo et al. (eds.)*	Advances in freshwater decapod systematics and biology

Editors' addresses:
Hironori Komatsu, Department of Zoology, National Museum of Nature and Science, 4-1-1 Amakubo, Tsukuba, Ibaraki 305-0005, Japan; e-mail: h-komatu@kahaku.go.jp
Junji Okuno, Coastal Branch of Natural History Museum and Institute, Chiba, 123 Yoshio, Katsuura, Chiba 299-5242, Japan; e-mail: okuno@chiba-muse.or.jp
Kouki Fukuoka, Research Center for Subtropical Fisheries, Seikai National Fisheries Research Institute, Fisheries Research Agency, 148-446 Fukai-Ohta, Ishigaki, Okinawa 907-0451, Japan; e-mail: fukuokak@fra.affrc.go.jp

Cover: Philyra taekoae Takeda, 1972, reproduced from Takeda & Nakasone (1991) with permission of the publisher.
Frontispiece: Crab in Kanji (Chinese) character, written with a calligraphy brush by Masatsune Takeda.

Studies on Eumalacostraca: a homage to Masatsune Takeda

By

Hironori Komatsu, Junji Okuno and Kouki Fukuoka
(Editors)

CRUSTACEANA MONOGRAPHS, 17

BRILL

LEIDEN • BOSTON

This book is printed on acid-free paper.

Library of Congress Cataloging-in-Publication Data

The Library of Congress Cataloging-in-Publication Data is available from the Publisher.

ISBN13: 978-90-04-202887

Koninklijke Brill NV incorporates the imprints BRILL, Hotei Publishing, IDC Publishers, Martinus Nijhoff Publishers and VSP.

PRINTED IN THE NETHERLANDS

CONTENTS

PREFACE

Professor Masatsune Takeda has devoted his life to the study of brachyuran taxonomy and has been a leader of Japanese carcinology for decades. He published a total of 323 papers and described 231 new taxa in a very active career spanning 45 years. He contributed to not only research but also education, and along the way, trained many young scientists including doctoral students. In addition to his professional career, he has also worked tirelessly to “diffuse” carcinology to amateurs, the general public and children through the writing of guidebooks, general scientific and pictorial books, newspaper columns, popular magazines, as well as helping TV programs and so on. His innumerable contributions have greatly influenced Japanese carcinology as a whole. These significant accomplishments have marked him out as one of the most influential crustacean workers of his generation.

A total of 40 of his friends and colleagues have contributed to this volume, submitting papers dealing with the taxonomy, larval biology, palaeontology and ecology of Brachyura, Caridea, Stenopodidea, Astacidea, Axiidea, Anomura, Amphipoda, Isopoda and Mysida, all that Takeda-sensei loved. Despite the trauma Japan endured as a result of the huge earthquake and Tsunami of 11th March 2011, our many esteemed Japanese contributors and referees have not waived their responsibilities; allowing the present volume to be completed on time. The editors wish to express their cordial thanks to their kind and sincere help. We also thank Peter K. L. Ng (National University of Singapore) and Sammy De Grave (Oxford University Museum of Natural History) who unselfishly helped to review the English throughout the volume and Danièle Guinot (Muséum national d'Histoire naturelle, Paris) who kindly translated the abstracts into French. It is a great pleasure for us to dedicate this special volume to Takeda-sensei in the year of his 70th birthday. ありがとう, 武田先生! (Thank you very much, Takeda-sensei!)

February 2012, *Guest Editors for the present volume*,
HIRONORI KOMATSU
JUNJI OKUNO
KOUKI FUKUOKA

PROFESSOR MASATSUNE TAKEDA

Prof. Masatsune Takeda was born in March 1942 at Hachioji, western Tokyo, the second son of a Japanese priest. He enjoyed collecting insects and freshwater crabs in the nearby forest in his boyhood and especially loved moths. He attended Yokohama National University, and originally intended to graduate to become a teacher of Biology in a junior high school. However, his life changed forever after he met the late Prof. Tune Sakai (1903-1986), the father of carcinology in Japan. His destiny was now intermingled with the taxonomy of Brachyura, and he subsequently registered at the graduate school of Kyushu University under the supervision of the late Prof. Sadayoshi Miyake (1908-1998), in the so-called "Miyake School". At the "Miyake School", he worked on his dissertation entitled "Studies on the Crustacea Brachyura from the Palau Islands". His classmates included an impressive lineup of decapod carcinologists, many of whom in time would become renowned internationally, Drs. Katsushi Sakai, Keiji Baba, Yasuhiko Miya, Ken-Ichi Hayashi, Yukio Nakasone, Yoshihisa Hashiguchi, Takahiro Fujino, and Yu Hsiang-Ping (fig. 2). His years at the graduate school were enormously productive, resulting in some 23 scientific papers before he graduated with his doctoral degree in 1971. His scientific life got off to a promising start, but in the final year of his doctoral course an unfortunate motorcycle accident happened. This resulted in him having a complex leg fracture which caused him to stay a long time in hospital, two weeks before his scheduled wedding and required a metal bolt which activates the security alarm each time he passes through airport scanners. During his long stay in hospital, his devoted fiancee, Taeko, remained at his side, who later had a son with him. Not surprisingly, he named a beautiful leucosiid crab after his wife — *Philyra taekoae* Takeda, 1972 (cover illustration). Today, he is also the proud grandfather of both a boy and a girl.

After recovering from his accident, he was awarded his doctoral degree and took his first job at Nihon University School of Medicine in 1971. This job did not last long — he was shortly after recruited to join the National Science Museum, Tokyo, as the first curator of Crustacea in 1973. He remained at the museum for the next 33 years, becoming its director in 1997. During this period, he took part in or led a number of field trips all over Japan and East Asia, and wrote a prodigious number of papers. He worked not only on

Fig. 1. Masatsune Takeda in the laboratory of the National Museum of Nature and Science, Tokyo, 12 July 2007. [Photo: K. Ichioka.]

brachyuran taxonomy, but also on their larvae, other crustaceans, fossils, and even micro-organisms. His powerful work was supported by cigarettes (non-smoking now), sweets, beer and his family (including his dog). He retired from the museum in 2006 at the mandatory retirement age of 64, but immediately moved to Teikyo Heisei University as a Professor of Biology. On March 2012 he reached his 70th birthday and finally retired from the University. Today, he continues to work at the National Museum of Nature and Science, Tsukuba, as an Honorary Curator. Regardless of his position and retirement, his love for crustaceans has never abated, and he continues to work as actively as ever on their taxonomy.

TEACHING

Takeda-sensei ("sensei" is an honorific term for teacher or senior scientist in Japan) was active not only in research but also in education. He helped train many young scientists, such as Kooichi Konishi, Yukio Hanamura, Masakazu

Fig. 2. Members of the "Miyake School" in the Zoological Laboratory of Kyushu University, 1966; upper, from left to right: Ken-Ichi Hayashi, Takahiro Fujino, Yasuhiko Miya, Masatsune Takeda and Yu Hsiang-Ping; bottom, from left to right: Ritsuko Kuraki (secretary), Teruaki Uchida (ex. Professor of Kyushu University), Yong Ock Lee. [Photo: K. Baba.]

Aoki, Tomoyuki Komai, Kazuyoshi Hashizume, Junji Okuno, Masayuki Osawa, and Kouki Fukuoka. He also worked as an adjunct lecturer at various universities on Systematic Zoology and Science History over a long period of time. From 1995 to 2003, he was also an adjunct professor at The University of Tokyo, helping to train a new generation of invertebrate taxonomists. He was a good teacher, encouraging his students by helpful and precise comments but always with a warm and gentle smile. Unfortunately, during this period, he developed diabetes. As a result, he lost a lot of weight and his hair turned white, with even his characteristic smile disappearing. Nevertheless, his strong fighting spirit helped him eventually to overcome his illness. Even during this testing time, he continued helping and advising his students. Under his supervision, six taxonomists graduated with doctoral degrees — Wataru Abe (Tardigrada), Tomomi Saito (Spongicolidae), Hironori Komatsu (Leucosiidae), Masako Mitsuhashi (Pontoniinae), Yuji Ise (Porifera), and Sayumi Hoshino (Porifera). In addition to his professional duties, he was also tireless in trying to "diffuse" carcinology to the general public including children through

Fig. 3. Masatsune Takeda on field trips. Top left, Rebun Island, northern Hokkaido, searching for freshwater crabs, 1991 [Photo: T. Kubodera]; top right, trawling on board of the FV *Seishin Maru*, Suruga Bay, 1993 [Photo: H. Saito]; bottom left, Nha Trang, Vietnam, 2002 [Photo: T. Kubodera]; bottom right, Iloilo Island, Philippines, 2005.

producing number of guidebooks, general scientific and pictorial books, newspaper columns, popular magazines, as well as helping various TV programs and so on.

HIS RESEARCH FIELD

Takeda-sensei has always focused his energy on alpha-level taxonomic work of crabs. This was intrinsically linked with one of the main purposes of the national museum, which was to inventorise the biodiversity of Japan. As part of the museum's long term project "Natural History Researches of the Japanese Islands" in the years 1967-2001, he surveyed almost all over the country (1973 to 2001), exploring Tsushima Islands, Tanegashima Island, Ogasawara Islands, Izu Islands, Kii Peninsula, Sea of Japan, Amami Islands, Soya Strait, eastern Hokkaido, Onagawa Bay, East China Sea, Seto Inland Sea, Kanto Plain and western Sagami Bay. In 1993, he and his colleagues started another project "Study on Deep-sea Fauna and Conservation of Deep-sea

Ecosystem", resulting in four-year long studies of Suruga Bay, Tosa Bay and Nansei Islands. He also led a project titled "Study on Environmental Changes in the Sagami Sea and Adjacent Coastal Area with Time Serial Comparison of Fauna and Flora" between 2001 and 2005. Throughout his long career, he also investigated the crustacean fauna of various East Asian countries like the Philippines, Malaysia, Thailand, Taiwan, Indonesia, China and Vietnam.

HIS IMPORTANT COLLABORATORS

Since Takeda-sensei has collaborated with so many researchers over the years, it is difficult to give a complete or fair summary. Some are highlighted below.

The late Sadayoshi Miyake, the founder of the "Miyake School", supervised his doctoral dissertation and published many papers with him on the taxonomy of all sorts of crabs, notably on xanthoids and pilumnoids. His old friend, Yukio Nakasone, a senior student of the "Miyake School", co-authored a number of papers on the taxonomy of Okinawan crabs. He also collaborated with Takao Yamaguchi on the morphology of the cheliped of fiddler crabs, whilst Yôji Kurata, a specialist on the fauna of the Ogasawara Islands, provided him with large numbers of crab specimens collected from the Ogasawara Islands, including from the gut contents of fishes. The late Seiji Nagai, a prolific amateur collector of crabs and shells, also gave Takeda-sensei many uncommon crabs he had dredged by himself from various places. Ienori Fujiyama and Kazuo Masubuchi are frequent collaborators on fossil crabs. Yôichi Tamura provided him with a good collection of cryptochirid crabs, revealing its diversity in Japan. Yukio Hanamura collaborated with him on the taxonomic study on shrimps, whilst Kooichi Konishi also collaborated mostly on the taxonomy of pinnotherids. Tomoyuki Komai, current-day a leading Japanese carcinologist on many decapod groups, wrote his first paper with Takeda-sensei and has since collaborated on the taxonomy of various shrimps and hermit crabs. Junji Okuno was his first "unofficial" student working on the systematics of hinge-beak shrimps and hermit crabs and Masahiro Marumura, a high school teacher in Wakayama Prefecture, is a good "crab hunter" at Kii Peninsula, and contributed to many important Japanese brachyuran discoveries through his close collaboration with Takeda-sensei. Wataru Abe was one of his first graduate students, but studying tardigrades, whilst Masako Mitsuhashi studied pontoniine shrimps with Takeda-sensei. Another student, Hironori Komatsu worked on leucosiid crabs. Tomomi Saito, a biologist based at Nagoya Aquarium, did his doctoral dissertation on the taxonomy, phylogeny and biology of spongicolid shrimps with Takeda-sensei, whilst Masayuki Osawa was a postdoctoral fellow at the museum, working with him on the taxonomy of hermit crabs. Masaaki Tanaka collaborated on limnological fauna and flora of microorganisms in the Imperial Palace, Tokyo.

Takeda-sensei's collaborations were not restricted to Japan. Bella Galil from the National Institute of Oceanography in Israel, collaborated with him on several occasions on the systematics of trapeziid, xanthoid, pilumnoid and calappid crabs. Marivene Manuel-Santos from the National Museum of the Philippines was a close collaborator on many groups of crabs, notably deep-water taxa from Balicasag. Richard ("Rick") Webber carried out a number of taxonomic studies with him on New Zealand crabs. Peter Ng from the National University of Singapore has remained a close ally for over two decades, and has actively collaborated with him on freshwater crabs as well as other several other groups. Their shared passion for crabs often resulted in extremely animated discussions whenever they met, discussions which the editors have witnessed with pleasure!

PUBLICATION

Takeda-sensei started his publication career in 1967 with a paper on some rare species of xanthid crabs. Following this, he so far has produced a total of 323 papers over his career span of 45 years. Two years stand out. In 1969, he wrote seven papers totaling 204 pages and treated 14 new taxa. The year 2000 was perhaps his most productive one, publishing 21 papers, 212 pages and four new taxa. Over those 45 years, he described 27 new genera and 204 new species, including seven fossil ones (table I). Most of them are crabs, but he also named 14 caridean, 18 anomuran, one amphipod and two tardigrade taxa. In spite of this enormous output, only 3 genera and 5 species (excluding what is done in this volume, see table II) have so far been named after him. A rather small amount considering his valuable contribution to carcinology. It is therefore our great pleasure that a further two new genera and 20 new species are named after Takeda-sensei in this volume.

HIRONORI KOMATSU
JUNJI OKUNO
KOUKI FUKUOKA

TABLE I
Taxa named by Masatsune Takeda

CARIDEA
Genus: 1
Bathystylodactylus Hanamura & Takeda, 1996

Species: 13
Palaemon ogasawaraensis Kato & Takeda, 1981
Paracrangon okutanii Ohé & Takeda, 1986
Sclerocrangon unidentata Komai & Takeda, 1989
Macrobrachium niphanae Shokita & Takeda, 1989
Rhynchocinetes conspiciocellus Okuno & Takeda, 1992a
Glyphocrangon hakuhoae Takeda & Hanamura, 1994
Bathystylodactylus inflatus Hanamura & Takeda, 1996
Glyphocrangon spinossisima Brand & Takeda, 1996
Coralliocaris tridens Mitsuhashi, Fujino & Takeda, 2001
Pandalopsis gibba Komai & Takeda, 2002
Lebbeus nudirostris Komai & Takeda, 2004a
Coralliocaris labyrintha Mitsuhashi & Takeda, 2008
Coralliocaris sandyi Mitsuhashi & Takeda, 2008

ANOMURA
Species: 18
Paralomis medipacifica Takeda, 1974b
Paralomis cristata Takeda & Ohta, 1979
Paralomis truncatispinosa Takeda & Miyake, 1980
Porcellanopagurus truncatifrons Takeda, 1981c
Paragalathea miyakoensis Takeda & Fujiyama, 1983 (fossil)
Paralomis shinkaimaruae Takeda, in Takeda & Hatanaka, 1984
Paralomis kyushupalauensis Takeda, 1985a
Porcellanopagurus nihonkaiensis Takeda, 1985b
Porcellanopagurus belauensis Suzuki & Takeda, 1987
Paralomis jamsteci Takeda & Hashimoto, 1990
Discorsopagurus cavicola Komai & Takeda, 1996
Lithodes ceramensis Takeda & Nagai, 2004
Catapaguroides iejimensis Osawa & Takeda, 2004
Catapaguroides kasei Osawa & Takeda, 2004
Bathynarius izuensis Komai & Takeda, 2004b
Bathypaguropsis carinatus Komai & Takeda, 2004b
Cestopagurus puniceus Komai & Takeda, 2005
Paralomis danida Takeda & Bussarawit, 2007

BRACHYURA
Genera: 26
Prosphorachaeus Takeda & Miyake, 1969b
Neommatocarcinus Takeda & Miyake, 1969d
Genkaia Miyake & Takeda, 1970a
Libystes Miyake & Takeda, 1970b
Zehntneria Takeda, 1972b = *Zehntneriana* Ng & Takeda, 2010

TABLE I
(Continued)

Glyptocarcinus Takeda, 1973a
Nanopilumnus Takeda, 1974c
Xenograpsus Takeda & Kurata, 1977a
Fizesereneia Takeda & Tamura, 1980b
Hiroia Takeda & Tamura, 1981c
Utinomia Takeda & Tamura, 1981c, preoccupied by *Utinomia* Tomlinson, 1963
Palmyria Galil & Takeda, 1986
Carpoporoides Takeda & Nagai, 1986 = *Hepatoporus* Serène, 1984
Gorgonariana Galil & Takeda, 1988
Lentilumnus Galil & Takeda, 1988
Serenolumnus Galil & Takeda, 1988
Xlumnus Galil & Takeda, 1988
Utinomiella Kropp & Takeda, 1988
Ovitamon Ng & Takeda, 1992c
Insulamon Ng & Takeda, 1992c
Mindoron Ng & Takeda, 1992c
Skelosophusa Ng & Takeda, 1994
Atoportunus Ng & Takeda, 2003
Nobiliella Komatsu & Takeda, 2003a
Zehntneriana Ng & Takeda, 2010
Pilomedaeus Takeda & Komatsu, 2011

Species: 172
Chlorodiella ohshimai Miyake & Takeda, 1967
Chlorodiella corallicola Miyake & Takeda, 1968
Pilodius serenei Miyake & Takeda, 1968 = *Pilodius granulatus* Stimpson, 1858
Pilumnus senahai Takeda & Miyake, 1968b
Pilumnus annamensis Takeda & Miyake, 1968b
Pilumnus ikedai Takeda & Miyake, 1968b
Pilumnus guinotae Takeda & Miyake, 1968b
Paractaeopsis quadriareolata (Takeda & Miyake, 1968c) [*Actaea*]
Vellodius etisoides (Takeda & Miyake, 1968c) [*Pilodius*]
Etisus sakaii Takeda & Miyake, 1968d
Achaeus boninensis Miyake & Takeda, 1969
Zalasius imajimai Takeda & Miyake, 1969b
Oncinopus angustifrons Takeda & Miyake, 1969b
Achaeus varians Takeda & Miyake, 1969b
Maja sakaii Takeda & Miyake, 1969b
Actumnus similis Takeda & Miyake, 1969c
Viaderiana incerta (Takeda & Miyake, 1969c) [*Parapilumnus*]
Nanopilumnus hondai (Takeda & Miyake, 1969c) [*Parapilumnus*]
Nanopilumnus coralliophilus (Takeda & Miyake, 1969c) [*Parapilumnus*]
Nectocarcinus bennetti Takeda & Miyake, 1969d
Microcassiope orientalis Takeda & Miyake, 1969e
Zehntneriana amakusae (Takeda & Miyake, 1969f) [*Litocheira*]
Genkaia gordonae Miyake & Takeda, 1970a

TABLE I
(Continued)

Xenophthalmus wolffi Takeda & Miyake, 1970a
Libystes lepidus Miyake & Takeda, 1970b
Pilumnus ohshimai Takeda & Miyake, 1970b
Pilumnus parapilumnoides Takeda & Miyake, 1970b
Cosmonotus genkaiae Takeda & Miyake, 1970c
Tunedromia yamashitai (Takeda & Miyake, 1970c) [*Petalomera*]
Lybia edmondsoni Takeda & Miyake, 1970d
Thalamita miyakei Takeda, 1971
Etisus odhneri Takeda, 1971
Camptandrium rathbunae Takeda, 1971 = *Takedellus ambonense* (Serène & Moosa, 1971)
Parasesarma palauense Takeda, 1971
Neorhynchoplax inermis Takeda & Miyake, 1971a
Halicarcinus hondai (Takeda & Miyake, 1971b) [*Rhynchoplax*]
Halicarcinus keijibabai (Takeda & Miyake, 1971b) [*Rhynchoplax*]
Scutumara miyakei (Nakamura & Takeda, 1972) [*Pseudograpsus*]
Philyra taekoae Takeda, 1972a
Zehntneriana miyakei (Takeda, 1972b) [*Zehntneria*]
Pilumnus habei Takeda & Miyake, 1972a
Cancer sakaii Takeda & Miyake, 1972c (unnecessary replacement name for *Cancer anaglyptus* (Balss, 1922) = *Platepistoma anaglyptum* Balss, 1922)
Pilumnus balssi Takeda & Miyake, 1972c
Stimdromia kosugei (Takeda & Miyake, 1972c) [*Petalomera*]
Ebalia cryptocnemoides Takeda & Miyake, 1972c
Ebalia sakaii Takeda & Miyake, 1972c
Pugettia marissinica Takeda & Miyake, 1972c
Typhlocarcinops genkaiae Takeda & Miyake, 1972c
Glyptocarcinus lophopus Takeda, 1973a
Sestrostoma toriumii (Takeda, 1974a) [*Acmaeopleura*]
Umalia horikoshii (Takeda, 1975a) [*Ranilia*]
Aethra seychellensis Takeda, 1975b
Xanthias rathbunae Takeda, 1976a
Dromidiopsis lethrinusae (Takeda & Kurata, 1976a) [*Sphaerodromia*]
Takedromia yoshidai (Takeda & Kurata, 1976b) [*Cryptodromia*]
Urnalana insularis (Takeda & Kurata, 1976b) [*Leucosia*]
Metadynomene devaneyi (Takeda, 1977a) [*Dynomene*]
Pseudophilyra dinops Takeda, 1977b = *Pseudophilyra tridentata* Miers, 1879
Pseudolambrus saishoi Takeda, 1977b
Ebalia humilis Takeda, 1977d
Praebebalia taeniata Takeda, 1977d
Osachila expansa Takeda, 1977d
Actumnus setosiareolatus Takeda, 1977d
Vellumnus pygmaeus (Takeda, 1977d) [*Planopilumnus*]
Actumnus marissinicus Takeda & Kim, 1977
Zehntneriana novaeinsulicola (Takeda & Kurata, 1977a) [*Zehntneria*]

TABLE I
(Continued)

Xenograpsus novaeinsularis Takeda & Kurata, 1977a
Grypachaeus tenuicollis Takeda, 1978b
Pseudocryptochirus ishigakiensis Takeda & Tamura, 1979
Ranidina teshimai Fujiyama & Takeda, 1980 (fossil)
Oxypleurodon coralliophilum (Takeda, 1980a) [*Sphenocarcinus*]
Calocarcinus habei Takeda, 1980a
Hypocolpus kurodai Takeda, 1980c
Lamoha williamsi (Takeda, 1980d) [*Hypsophrys*]
Calocarcinus lewinsohni Takeda & Galil, 1980
Fizesereneia ishikawai Takeda & Tamura, 1980b
Cymonomus umitakae Takeda, 1981b
Dacryomaia japonica (Takeda & Tamura, 1981b) [*Favicola*]
Sundathelphusa cavernicola (Takeda, 1983b) [*Archipelothelphusa*]
Homolopsis hachiyai Takeda & Fujiyama, 1983 (fossil)
Necrocarcinus undecimtuberculata Takeda & Fujiyama, 1983 (fossil)
Eupilumnus kiiensis (Takeda & Nagai, 1983) [*Globopilumnus*]
Lithoscaptus grandis (Takeda & Tamura, 1983) [*Cryptochirus*]
Cryptochirus planus (Takeda & Tamura, 1983) [*Favicola*]
Ptychognathus capillidigitatus Takeda, 1984
Achaeus nomurai Takeda & Fujiyama, 1984 (fossil)
Lophoplax sextuberculata Takeda & Kurata, 1984
Tymolus ingens Takeda & Tomida, 1984 (fossil)
Tymolus itoigawai Takeda & Tomida, 1984 (fossil)
Hepatoporus distinctus (Takeda & Nagai, 1986) [*Carpoporoides*]
Varuna yui Hwang & Takeda, 1986
Miersiella cavifrons Takeda, 1989c
Pedroplax megalops (Takeda, 1989c) [*Psopheticus*]
Calappa quadrimaculata Takeda & Shikatani, 1990
Cymonomus hakuhoae Takeda & Moosa, 1990
Dairoides seafdeci Takeda & Ananpongsuk, 1991
Macrophthalmus (*Macrophthalmus*) *serenei* Takeda & Komai, 1991
Philyra nishihirai Takeda & Nakasone, 1991
Geosesarma protos Ng & Takeda, 1992a
Johora hoiseni Ng & Takeda, 1992b
Parathelphusa malaysiana Ng & Takeda, 1992b
Ovitamon arcanum Ng & Takeda, 1992c
Ovitamon tomaculum Ng & Takeda, 1992c
Insulamon unicorn Ng & Takeda, 1992c
Mindoron pala Ng & Takeda, 1992c
Isolapotamon spatha Ng & Takeda, 1992c
Parathelphusa saginata Ng & Takeda, 1993a
Parathelphusa rasilis Ng & Takeda, 1993a
Parathelphusa nana Ng & Takeda, 1993a
Parathelphusa balabac Ng & Takeda, 1993a

TABLE I
(Continued)

Parathelphusa parma Ng & Takeda, 1993a
Parathelphusa mindoro Ng & Takeda, 1993a
Carupa ohashii Takeda, 1993b
Ostracotheres tomentipes Takeda & Konishi, 1994
Neorhynchoplax okinawaensis (Nakasone & Takeda, 1994) [*Elamenopsis*]
Skelosophusa prolixa Ng & Takeda, 1994
Skelosophusa eumeces Ng & Takeda, 1994
Raylilia uenoi (Takeda, 1995b) [*Arcania*]
Drachiella caelata Takeda & Tachikawa, 1995
Atergatis interruptus Takeda & Marumura, 1997b
Lipkemera corallina (Takeda & Marumura, 1997b) [*Meriola*]
Lamarckopilumnus izuogasawarensis (Takeda & Ng, 1997) [*Pilumnus*]
Nursia alata Komatsu & Takeda, 1999
Clistocoeloma amamaparense Rahayu & Takeda, 2000
Cryptocnemus trigonus Komatsu & Takeda, 2000
Gandalfus yunohana (Takeda, Hashimoto & Ohta, 2000) [*Austinograea*]
Rochinia kotakae Takeda, 2001a
Pilumnopeus riui Takeda, 2001b
Merocryptoides ohtsukai Komatsu & Takeda, 2001a
Merocryptoides peteri Komatsu & Takeda, 2001a
Nursia guinotae Komatsu & Takeda, 2001b
Sundathelphusa hades Takeda & Ng, 2001
Pseudactaea multiareolata Takeda & Marumura, 2002
Atoportunus gustavi Ng & Takeda, 2003
Atoportunus pluto Ng & Takeda, 2003
Atoportunus dolichopus Takeda, 2003
Viaderiana kasei Takeda & Manuel, 2003
Goneplacoides marivenae (Komatsu & Takeda, 2004) [*Goneplax*]
Goneplax megalops Komatsu & Takeda, 2004 = *Goneplacoides marivenae* (Komatsu & Takeda, 2004)
Mursia baconaua Galil & Takeda, 2004
Mursia buwaya Galil & Takeda, 2004
Mursia mameleu Galil & Takeda, 2004
Mursia diwata Galil & Takeda, 2004
Arcania marinduquensis Komatsu, Manuel & Takeda, 2004
Mathildella kyushupalauensis Takeda & Watabe, 2004
Rochinia daiyuae Takeda & Komatsu, 2005
Mursia orientalia Takeda & Galil, 2005
Raylilia intermedia Komatsu, Manuel & Takeda, 2005
Tokoyo trilobata Komatsu, Manuel & Takeda, 2005 = *Tokoyo eburnea* (Alcock, 1896)
Achaeus kermadecensis Webber & Takeda, 2005
Heteronucia toyoshioae Komatsu & Takeda, 2005
Heteronucia granulata Komatsu & Takeda, 2005

TABLE I
(Continued)

Caphyra acheronae Takeda & Webber, 2006
Liomera yaldwyni Takeda & Webber, 2006
Platypodia delli Takeda & Webber, 2006
Xanthias dawsoni Takeda & Webber, 2006
Actumnus griffini Takeda & Webber, 2006
Ebalia webberi Komatsu & Takeda, 2007
Pseudolambrus constrictus Takeda & Webber, 2007
Rochinia planirostris Takeda, 2009
Pinnixa lata Komatsu & Takeda, 2009a
Heteronucia fujitai Komatsu & Takeda, 2009b
Huenia nagaii Takeda & Marumura, 2010
Huenia toyoshioae Takeda & Marumura, 2010
Notonyx angulatus Naruse & Takeda, 2010
Catoptrus madingondonensis Takeda, 2010
Singhaplax danielae Takeda & Komatsu, 2010a
Microgoneplax guinotae Takeda & Komatsu, 2010a
Gaillardiellus holthuisi Takeda & Komatsu, 2010b
Pilomedaeus okutanii Takeda & Komatsu, 2011
Meractaea takunan Komatsu & Takeda, 2011

AMPHIPODA
Species: 1
Neoxenodice hoshiaii Takeuchi & Takeda, 1992

TARDIGRADA
Species: 2
Pseudechiniscus asper Abe, Utsugi & Takeda, 1998
Calcarobiotus imperialis Abe & Takeda, 2000a

TABLE II
Taxa named after Masatsune Takeda

Genera: 3
Takedana Davie, 1989
Takedromia McLay, 1993
Takedellus C. G. S. Tan & Ng, 1999
[+ 2 new genera in this volume]

Species: 5
Tiarinia takedai Griffin & Tranter, 1986
Typhlocarcinops takedai Ng, 1987
Pilumnus takedai Ng, 1988
Gonodactylus takedai Moosa, 1989 = *Gonodactylaceus falcatus* (Forskål, 1775)
Tridentella takedai Nunomura, 2006
[+ 20 new species in this volume]

MASATSUNE TAKEDA BIBLIOGRAPHY

1967

MIYAKE, S. & M. TAKEDA, 1967. On some rare xanthid crabs from the Ryukyu Islands, with description of a new species. Journal of the Faculty of Agriculture, Kyushu University, **14**(2): 293-302.

1968

BABA, K., Y. NAKASONE & M. TAKEDA, 1968. Two species of *Microprosthema* found on the coral reefs of the Ryukyu Islands (Stenopodidae, Crustacea). OHMU, Occasional Papers of Zoological Laboratory, Faculty of Agriculture, Kyushu University, **1**(8): 173-181.

MIYAKE, S. & M. TAKEDA, 1968. Two new species of xanthid crabs from the Palau Islands. Journal of the Faculty of Agriculture, Kyushu University, **14**(3): 389-398.

TAKEDA, M. & S. MIYAKE, 1968a. First zoea of two pilumnid crabs of the family Xanthidae. Bulletin of the Faculty of Agriculture, Kyushu University, **23**(3): 127-133. [In Japanese.]

TAKEDA, M. & S. MIYAKE, 1968b. Pilumnid crabs of the family Xanthidae from the West Pacific. I. Twenty-three species of the genus *Pilumnus*, with description of four new species. OHMU, Occasional Papers of Zoological Laboratory, Faculty of Agriculture, Kyushu University, **1**(1): 1-60, pls. 1-4.

TAKEDA, M. & S. MIYAKE, 1968c. Two new xanthid crabs inhabiting coral reefs of the Ryukyu Islands. OHMU, Occasional Papers of Zoological Laboratory, Faculty of Agriculture, Kyushu University, **1**(8): 183-189, pl. 8.

TAKEDA, M. & S. MIYAKE, 1968d. A new xanthid crab of the genus *Etisus* from the Palau Islands. OHMU, Occasional Papers of Zoological Laboratory, Faculty of Agriculture, Kyushu University, **1**(11): 201-210.

TAKEDA, M. & S. MIYAKE, 1968e. Crabs from the East China Sea. I. Corystoidea and Brachygnatha Brachyrhyncha. Journal of the Faculty of Agriculture, Kyushu University, **14**(4): 541-582, pl. 6.

TAKEDA, M. & S. MIYAKE, 1968f. Six unrecorded xanthid crabs from the Ryukyu Islands preserved in the Zoological Laboratory, Kyushu University. Biological Magazine Okinawa, **5**: 1-10.

1969

MIYAKE, S. & M. TAKEDA, 1969. On two majid crabs of the genus *Achaeus* from the Ogasawara Islands (Crustacea, Brachyura). Proceeding of the Japanese Society of Systematic Zoology, **5**: 22-32.

TAKEDA, M. & S. MIYAKE, 1969a. Crabs from the East China Sea. II. Addition to Brachygnatha Brachyrhyncha. Journal of the Faculty of Agriculture, Kyushu University, **15**(4): 449-468.

TAKEDA, M. & S. MIYAKE, 1969b. Crabs from the East China Sea. III. Brachygnatha Oxyrhyncha. Journal of the Faculty of Agriculture, Kyushu University, **15**(4): 469-521, pls. 17, 18.

TAKEDA, M. & S. MIYAKE, 1969c. Pilumnid crabs of the family Xanthidae from the West Pacific. II. Twenty-one species of four genera, with descriptions of four new species. OHMU, Occasional Papers of Zoological Laboratory, Faculty of Agriculture, Kyushu University, **2**(7): 93-156.

TAKEDA, M. & S. MIYAKE, 1969d. A small collection of crabs from New Zealand. OHMU, Occasional Papers of Zoological Laboratory, Faculty of Agriculture, Kyushu University, **2**(8): 157-193, pls. 1-3.

TAKEDA, M. & S. MIYAKE, 1969e. On two species of the family Xanthidae (Crustacea, Brachyura) from southern Japan. OHMU, Occasional Papers of Zoological Laboratory, Faculty of Agriculture, Kyushu University, **2**(9): 195-206.

TAKEDA, M. & S. MIYAKE, 1969f. A new species of the family Goneplacidae (Crustacea, Brachyura) from the Amakusa Islands, Kyushu. Publications from the Amakusa Marine Biological Laboratory, Kyushu University, **2**(1): 9-15.

1970

MIYAKE, S. & M. TAKEDA, 1970a. A remarkable species of the Dromiacea (Crustacea, Decapoda) from the Tsushima Islands, Japan. OHMU, Occasional Papers of Zoological Laboratory, Faculty of Agriculture, Kyushu University, **3**(3): 19-28.

MIYAKE, S. & M. TAKEDA, 1970b. A new portunid crab of the genus *Libystes* from the Ogasawara Islands, with note on *L. villosus* Rathbun from the Ryukyu Islands. OHMU, Occasional Papers of Zoological Laboratory, Faculty of Agriculture, Kyushu University, **3**(4): 29-36.

TAKEDA, M. & S. MIYAKE, 1970a. A new free-living pinnotherid crab (Crustacea, Brachyura) from the Iranian Gulf. OHMU, Occasional Papers of Zoological Laboratory, Faculty of Agriculture, Kyushu University, **3**(2): 11-18.

TAKEDA, M. & S. MIYAKE, 1970b. Pilumnid crabs of the family Xanthidae from the West Pacific. III. Descriptions of two new species of the genus *Pilumnus*. OHMU, Occasional Papers of Zoological Laboratory, Faculty of Agriculture, Kyushu University, **3**(5): 37-44, pl. 1.

TAKEDA, M. & S. MIYAKE, 1970c. Crabs from the East China Sea. IV. Gymnopleura, Dromiacea and Oxystomata. Journal of the Faculty of Agriculture, Kyushu University, **16**(3): 193-235, pl. 1.

TAKEDA, M. & S. MIYAKE, 1970d. *Lybia edmondsoni* sp. nov., a new anemone crab from the Hawaiian Islands. Proceeding of the Japanese Society of Systematic Zoology, **6**: 11-15.

1971

TAKEDA, M., 1971. New and rare crabs from the Palau Islands. Micronesica, **7**: 185-213.

TAKEDA, M. & S. MIYAKE, 1971a. A new freshwater crab of the family Hymenosomatidae from the Palau Islands. Researches on Crustacea, **4/5**: 157-163.

TAKEDA, M. & S. MIYAKE, 1971b. Two new hymenosomatid crabs of the genus *Rhynchoplax* from the West and South Pacific. Researches on Crustacea, **4/5**: 164-172.

1972

NAKAMURA, I. T. & M. TAKEDA, 1972. A new species of the Grapsidae (Crustacea, Brachyura) from Bali Island in the collection of the National Science Museum, Tokyo. Bulletin of the National Science Museum, Tokyo, **15**(3): 435-439.

TAKEDA, M., 1972a. A new species of the Leucosiidae (Crustacea, Brachyura) from Amami-Oshima. Bulletin of the Biogeographical Society of Japan, **28**(1): 1-4.

TAKEDA, M., 1972b. Systematic status of *Ceratoplax villosa* Zehntner and some related species (Crustacea, Decapoda, Brachyura). Proceedings of the Japanese Society of Systematic Zoology, **8**: 34-41.

TAKEDA, M., 1972c. Further notes on the unrecorded xanthid crabs from the Ryukyu Islands. Biological Magazine of Okinawa, **9**: 15-24.

TAKEDA, M. & S. MIYAKE, 1972a. Pilumnid crabs of the family Xanthidae from the West Pacific. IV. A new *Pilumnus* in the collection of the National Science Museum. OHMU, Occasional Papers of Zoological Laboratory, Faculty of Agriculture, Kyushu University, **3**(7): 57-62, pl. 2.

TAKEDA, M. & S. MIYAKE, 1972b. Crabs from the East China Sea. V. A remaining collection. OHMU, Occasional Papers of Zoological Laboratory, Faculty of Agriculture, Kyushu University, **3**(8): 63-90, pl. 3.

TAKEDA, M. & S. MIYAKE, 1972c. New crabs from the sea around the Tsushima Islands. Bulletin of the National Science Museum, Tokyo, **15**(2): 253-265.

1973

TAKEDA, M., 1973a. A new genus and a new species of the Parthenopidae from the sea off the Ogasawara Islands (Crustacea, Brachyura). Bulletin of the National Science Museum, Tokyo, **16**(1): 31-36.

TAKEDA, M., 1973b. Crabs from the sea around the Tsushima Islands. Bulletin of the Biogeographical Society of Japan, **29**(3): 9-16. [In Japanese with English abstract.]

TAKEDA, M., 1973c. Report on the crabs from the sea around the Tsushima Islands collected by the research vessel "Genkai" for the trustees of the National Science Museum, Tokyo. Bulletin of the Liberal Arts & Science Course, Nihon University School of Medicine, **1**: 17-68.

TAKEDA, M., 1973d. Studies on the Crustacea Brachyura of the Palau Islands, I. Dromiidae, Dynomenidae, Calappidae, Leucosiidae, Hymenosomatidae, Majidae and Parthenopidae. Bulletin of the Liberal Arts & Science Course, Nihon University School of Medicine, **1**: 75-122, pls. 2, 3.

TAKEDA, M., 1973e. A record of a fiddler crab, *Uca tetragonon* (Herbst), from the Ryukyu Islands. Biological Magazine Okinawa, **10**: 7-11. [In Japanese with English abstract.]

TAKEDA, M. & H. HAYASHI, 1973. On a small collection of crabs from the Palau Islands. Bulletin of the Liberal Arts & Science Course, Nihon University School of Medicine, **1**: 69-74, pl. 1.

TAKEDA, M. & T. YAMAGUCHI, 1973. Occurrences of abnormal males in a fiddler crab, *Uca marionis* (Desmarest) with notes on asymmetry of chelipeds. Proceeding of the Japanese Society of Systematic Zoology, **9**: 13-20. [In Japanese with English abstract.]

YAMAGUCHI, T. & M. TAKEDA, 1973a. Dimorphism of chelipeds and systematic status of a fiddler crab, *Uca marionis* (Desmarest). Proceeding of the Japanese Society of Systematic Zoology, **9**: 7-12. [In Japanese with English abstract.]

YAMAGUCHI, T. & M. TAKEDA, 1973b. Notes on a fiddler crab, *Uca dussumieri* (H. Milne Edwards), with special reference to dimorphism of chelipeds. Biological Magazine Okinawa, **10**: 12-16. [In Japanese with English abstract.]

1974

SUZUKI, K. & M. TAKEDA, 1974. On a parthenopid crab, *Zebrida adamsii* on the sea urchins from Suruga Bay, with a special reference to their parasitic relations. Bulletin of the National Science Museum, Tokyo, **17**(4): 287-296, pl. 1.

TAKEDA, M., 1974a. Accounts of some crabs from Mutsu Bay, with description of a new grapsid from Onagawa Bay. Bulletin of the Marine Biological Station of Asamushi, Tohoku University, **15**(1): 13-21.

TAKEDA, M., 1974b. On three species of Lithodidae (Crustacea, Anomura) from the central Pacific. Bulletin of the National Science Museum, Tokyo, **17**(3): 205-214, pls. 1-3.

TAKEDA, M., 1974c. Pilumnid crabs of the family Xanthidae from the West Pacific V. Definition of a new genus, with description of its type-species. Bulletin of the National Science Museum, Tokyo, **17**(3): 215-219.

TAKEDA, M. & Y. KOYAMA, 1974. On some rare crabs from Kii Province. Researches on Crustacea, **6**: 103-121, pls. 10-12. [In Japanese and English.]

TAKEDA, M. & S. SHIMAZAKI, 1974. Studies on the Crustacea Brachyura of the Palau Islands, II. Atelecyclidae, Portunidae, Goneplacidae, Pinnotheridae and Palicidae. Bulletin of the Liberal Arts & Science Course, Nihon University School of Medicine, **2**: 41-79, pls. 1-4.

1975

IMAJIMA, M. & M. TAKEDA, 1975. [Benthic animals of Kagoshima Bay. Perticularly Polychaeta and Crustacea]. [Reports of Kagoshima Fisheries Experimental Stations]: 350-370. [In Japanese.]

TAKEDA, M., 1975a. Crabs from the East China Sea, VI. A collection from off the Danjo Islands made by the R/V Hakuhô Maru Cruise KH-74-3. Bulletin of the National Science Museum, Tokyo, (A) **1**(3): 137-156, pls. 1-3.

TAKEDA, M., 1975b. Preliminary description of a new crab of the Parthenopidae from the Seychelles Bank. Bulletin of the National Science Museum, Tokyo, (A) **1**(4): 199-202, pl. 1.

1976

TAKEDA, M., 1976a. Studies on the Crustacea Brachyura of the Palau Islands, III. Xanthidae (1). Researches on Crustacea, **7**: 69-99, pls. 9-11.

TAKEDA, M., 1976b. Littoral and inshore decapod crustaceans of Tanega-shima Island, southwest Japan. Memoirs of the National Science Museum, Tokyo, **9**: 151-161. [In Japanese and English abstract.]

TAKEDA, M. & Y. KURATA, 1976a. Crabs of the Ogasawara Islands, II. First report on the species obtained from stomachs of fishes. Researches on Crustacea, **7**: 116-137. [In Japanese and English.]

TAKEDA, M. & Y. KURATA, 1976b. Crabs of the Ogasawara Islands, III. Some species collected by coral fishing boats. Bulletin of the National Science Museum, Tokyo, (A) **2**(1): 19-32, pls. 1, 2.

TAKEDA, M., M. MIKI, M. ASAI & K. SUZUKI, 1976. *Halicarcinus orientalis* Sakai (Crustacea, Hymenosomatidae) associated with *Ophiocoma brevipes* Peters (Ophiuroidea, Ophiocomidae) from Hachijo Island. Bulletin of the National Science Museum, Tokyo, (A) **2**(1): 33-38, pl. 1.

TAKEDA, M. & S. MIYAKE, 1976. Crabs of the Ogasawara Islands, I. List of the known species. Researches on Crustacea, **7**: 101-115. [In Japanese with English abstract.]

TAKEDA, M. & N. NUNOMURA, 1976. Crabs collected by the Melanesia Expedition of the Osaka Museum of Natural History, 1958. Bulletin of the Osaka Museum of Natural History, **30**: 61-92.

TAKEDA, M., K. OKAMOTO & T. FUKUDA, 1976. *Huenia brevifrons* Ward (Decapoda, Majidae) attached to calcareous green alga, *Halimeda*, from the Ryukyu Islands. Bulletin of the National Science Museum, Tokyo, (A) **2**(2): 103-108, pl. 1.

YAMAGUCHI, T., M. TAKEDA & K. TOKUDOME, 1976. A list of crabs collected in the vicinity of the Aitsu Marine Biological Station and a preliminary report on the cheliped asymmetry of the crabs. Calanus, **5**: 31-46.

1977

FUJINO, T. & M. TAKEDA, 1977. *Levicaris mammillata* (Edmondson), a gnathophyllid shrimp associated with slate-pencil sea urchin, *Heterocentrotus mammillatus* (Linnaeus), from the Ogasawara and Ryukyu Islands. Bulletin of the National Science Museum, Tokyo, (A) **3**(3): 131-140.

TAKEDA, M., 1977a. Two interesting crabs from Hawaii. Pacific Science, **31**(1): 31-38.

TAKEDA, M., 1977b. Crabs from shallow waters off Mage-jima Island, Southwest Japan. Bulletin of the National Science Museum, Tokyo, (A) **3**(2): 73-89.

TAKEDA, M., 1977c. Rediscovery of *Lophoplax sculpta* (Stimpson, 1858) (Crustacea, Decapoda, Brachyura). Zoological Magazine, **86**(2): 120-123.

TAKEDA, M., 1977d. Crabs of the Ogasawara Islands, V. A collection made by dredging. Memoirs of the National Science Museum, Tokyo, **10**: 113-140, pls. 12-17.

TAKEDA, M. & H. S. KIM, 1977. A new *Actumnus* (Crustacea; Decapoda; Brachyura) from Jeju Island, Korea. Korean Journal of Zoology, **20**(3): 135-139.

TAKEDA, M. & Y. KURATA, 1977a. Crabs of the Ogasawara Islands, IV. A collection made at the new volcanic island, Nishino-shima-shinto, in 1975. Bulletin of the National Science Museum, Tokyo, (A) **3**(2): 91-111.

TAKEDA, M. & Y. KURATA, 1977b. Crabs of the Ogasawara Islands, VI. Second report on the species obtained from stomachs of fishes. Memoirs of the National Science Museum, Tokyo, **10**: 141-145.

1978

SABA, M., M. TAKEDA & Y. NAKASONE, 1978a. Larval development of *Baptozius vinosus* (H. Milne Edwards) (Crustacea, Brachyura, Xanthidae). Proceeding of the Japanese Society of Systematic Zoology, **14**: 29-42.

SABA, M., M. TAKEDA & Y. NAKASONE, 1978b. Larval development of *Epixanthus dentatus* (White) (Brachyura, Xanthidae). Bulletin of the National Science Museum, Tokyo, (A) **4**(3): 151-161.

TAKEDA, M., 1978a. Soldier crabs from Australia and Japan. Bulletin of the National Science Museum, Tokyo, (A) **4**(1): 31-38.

TAKEDA, M., 1978b. A new spider crab from off Iriomote Island, the Ryukyu Islands. Bulletin of the National Science Museum, Tokyo, (A) **4**(2): 117-121.

TAKEDA, M., 1978c. Biogeographical notes on the crabs obtained by dredging at the Sea around Nii-jima and O-shima, Izu Islands. Memoirs of the National Science Museum, Tokyo, **11**: 73-80. [In Japanese with English abstract.]

1979

HORIKOSHI, M., E. TSUCHIDA, M. IMAJIMA, M. TAKEDA, S. GAMO & S. OHTA, 1979. Benthic invertebrates recorded from Otsuchi Bay and the adjacent Sanriku Coast — primary catalogue of fauna. Otsuchi Marine Rsearch Center Report, **5**: 37-85. [In Japanese.]

TAKEDA, M., 1979a. Systematic and biogeographic notes on the crabs obtained by dredging at the sea around Cape Shionomisaki, Kii Peninsula. Memoirs of the National Science Museum, Tokyo, **12**: 151-157. [In Japanese with English abstract.]

TAKEDA, M., 1979b. Generic and specific validity of *Glyptocarcinus lophopus* Takeda (Crustacea, Brachyura). Proceedings of the Japanese Society of Systematic Zoology, **17**: 68-72.

TAKEDA, M. & S. NAGAI, 1979. Occurrence of a majid crab, *Sphenocarcinus auritus* Rathbun, in Tosa Bay. Nankiseibutu, **21**(1): 18. [In Japanese with English abstract.]

TAKEDA, M. & S. OHTA, 1979. A new species of the Lithodidae (Crustacea, Anomura) from Suruga Bay, Central Japan. Bulletin of the National Science Museum, Tokyo, (A) **5**(3): 195-200, pls. 1-3.

TAKEDA, M. & H. SUGA, 1979. Feeding habits of box crabs, *Calappa*. Researches on Crustacea, **9**: 43-46, pl. 1. [In Japanese with English abstract.]

TAKEDA, M. & Y. TAMURA, 1979. Coral-inhabiting crabs of the family Hapalocarcinidae from Japan. I. Three species obtained from mushroom coral, *Fungia*. Bulletin of the National Science Museum, Tokyo, (A) **5**(3): 183-194, pls. 1-7.

1980

FUJIYAMA, I. & M. TAKEDA, 1980. A fossil raninid crab from the Poronai Formation, Hokkaido, Japan. In: H. IGO & H. NODA (eds.), Professor Saburo Kanno Memorial Volume: 339-342, pls. 39, 40. (Memorial Association of Professor Saburo Kanno's Retirement, Institute of Geoscience, the University of Tsukuba, Tsukuba).

GAMO, S., Y. SUGIURA, M. TAKEDA, H. MINEGISHI & S. IRIMURA, 1980. Notes on some invertebrates (hydromedusae, tubellarians, decapod and some lower crustaceans, and ophiuroids) from Otsuchi Bay and its vicinity. Otsuchi Marine Research Center Report, **6**: 25-31. [In Japanese.]

TAKEDA, M., 1980a. Two new crabs associated with precious coral from the central Pacific. Bulletin of the National Science Museum, Tokyo, (A) **6**(2): 71-76.

TAKEDA, M., 1980b. *Pilumnus planus* Edmondson and *Leptodius leptodon* Forest & Guinot as synonyms of *Forestia depressa* (White) and *Leptodius davaoensis* Ward (Decapoda, Brachyura). Crustaceana, **39**(3): 318-320.

TAKEDA, M., 1980c. A new xanthid crab from the Ryukyu Islands. Bulletin of the Biogeographical Society of Japan, **35**(3): 39-44.

TAKEDA, M., 1980d. A small collection of crabs from the Kyushu-Palau Submarine Ridge, with a description of a new species of the Homolidae. Micronesica, **16**(2): 279-287.

TAKEDA, M. & B. S. GALIL, 1980. A new species of *Calocarcinus* (Crustacea, Brachyura) from the submarine bank off the Izu Islands, central Japan. Bulletin of the National Science Museum, Tokyo, (A) **6**(4): 191-194.

TAKEDA, M. & S. MIYAKE, 1980. A new species of *Paralomis* (Crustacea, Anomura) from the East China Sea. Annotationes Zoologicae Japonenses, **53**(1): 42-45.

TAKEDA, M. & Y. TAMURA, 1980a. Coral-inhabiting crabs of the family Hapalocarcinidae from Japan. II. Genus *Pseudohapalocarcinus*. Proceedings of the Japanese Society of Systematic Zoology, **18**: 54-59, pls. 2-5.

TAKEDA, M. & Y. TAMURA, 1980b. Coral-inhabiting crabs of the family Hapalocarcinidae from Japan. III. New genus *Fizesereneia*. Bulletin of the National Science Museum, Tokyo, (A) **6**(3): 137-146.

TAKEDA, M. & Y. TAMURA, 1980c. Coral-inhabiting crabs of the family Hapalocarcinidae from Japan. IV. Genus *Neotroglocarcinus*. Bulletin of the National Science Museum, Tokyo, (A) **6**(3): 147-151.

TAKEDA, M. & Y. TAMURA, 1980d. Coral-inhabiting crabs of the family Hapalocarcinidae from Japan. V. Genus *Cryptochirus*. Researches on Crustacea, **10**: 45-56, pls. 2-4.

1981

KATO, H. & M. TAKEDA, 1981. A new shrimp of the genus *Palaemon* (Crustacea, Decapoda) from the Ogasawara Islands. Bulletin of the National Science Museum, Tokyo, (A) **7**(3): 101-109.

TAKEDA, M., 1981a. *Macrophthalmus* (Crustacea, Brachyura, Ocypodidae) from the Ryukyu Island. In: T. YAMAGUCHI (ed.), Ecological Studies of Coastal Marine and Freshwater Crabs: 69-77. [In Japanese with English abstract.] (Ministry of Education, Tokyo).

TAKEDA, M., 1981b. A new crab of the genus *Cymonomus* (Crustacea: Brachyura) from off Boso Peninsula, central Japan. Researches on Crustacea, **11**: 36-40.

TAKEDA, M., 1981c. A new hermit crab of the genus *Porcellanopagurus* from the Ogasawara Islands. Bulletin of the Biogeographical Society of Japan, **36**(2): 8-13.

TAKEDA, M. & Y. TAMURA, 1981a. Coral-inhabiting crabs of the family Hapalocarcinidae from Japan. VI. Genus *Hapalocarcinus*. In: T. YAMAGUCHI (ed.), Ecological Studies of Coastal Marine and Freshwater Crabs: 57-62, pls. 1-3. [In Japanese with English abstract.] (Ministry of Education, Tokyo).

TAKEDA, M. & Y. TAMURA, 1981b. Coral-inhabiting crabs of the family Hapalocarcinidae from Japan. VII. Genus *Favicola*. Researches on Crustacea, **11**: 41-50, pls. 1-3.

TAKEDA, M. & Y. TAMURA, 1981c. Coral-inhabiting crabs of the family Hapalocarcinidae from Japan. VIII. Genus *Pseudocryptochirus* and two new genera. Bulletin of the Biogeographical Society of Japan, **36**(3): 14-27, pls. 1-4.

1982

TAKEDA, M., 1982a. Keys to the Japanese and foreign crustaceans fully illustrated in colors. 284 pp. [In Japanese.] (Hokuryukan, Tokyo).

TAKEDA, M., 1982b. Biogeographical notes on the crabs obtained by dredging off the southeast coast of the Izu Peninsula, Central Japan. Bulletin of the Biogeographical Society of Japan, **37**(4): 15-21. [In Japanese with English abstract.]

TAKEDA, M., 1982c. Two land-dwelling crabs from Minami-Iwojima Island. In: NATURE CONSERVATION BUREAU, ENVIRONMENT AGENCY OF JAPAN (ed.), Conservation Reports of the Minami-Iwojima Wilderness Area, Tokyo, Japan: 379-382. [In Japanese with English abstract.] (Japan Wildlife Research Center, Tokyo).

TAKEDA, M. & P. MASAHITO, 1982. *Sympasiphaea imperialis* Terao as a junior synonym of *Glyphus marsupialis* Filhol (Crustacea, Decapoda, Macrura). Memoirs of the National Science Museum, Tokyo, **15**: 181-185. [In Japanese with English abstract.]

1983

TAKEDA, M., 1983a. Crustaceans. Crustaceans and mollusks trawled off Suriname and French Guiana: 19-185. (Japan Marine Fishery Resource Research Center, Tokyo).

TAKEDA, M., 1983b. A new cavernicolous crab from Bohol, the Philippines. Bulletin of the National Science Museum, Tokyo, (A) **9**(4): 169-173.

TAKEDA, M. & I. FUJIYAMA, 1983. Three decapod crustaceans from the Lower Cretaceous Miyako Group, northern Japan. Bulletin of the National Science Museum, Tokyo, (C) **9**(4): 129-136.

TAKEDA, M. & K. IWASAKI, 1983. Remarkable crabs from the Ryukyu Islands (1). Bulletin of the Biogeographical Society of Japan, **38**(9): 87-91.

TAKEDA, M. & S. NAGAI, 1983. Description of a new crab from Kushimoto, the Kii Peninsula, Southwest Japan. Bulletin of the National Science Museum, Tokyo, (A) **9**(2): 45-49.

TAKEDA, M. & Y. TAMURA, 1983. Coral-inhabiting crabs of the family Hapalocarcinidae from Japan. IX. A small collection made at Kushimoto and Koza, the Kii Peninsula. Bulletin of the National Science Museum, Tokyo, (A) **9**(1): 1-12, pls. 1-4.

1984

TAKEDA, M., 1984. A new crab of the family Grapsidae from Japan. Bulletin of the National Science Museum, Tokyo, (A) **10**(3): 117-120.

TAKEDA, M. & I. FUJIYAMA, 1984. A new majid crab from the Miocene Matsuyama Group, Saitama Prefecture, central Japan. Bulletin of the National Science Museum, Tokyo, (C) **10**(2): 49-53.

TAKEDA, M. & H. HATANAKA, 1984. Records of decapod crustaceans from the southwestern Atlantic collected by the Japanese Fisheries Research trawlers. Bulletin of the National Science Museum, Tokyo, (A) **10**(1): 7-24.

TAKEDA, M., K. HIRAMOTO & Y. SUZUKI, 1984. Additional material of *Paralomis cristata* Takeda et Ohta (Crustacea, Decapoda) from Suruga Bay, Japan. Bulletin of the Biogeographical Society of Japan, **39**(5): 27-31.

TAKEDA, M. & Y. KURATA, 1984. Crabs of the Ogasawara Islands. VII. Third report on the species obtained from stomachs of fishes. Bulletin of the National Science Museum, Tokyo, (A) **10**(4): 195-202.

TAKEDA, M., M. OISHI & I. FUJIYAMA, 1984. A record of Pliocene cancrid crab, *Cancer minutoserratus* Nagao, from the Yushima Formation, Iwate Prefecture, with a note on subgeneric diversity of *Cancer* in Japan. Bulletin of the Iwate Prefectural Museum, **2**: 157-163.

TAKEDA, M. & S. TOMIDA, 1984. Two new fossil crabs of the Tymolidae from the Miocene Mizunami Group, central Japan. Bulletin of the Mizunami Fossil Museum, **12**: 39-48, pl. 13.

1985

GALIL, B. & M. TAKEDA, 1985. Crabs of the genus *Quadrella* (Crustacea, Decapoda, Trapeziidae) from Japanese waters. Bulletin of the National Science Museum, Tokyo, (A) **11**(4): 197-207.

TAKEDA, M., 1985a. A new species of *Paralomis*, the crab-shaped Anomura (Crustacea, Decapoda), from the Kyushu-Palau submarine ridge. Bulletin of the National Science Museum, Tokyo, (A) **11**(3): 137-140.

TAKEDA, M., 1985b. Occurrence of a new hermit crab of the genus *Porcellanopagurus* in the Sea of Japan. Memoirs of the National Science Museum, Tokyo, **18**: 141-144.

TAKEDA, M., 1985c. Record of a male *Genkaia gordonae* Miyake et Takeda from Japan (Crustacea: Decapoda: Brachyura). Special Publication of the Mukaishima Marine Biological Station, **1985**: 97-100.

TAKEDA, M. & K. MASUBUCHI, 1985a. On a fossil *Cancer japonicus* from the Iimuro Formation (Kazusa Group) in the Tama River. Annual report of the Kawasaki Municipal Science Museum for Youth, **2**: 27. [In Japanese.] [Dated 1984.]

TAKEDA, M. & K. MASUBUCHI, 1985b. Crab fossils of *Ovalipes puncatus* (de Haan) and *Carcinoplax longimana* (de Haan) from the Iimuro Formation (Kazusa Group) in the Tama River. Annual Report of the Kawasaki Municipal Science Museum for Youth, **3**: 35-38. [In Japanese.]

TAKEDA, M. & Y. TAMURA, 1985. Coral-inhabiting crabs of the family Hapalocarcinidae from Japan. X. Collections from Hachijo Island in the Izu Islands. Bulletin of the National Science Museum, Tokyo, (A) **11**(2): 99-108, pls. 1-3.

TAKEDA, M. & F. YANAGISAWA, 1985. Rediscovery of two *Carcinoplax* crabs in the Kumanonada Sea, east of the Kii Peninsula, central Japan, with notes on distribution of the related species. Bulletin of the Biogeographical Society of Japan, **40**(8): 57-62. [In Japanese with English abstract.]

1986

GALIL, B. & M. TAKEDA, 1986. Resurrection of the genus *Jonesius* and establishment of a new genus: commensal crabs associated with corals from the Indo-Pacific Ocean. Bulletin of the National Science Museum, Tokyo, (A) **12**(4): 163-171.

HWANG, J.-J. & M. TAKEDA, 1986. A new freshwater crab of the family Grapsidae from Taiwan. Proceedings of the Japanese Society of Systematic Zoology, **33**: 11-18.

OHÉ, M. & M. TAKEDA, 1986. A new deep-sea shrimp of the genus *Paracrangon* from central Japan. Bulletin of the National Science Museum, Tokyo, (A) **12**(2): 75-81.

TAKEDA, M., 1986a. Crustacea. In: I. NAKAMURA (ed.), Important Fishes Trawled off Patagonia: 318-343. [In Japanese and English.] (Japan Marine Fishery Resource Research Center, Tokyo).

TAKEDA, M., 1986b. *Carpoporoides* Takeda et Nagai, 1986, a synonym of *Hepatoporus* Serène, 1984 (Crustacea, Decapoda, Brachyura). Proceedings of the Japanese Society of Systematic Zoology, **33**: 51.

TAKEDA, M., Y. MIZUNO & M. YAMAOKA, 1986. Some fossil crustaceans from the Miocene Morozaki Group in the Chita Peninsula, central Japan. Publication of the Tokai Fossil Society, **28**: 12-22. [In Japanese.]

TAKEDA, M. & S. NAGAI, 1986. Establishment of a new genus on two Japanese crabs. Zoological Science, **3**(3): 547-550.

TAKEDA, M. & Y. TAMURA, 1986. Coral-inhabiting crabs of the family Hapalocarcinidae from Japan. XI. Biogeographical distribution. Bulletin of the Biogeographical Society of Japan, **41**(8): 61-70. [In Japanese with English abstract.]

1987

HANAMURA, Y. & M. TAKEDA, 1987. Family Pandalidae (Crustacea, Decapoda, Caridea) collected by the RV "Soela" from the northwest Australian shelf. Bulletin of the National Science Museum, Tokyo, (A) **13**(3): 103-121.

MUKAI, H. & M. TAKEDA, 1987. A giant amphipod Crustacea from the Miocene Morozaki Group in the Chita Peninsula, central Japan. Bulletin of the National Science Museum, Tokyo, (C) **13**(1): 35-39.

SUZUKI, H. & M. TAKEDA, 1987. Occurrence of a new hermit crab of the genus *Porcellanopagurus* (Decapoda, Paguridae) in the sea adjacent to the Palau Islands. Proceedings of the Japanese Society of Systematic Zoology, **36**: 17-24.

TAKEDA, M., 1987a. Freshwater crabs from Peninsular Malaysia, with notes on some species from the Philippines. In: K. KAWASHIMA (ed.), Paragonimus Research Report 1. *Paragonimus* in Asia: biology, genetic variation and speciation: 89-106. (Shunpposha Photographic Printing Co., Fukuoka).

TAKEDA, M., 1987b. *Oregonia mutsuensis* Yokoya, 1928, as a synonym of *O. gracilis* Dana, 1851. Memoirs of the National Science Museum, Tokyo, **20**: 133-136.

YAMAGUCHI, T., K. HARADA, M. TAKEDA & T. KIKUCHI, 1987. Crab fauna of the Amakusa Islands. Calanus, **10**: 1-70. [In English and Japanese.]

1988

GALIL, B. & M. TAKEDA, 1988. A revision of the genus *Glabropilumnus* (Crustacea, Decapoda, Brachyura). Bulletin of the National Science Museum, Tokyo, (A) **14**(2): 67-90.

KROPP, R. K. & M. TAKEDA, 1988. *Utinomiella*, a replacement name for *Utinomia* Takeda et Tamura, 1981 (Crustacea, Decapoda), *non* Tomlinson, 1963 (Crustacea, Acrothoracica). Bulletin of the Biogeographical Society of Japan, **43**(5): 29.

MASUBUCHI, K. & M. TAKEDA, 1988. A crab fossil of *Macrophthalmus japonicus* (de Haan) from the early Pleistocene Renkôji Formation in the Tama Hill, central Japan. Journal of the Hiraoka Environmental Science Laboratory, **1**(1): 13-18. [In Japanese with English abstract.]

QUINTANA, R. & M. TAKEDA, 1988. On the megalopa and first crab stages of a species of the family Xanthidae (Crustacea, Brachyura) from Japanese waters. Bulletin of the National Science Museum, Tokyo, (A) **14**(1): 27-33.

TAKEDA, M. & K. KONISHI, 1988. Redescription of *Pinnotheres alcocki* Rathbun, 1909, a commensal pea crab new to Japan. Bulletin of the National Science Museum, Tokyo, (A) **14**(3): 137-144.

1989

FUJII, M., M. TAKEDA, M. KAYAMA, K. KAYAMA, T. KAYAMA & I. MITANI, 1989. Some biological aspects of Japanese lobster, *Metanephrops japonicus*, in Sagami Bay — IV. Egg, newly hatched larva and juvenile. Bulletin of the Kanagawa Prefectural Fisheries Experiment Station, **10**: 21-25, pls. 1-5. [In Japanese.]

HUANG, J.-F., H.-P. YU & M. TAKEDA, 1989. Fiddler crabs (Crustacea: Decapoda: Ocypodidae) of Taiwan. Bulletin of the Institute of Zoology, Academia Sinica, **28**(3): 191-209.

KOMAI, T. & M. TAKEDA, 1989. *Sclerocrangon unidentata*, a new crangonid shrimp from the Pacific coast of Honshu, Japan (Crustacea: Decapoda). Bulletin of the Biogeographical Society of Japan, **44**: 77-84.

NAIYANETR, P. & M. TAKEDA, 1989. Thai freshwater crabs collected during the field survey in 1987. In: K. KAWASHIMA (ed.), Paragonimus Research Report 2. *Paragonimus* in Asia: biology, genetic variation and speciation: 114-121. (Shunpposha Photographic Printing Co., Fukuoka).

SHOKITA, S. & M. TAKEDA, 1989. A new freshwater prawn of the genus *Macrobrachium* (Decapoda, Caridea, Palaemonidae) from Thailand. Bulletin of the National Science Museum, Tokyo, (A) **15**(3): 147-154, pl. 1.

TAKEDA, M., 1989a. Biogeographical distribution of crabs from the Palau Islands. In: H. OHBA, I. HAYAMI & K. MOCHIZUKI (eds.), Current aspects of biogeography in west Pacific and east Asian regions. The University Museum, The University of Tokyo, Nature and Culture, No. 1: 45-56. (The University Museum, The University of Tokyo, Tokyo).

TAKEDA, M., 1989b. Three freshwater crabs and one crayfish from China as intermediate host of lung flukes. In: K. KAWASHIMA (ed.), Paragonimus Research Report 2. *Paragonimus* in Asia: biology, genetic variation and speciation: 60-63. (Shunpposha Photographic Printing Co., Fukuoka).

TAKEDA, M., 1989c. Shallow-water crabs from the Oshima Passage between Amami-Oshima and Kakeroma-jima Islands, the northern Ryukyu Islands. Memoirs of the National Science Museum, Tokyo, **22**: 135-184, pl. 4.

TAKEDA, M. & K. MASUBUCHI, 1989. A crab fossil of *Nanocassiope alcocki* (Rathbun) hitherto known as *N. granulipes* (Sakai) from the Middle Pleistocene Ninomiya Formation in Kanagawa Prefecture, Japan. Journal of the Hiraoka Environmental Science Laboratory, **2**: 51-54. [In Japanese with English abstract.]

TAKEUCHI, I., M. TAKEDA & K. TAKESHITA, 1989. Redescription of the bathyal caprellid, *Caprella ungulina* Mayer, 1903 (Crustacea, Amphipoda) from the North Pacific. Bulletin of the National Science Museum, Tokyo, (A) **15**(1): 19-28.

1990

HUANG, J.-F., H.-P. YU & M. TAKEDA, 1990. Occurrence of the giant spider crab, *Macrocheira kaempferi* (Temminck, 1836) (Crustacea: Decapoda: Majidae) in Taiwan. Bulletin of the Institute of Zoology, Academia Sinica, **29**(3): 207-212.

MANUEL, M. R. & M. TAKEDA, 1990. Rediscovery of *Lophopilumnus dilatipes* (Adams et White, 1848), a pilumnid crab from the Philippines. Bulletin of the Biogeographical Society of Japan, **45**(8): 61-64.

TAKEDA, M., 1990. Crustacea. In: K. AMAOKA, K. MATSUURA, T. INADA, M. TAKEDA, H. HATANAKA & K. OKADA (eds.), Fishes collected by the R/V Shinkai Maru around New Zealand: 352-376. [In Japanese and English.] (Japan Marine Fishery Resource Research Center, Tokyo).

TAKEDA, M. & J. HASHIMOTO, 1990. A new species of the genus *Paralomis* (Crustacea, Decapoda, Lithodidae) from the Minami-Ensei Knoll in the mid-Okinawa Trough. Bulletin of the National Science Museum, Tokyo, (A) **16**(2): 79-88, pl. 1.

TAKEDA, M. & H. HAYASHI, 1990. Biogeographical notes on the anomuran and brachyuran crustaceans from the Pacific coast of eastern Hokkaido, northern Japan. Bulletin of the Biogeographical Society of Japan, **45**(7): 51-59. [In Japanese with English abstract.]

TAKEDA, M. & M. K. MOOSA, 1990. A small collection of deep-sea crabs from the Florès Sea. Indo-Malayan Zoology, **6**(1): 53-72.

TAKEDA, M. & N. SHIKATANI, 1990. Crabs of the genus *Calappa* from the Ryukyu Islands, with description of a new species. Zoological Science, **7**(3): 477-484.

1991

SHOKITA, S., M. TAKEDA, S. SITTILERT & T. POLPAKDEE, 1991. Abbreviated larval development of a fresh-water prawn, *Macrobrachium niphanae* Shokita and Takeda (Decapoda: Palaemonidae), from Thailand. Journal of Crustacean Biology, **11**(1): 90-102.

TAKEDA, M. & S. ANANPONGSUK, 1991. A new deep-sea crab from the Andaman Sea off Thailand. Bulletin of the National Science Museum, Tokyo, (A) **17**(2): 93-100.

TAKEDA, M. & T. KOMAI, 1991. Japanese species of the *Macrophthalmus telescopicus* complex (Crustacea: Decapoda: Brachyura: Ocypodidae). Bulletin of the National Science Museum, Tokyo, (A) **17**(4): 165-171.

TAKEDA, M. & K. KONISHI, 1991. Pinnotherid crabs from Otsuchi Bay, Iwate Prefecture, northeastern Japan. Otsuchi Marine Research Center Report, **17**: 29-39. [In Japanese with English abstract.]

TAKEDA, M., S. MARUYAMA & K. KONISHI, 1991. Taxonomical and biogeographical notes on *Sakaina yokoyai* (Glassell) from northern Japan (Crustacea: Decapoda: Pinnotheridae). Bulletin of the Biogeographical Society of Japan, **46**(9): 89-95.

TAKEDA, M. & Y. NAKASONE, 1991. Three leucosiid crabs of the genus *Philyra* from Okinawa, the Ryukyu Islands, with description of a new species. Bulletin of the National Science Museum, Tokyo, (A) **17**(1): 19-24.

TAKEDA, M. & T. YASUHARA, 1991. Remarkable abnormalities found in giant spider crab, *Macrocheira kaempferi* (Temminck), and spiny lobster, *Panulirus japonicus* (von Siebold), from Japan. Researches on Crustacea, **20**: 57-62. [In Japanese with English abstract.]

1992

HUANG, J.-F., H.-P. YU & M. TAKEDA, 1992. A review of the ocypodid and mictyrid crabs (Crustacea: Decapoda: Brachyura) in Taiwan. Bulletin of the Institute of Zoology, Academia Sinica, **31**(3): 141-161.

NG, P. K. L. & M. TAKEDA, 1992a. A new freshwater crab of the genus *Geosesarma* de Man, 1892 from the Philippines (Crustacea, Brachyura, Grapsidae). Proceedings of the Japanese Society of Systematic Zoology, **47**: 29-32.

NG, P. K. L. & M. TAKEDA, 1992b. On some freshwater crabs (Crustacea: Brachyura: Potamidae, Parathelphusidae and Grapsidae) from Peninsular Malaysia. Bulletin of the National Science Museum, Tokyo, (A) **18**(3): 103-116.

NG, P. K. L. & M. TAKEDA, 1992c. The freshwater crab fauna (Crustacea, Brachyura) of the Philippines. I. The family Potamidae Ortmann, 1896. Bulletin of the National Science Museum, Tokyo, (A) **18**(4): 149-166.

OKUNO, J. & M. TAKEDA, 1992a. Description of a new hinge-beak shrimp, *Rhynchocinetes conspiciocellus*, from southern Japan, with designation of the lectotype of *R. uritai* Kubo, 1942. Bulletin of the National Science Museum, Tokyo, (A) **18**(2): 63-72.

OKUNO, J. & M. TAKEDA, 1992b. Distinction between two hinge-beak shrimps, *Rhynchocinetes durbanensis* Gordon and *R. uritai* Kubo (Family Rhynchocinetidae). Revue française d'aquariologie et herpetologie, **19**(3): 85-90.

TAKEDA, M. & T. MIYAUCHI, 1992. Anomuran and brachyuran crustaceans from the Soya Strait, northern Hokkaido. Memoirs of the National Science Museum, Tokyo, **25**: 143-153. [In Japanese with English abstract.]

TAKEUCHI, I. & M. TAKEDA, 1992. Three species of amphipod crustaceans collected from Breid and Lützow-Holm Bays, Antarctica, during the JARE-26 Cruise. Proceedings of the NIPR Symposium on Polar Biology, **5**: 65-82.

1993

NG, P. K. L. & M. TAKEDA, 1993a. The freshwater crab fauna (Crustacea, Brachyura) of the Philippines. II. The genus *Parathelphusa* H. Milne Edwards, 1853 (family Parathelphusidae). Bulletin of the National Science Museum, Tokyo, (A) **19**(1): 1-19.

NG, P. K. L. & M. TAKEDA, 1993b. The freshwater crab fauna (Crustacea, Brachyura) of the Philippines. III. The identity of *Telphusa cumingii* Miers, 1884, and its placement in the genus *Ovitamon* Ng et Takeda, 1992 (family Potamidae). Bulletin of the National Science Museum, Tokyo, (A) **19**(3): 111-116.

TAKEDA, M., 1993a. Four deep-sea species of anomuran crustaceans from Pacific coast of eastern Hokkaido. Memoirs of the National Science Museum, Tokyo, **26**: 95-100. [In Japanese with English abstract.]

TAKEDA, M., 1993b. A new swimming crab of the genus *Carupa* from submarine caves in the Ryukyu Islands. Bulletin of the National Science Museum, Tokyo, (A) **19**(4): 145-150.

TAKEDA, M., H. TAKEUCHI & H. SUGANUMA, 1993. Occurrence of *Xenograpsus novaeinsularis* Takeda et Kurata (Crustacea: Decapoda: Brachyura) in the Tokara and Iwo Islands. Natural Environmental Science Research, **6**: 59-64. [In Japanese with English abstract.]

1994

BRAND, T. M. & M. TAKEDA, 1994. New records of the hippolytid shrimp, *Koror misticius* Clark, 1989, from submarine caves of the Ryukyu Islands. Bulletin of the National Science Museum, Tokyo, (A) **20**(2): 77-81.

NAKASONE, Y. & M. TAKEDA, 1994. A new hymenosomatid crab, *Elamenopsis okinawaensis*, n. sp. (Crustacea: Hymenosomatidae), from Okinawa, the Ryukyu Islands, Japan. Pacific Science, **48**(2): 158-160.

NG, P. K. L. & M. TAKEDA, 1994. *Skelosophusa* (Crustacea, Decapoda, Brachyura), a new genus of potamonautid freshwater crab from Madagascar, with descriptions of two new species. Bulletin of the National Science Museum, Tokyo, (A) **20**(4): 161-172.

TAKEDA, M., Y. FUKUI, K. WADA & A. ASAKURA, 1994. Brachyura (Crustacea: Decapoda) from the northern Mariana Islands, Micronesia. In: A. ASAKURA & T. FURUKI (eds.), Biological Expedition to the northern Mariana Islands, Micronesia. Natural History Research, (Special Issue) **1**: 285-290.

TAKEDA, M. & Y. HANAMURA, 1994. Deep-sea shrimps and lobsters from the Flores Sea collected by the R.V. *Hakuho-Maru* during KH-85-1 cruise. Bulletin of the National Science Museum, Tokyo, (A) **20**(1): 1-37.

TAKEDA, M. & K. KONISHI, 1994. A new commensal crab of the family Pinnotheridae from Okinawa, the Ryukyu Islands. Proceedings of the Japanese Society of Systematic Zoology, **50**: 20-23.

TAKEDA, M. & M. MARUMURA, 1994. Rare crabs from the west coast of the Kii Peninsula, central Japan (I). Nankiseibutu, **36**(1): 26-30. [In Japanese with English abstract.]

1995

SAKAI, K. & M. TAKEDA, 1995. New records of two species of decapod crustaceans from Amami-Oshima Island, the northern Ryukyu Islands, Japan. Bulletin of the National Science Museum, Tokyo, (A) **21**(4): 203-210.

TAKEDA, M., 1995a. Geographical notes on the crabs from Onagawa Bay and its adjacent waters, northwest Honshu, Japan. Memoirs of the National Science Museum, Tokyo, **28**: 135-145. [In Japanese with English abstract.]

TAKEDA, M., 1995b. A new leucosiid crab, *Arcania uenoi*, from the Ryukyu Islands. Special Bulletin of the Japanese Society of Coleopterology, **4**: 151-155.

TAKEDA, M. & M. MARUMURA, 1995a. Rare crabs from the west coast of the Kii Peninsula, central Japan (II). Nankiseibutu, **37**(1): 1-7. [In Japanese with English abstract.]

TAKEDA, M. & M. MARUMURA, 1995b. Emendatory notes on *Lophoplax sextuberculata* Takeda et Kurata (Crustacea, Decapoda, Brachyura). Bulletin of the National Science Museum, Tokyo, (A) **21**(2): 87-91.

TAKEDA, M. & H. TACHIKAWA, 1995. Crabs of the Ogasawara Islands. VIII. Two species of the family Leucosiidae. Bulletin of the National Science Museum, Tokyo, (A) **21**(4): 211-218.

1996

BRAND, T. M. & M. TAKEDA, 1996. *Glyphocrangon* shrimps from the western Pacific collected by the R/V *Hakuho-Maru* during the KH-72-1 cruise. Bulletin of the National Science Museum, Tokyo, (A) **22**(4): 263-281.

HANAMURA, Y. & M. TAKEDA, 1996. Establishment of a new genus *Bathystylodactylus* (Crustacea: Decapoda: Stylodactylidae), with description of a new species from northwestern Pacific. Zoological Science, **13**(6): 929-934.

KOMAI, T. & M. TAKEDA, 1996. A new hermit crab of the genus *Discorsopagurus* (Crustacea: Anomura: Paguridae) from Japan. Species Diversity, **1**(1): 75-85.

KOMAI, T., I. TAKEUCHI & M. TAKEDA, 1996. Deep-sea shrimp (Crustacea: Decapoda: Caridea) from the Antarctic Sea collected during the JARE-35 cruise. Proceedings of the NIPR Symposium on Polar Biology, **9**: 179-206.

TAKEDA, M. & M. MARUMURA, 1996. Rare crabs from the west coast of the Kii Peninsula, central Japan (III). Nankiseibutu, **38**(1): 4-10. [In Japanese with English abstract.]

1997

AIBA, H., T. YAMAGUCHI, M. TAKEDA & T. KAWABE, 1997. Taphonomical approaches to the fossil cancrid crabs of the Pleistocene Higashiyatsu Formation of Japan. Benthos Research, **52**(1): 1-8.

TAKEDA, M., 1997. Deep-sea decapod crustacean fauna of Suruga Bay, central Japan. National Science Museum Monographs, **12**: 229-255, pls. 1-5.

TAKEDA, M. & M. MARUMURA, 1997a. Rare crabs from the west coast of the Kii Peninsula, central Japan (IV). Nankiseibutu, **39**(1): 15-20. [In Japanese with English abstract.]

TAKEDA, M. & M. MARUMURA, 1997b. Two new crabs of the family Xanthidae from Japan. Bulletin of the National Science Museum, Tokyo, (A) **23**(2): 97-106.

TAKEDA, M. & P. K. L. NG, 1997. The Indo-Pacific Pilumnidae XI. An unusual new species of *Pilumnus* (Crustacea, Decapoda, Brachyura) from Japan. Bulletin of the National Science Museum, Tokyo, (A) **23**(4): 185-190.

1998

ABE, W., K. UTSUGI & M. TAKEDA, 1998. *Pseudechiniscus asper*, a new Tardigrada (Heterotardigrada: Echiniscidae) from Hokkaido, northern Japan. Proceedings of the Biological Society of Washington, **111**(4): 843-848.

MITSUHASHI, M. & M. TAKEDA, 1998. *Coralliocaris taiwanensis* Fujino & Miyake, 1972 (Crustacea, Decapoda, Natantia, Palaemonidae), a pontoniine shrimp associated with coral, as a senior synonym of *C. pavonae* Bruce, 1972. Bulletin of the National Science Museum, Tokyo, (A) **24**(1): 1-6.

NG, P. K. L. & M. TAKEDA, 1998. *Aethra edentata* Edmondson, 1951 (Crustacea, Decapoda, Brachyura, Parthenopidae), new records for the Izu-Mariana Arch. Bulletin of the National Science Museum, Tokyo, (A) **24**(1): 7-10.

TAKEDA, M., 1998. Crabs collected from submarine caves in the Palau Islands. Natural Environmental Science Research, **11**: 43-47.

TAKEDA, M. & T. KUBODERA, 1998. A small collection of crabs from the East China Sea. Memoirs of the National Science Museum, Tokyo, **31**: 211-222.

YONEYAMA, S. & M. TAKEDA, 1998. Phyllosoma and nisto stage larvae of slipper lobster, *Parribacus*, from the Izu-Kazan Islands, southern Japan. Bulletin of the National Science Museum, Tokyo, (A) **24**(3): 167-175.

1999

KO, H. S. & M. TAKEDA, 1999. New records of three xanthid crabs (Decapoda: Brachyura: Xanthidae) in Korea. Korean Journal of Systematic Zoology, **15**(1): 75-82.

KOMATSU, H. & M. TAKEDA, 1999. A new leucosiid crab of the genus *Nursia* from the Ryukyu Islands. Bulletin of the National Science Museum, Tokyo, (A) **25**(1): 59-64.

NG, P. K. L. & M. TAKEDA, 1999. Notes on the taxonomy of two species of swimming crabs, *Charybdis smithii* McLeay, 1838, and *C. omanensis* Leene, 1938 (Crustacea: Decapoda: Brachyura: Portunidae) from the Indian Ocean. Journal of South Asian Natural History, **4**(1): 71-81.

TAKEDA, M. & S. GAMO, 1999. Record of a penicillate shore crab, *Hemigrapsus penicillatus* (de Haan), having a secondary individual on the carapace. Bulletin of the National Science Museum, Tokyo, (A) **25**(3): 225-228.

2000

ABE, W., M. ITO & M. TAKEDA, 2000. First record of *Echiniscus hoonsooi* (Tardigrada: Echiniscidae) from Japan. Species Diversity, **5**(2): 103-110.

ABE, W. & M. TAKEDA, 2000a. A new *Calcarobiotus* (Tardigrada: Macrobiotidae) from the Imperial Palace of Japan. Zoological Science, **17**(2): 259-263.

ABE, W. & M. TAKEDA, 2000b. A new record of *Cornechiniscus madagascarensis* Maucci, 1993 (Tardigrada: Echiniscidae) from India. Proceedings of the Biological Society of Washington, **113**(2): 480-485.

ABE, W. & M. TAKEDA, 2000c. Tardigrades from the Imperial Palace, Tokyo. Memoirs of the National Science Museum, Tokyo, **35**: 165-177.

CHOW, S., M. OKAZAKI, M. TAKEDA & T. KUBOTA, 2000. A rare abyssal shrimp, *Galatheocaris abyssalis*, found in the stomach of a lancetfish. Crustaceana, **73**(2): 243-246.

KO, H. S. & M. TAKEDA, 2000. New records of three xanthoid crabs (Decapoda, Brachyura) collected from Chejudo Island in Korea. Korean Journal of Systematic Zoology, **16**(1): 31-37.

KOMATSU, H. & M. TAKEDA, 2000. Leucosiid crabs (Crustacea: Decapoda: Brachyura) from the Osumi Islands, southwest Japan, with description of a new species of *Cryptocnemus*. Species Diversity, **5**(3): 267-283.

MITSUI, Y. & M. TAKEDA, 2000. *Eudorella truncatula* (Bate, 1856), a cumacean Crustacea new to Japan. Memoirs of the National Science Museum, Tokyo, **33**: 127-133.

RAHAYU, D. L. & M. TAKEDA, 2000. A new species of the genus *Clistocoeloma* (Crustacea: Decapoda: Grapsidae) from Irian Jaya, Indonesia. Bulletin of the National Science Museum, Tokyo, (A) **26**(2): 35-41.

TAKEDA, M., 2000a. Two species of crabs from an anchialine cave in Balicasag Island, the Philippines. Omon Ronso, **50**: 59-66.

TAKEDA, M., 2000b. Macrurans from the inside moats of the Imperial Palace, Tokyo. Memoirs of the National Science Museum, Tokyo, **35**: 69-74. [In Japanese with English abstract.]

TAKEDA, M., J. HASHIMOTO & S. OHTA, 2000. A new species of the family Bythograeidae (Crustacea, Decapoda, Brachyura) from the hydrothermal vents along volcanic front of the Philippine Sea Plate. Bulletin of the National Science Museum, Tokyo, (A) **26**(4): 159-172.

TAKEDA, M. & M. R. MANUEL, 2000. Taxonomic accounts of some rare crabs new to the Philippines. In: K. MATSUURA (ed.), Proceedings of the First and Second Symposia on Collection Building and Natural History Studies in Asia. National Science Museum Monographs, **18**: 149-162.

TAKEDA, M. & M. MARUMURA, 2000. Rare crabs from the west coast of the Kii Peninsula, central Japan (V). Nankiseibutu, **42**(1): 35-38. [In Japanese with English abstract.]

TAKEDA, M. & PRINCE MASAHITO, 2000. Systematic notes on the pinnotherid crabs of the genus *Pinnaxodes*. Bulletin of the National Science Museum, Tokyo, (A) **26**(3): 99-112.

TAKEDA, M., K. MATSUURA, S. NOMURA, M. OWADA, M. TOMOKUNI & A. SHINOHARA, 2000. Flora and fauna of the Imperial Palace, Tokyo. II. Fauna, excluding insects. Memoirs of the National Science Museum, Tokyo, **35**: 1-5. [In Japanese.]

TAKEDA, M., M. OWADA, M. TOMOKUNI, A. SHINOHARA, S. NOMURA & K. MATSUURA, 2000. Flora and fauna of the Imperial Palace, Tokyo. III. Insect fauna. Memoirs of the National Science Museum, Tokyo, **36**: 1-5. [In Japanese.]

TAKEDA, M., D. L. RAHAYU & I. ASWANDY, 2000. Prawns and crabs. In: K. MATSUURA, O. KURNAEN SUMADHIHARGA & K. TSUKAMOTO (eds.), Field Guide to Lombok Island: Identification Guide to Marine Organisms in Seagrass Beds of Lombok Island, Indonesia: 54-96. (Ocean Research Institute, University of Tokyo, Tokyo).

TAKEDA, M., K. SAKAI, S. SHINOMIYA & H. SUGA, 2000. Crabs from the Seto Inland Sea, southern Japan. Memoirs of the National Science Museum, Tokyo, **33**: 135-144. [In Japanese with English abstract.]

TAKEDA, M., T. SHIMAZU, S. URAWA, J. ARAKI, T. KURAMOCHI & M. MACHIDA, 2000. Ectoparasitic crustaceans of shrimps and fishes from the inside moats the Imperial Palace, Tokyo. Memoirs of the National Science Museum, Tokyo, **35**: 75-78. [In Japanese with English abstract.]

TANAKA, M. & M. TAKEDA, 2000. Protozoans, rotifers, cladocerans and copepods from the inside moats of the Imperial Palace, Tokyo. Memoirs of the National Science Museum, Tokyo, **35**: 234-244, pls. 1-16. [In Japanese with English abstract.]

2001

IWASAKI, N., M. TAKEDA, S. NISHIDA, M. MURANO, C. VALLET & M. IWAI, 2001. Bathybenthic fauna of the Nankai Trough off Muroto observed using a deep seafloor observatory 2. Crustacea. JAMSTEC Journal of Deep Sea Research, **18**: 217-224. [In Japanese with English abstract.]

KOMATSU, H. & M. TAKEDA, 2001a. On *Merocryptoides*, a leucosiid crab genus (Crustacea: Decapoda: Brachyura) endemic to Japan, with description of two new species. Zoological Science, **18**(7): 993-1002.

KOMATSU, H. & M. TAKEDA, 2001b. A new leucosiid crab of the genus *Nursia* Leach, 1817 from Vietnam (Crustacea: Decapoda: Brachyura), with redescription of *N. mimetica* Nobili, 1906. Proceedings of the Biological Society of Washington, **114**(3): 599-604.

MITSUHASHI, M., T. FUJINO & M. TAKEDA, 2001. A new pontoniine shrimp of the genus *Coralliocaris* Stimpson, 1860 (Crustacea: Decapoda: Palaemonidae) from the Ryukyu Islands. Proceedings of the Biological Society of Washington, **114**(4): 944-950.

SAITO, H., T. FUJITA, H. SAKAJI & M. TAKEDA, 2001. An outline of "Research on Deep-sea Fauna and Pollutants in Tosa Bay, 1997-2000". National Science Museum Monographs, **20**: 1-9.

SAITO, T., I. UCHIDA & M. TAKEDA, 2001. Pair formation in *Spongicola japonica* (Crustacea: Stenopodidea: Spongicolidae), a shrimp associated with deep-sea hexactinellid sponges. Journal of the Marine Biological Association of the United Kingdom, **81**(5): 789-797.

TAKEDA, M., 2001a. Annotated list of crabs from Tosa Bay, southwest Japan, collected by the R/V *Kotaka Maru* during the years 1997-2000. In: T. FUJITA, H. SAITO & M. TAKEDA (eds.), Deep-Sea Fauna and Pollutants in Tosa Bay. National Science Museum Monographs, **20**: 217-262.

TAKEDA, M., 2001b. On two estuarine crabs from Hainan Island, South China Sea. In: K. MATSUURA (ed.), Marine Fauna of the Shallow Waters around Hainan Island, South China Sea. National Science Museum Monographs, **21**: 87-93.

TAKEDA, M., T. KURAMOCHI, I. NISHIUMI, S. AKIYAMA, M. HIGUCHI, K. YOKOYAMA, K. UEMURA & Y. MIZOGUCHI, 2001. Preface to the natural history researches of the Kanto Plain, its northern mountains and their adjoining areas. Memoirs of the National Science Museum, Tokyo, **37**: 1-6.

TAKEDA, M. & P. K. L. NG, 2001. The freshwater crab fauna (Crustacea, Brachyura) of the Philippines. VI. A new cavernicolous crab from Mindanao. Zoological Science, **18**(8): 1123-1127.

2002

ITANI, G., P. J. F. DAVIE & M. TAKEDA, 2002. Taxonomic notes on *Acmaeopleura balssi* Shen, 1932 and *A. toriumii* Takeda, 1974 (Crustacea, Brachyura, Grapsidae) from Japanese waters. Bulletin of the National Science Museum, Tokyo, (A) **28**(1): 43-50.

KOMAI, T. & M. TAKEDA, 2002. A new deep-water shrimp of the genus *Pandalopsis* (Decapoda, Caridea, Pandalidae) from Sagami Bay, Japan. Bulletin of the National Science Museum, Tokyo, (A) **28**(2): 91-100.

SAITO, T., I. UCHIDA & M. TAKEDA, 2002. Skeletal growth of the deep-sea hexactinellid sponge *Euplectella oweni*, and host selection by the symbiotic shrimp *Spongicola japonica* (Crustacea: Decapoda: Spongicolidae). Journal of Zoology, **258**(4): 521-529.

TAKEDA, M. & M. MARUMURA, 2002. The genus *Pseudactea* Serène, 1962 (Crustacea, Decapoda), with description of a new species from central Japan. Bulletin of the National Science Museum, Tokyo, (A) **28**(2): 101-107.

TAKEDA, M. & R. UESHIMA, 2002. Shallow-water crabs from western part of the Sagami Sea, central Japan. Memoirs of the National Science Museum, Tokyo, **38**: 161-168.

2003

KOMATSU, H. & M. TAKEDA, 2003a. A new genus of leucosiid crabs (Crustacea, Decapoda, Brachyura) from the Red Sea. Zoosystema, **25**(3): 413-423.

KOMATSU, H. & M. TAKEDA, 2003b. First zoea of the deep-water spider crab, *Rochinia debilis* Rathbun, 1932 (Crustacea, Decapoda, Majidae) from Sagami Bay, central Japan. Bulletin of the National Science Museum, Tokyo, (A) **29**(4): 197-203.

NG, P. K. L. & M. TAKEDA, 2003. *Atoportunus*, a remarkable new genus of cryptic swimming crab (Crustacea; Decapoda; Brachyura: Portunidae), with descriptions of two new species from the Indo-West Pacific. Micronesica, **35-36**: 417-430.

SAITO, T. & M. TAKEDA, 2003. Phylogeny of the family Spongicolidae (Crustacea: Stenopodidea): evolutionary trend from shallow-water free-living to deep-water sponge-associated habitat. Journal of the Marine Biological Association of the United Kingdom, **83**(1): 119-131.

TAKEDA, M., 2003. *Atoportunus dolichopus*, a new cavernicolous crab of the family Portunidae (Crustacea: Decapoda) from the Ryukyu Islands. Bulletin of the National Science Museum, Tokyo, (A) **29**(3): 141-146.

TAKEDA, M. & M. R. MANUEL, 2003. A new crab of the family Pilumnidae, *Viaderiana kasei* from the Philippines. Bulletin of the National Science Museum, Tokyo, (A) **29**(4): 205-211.

2004

GALIL, B. S. & M. TAKEDA, 2004. On a collection of *Mursia* (Crustacea, Decapoda, Brachyura, Calappidae) from Balicasag Island, Philippines. Bulletin of the National Science Museum, Tokyo, (A) **30**(1): 23-35.

HOSHINO, S., M. TAKEDA & Y. WATANABE, 2004. Systematic status of *Halichondria japonica* (Kadota) (Demospongiae, Halichondrida) from Japan. Bollettino dei Musei e degli Istituti biologici dell' Universita di Genova, **68**: 373-379.

ISE, Y., M. TAKEDA & Y. WATANABE, 2004. Psammobiontic Clionaidae (Demospongiae: Hydromerida) in lagoons of the Ryukyu Islands, southwestern Japan. Bollettino dei Musei e degli Istituti biologici dell' Universita di Genova, **68**: 381-389.

KOMAI, T. & M. TAKEDA, 2004a. A new hippolytid shrimp of the genus *Lebbeus* White, 1847 from the Sagami-Nada Sea, central Japan, with further records of two little-known species (Crustacea: Decapoda: Caridea). Bulletin of the National Science Museum, Tokyo, (A) **30**(2): 77-86.

KOMAI, T. & M. TAKEDA, 2004b. Two new deepwater species of hermit crabs (Crustacea: Decapoda: Anomura: Paguroidea) from Japan. Bulletin of the National Science Museum, Tokyo, (A) **30**(3): 113-127.

KOMATSU, H., M. R. MANUEL & M. TAKEDA, 2004. Some rare leucosiid crabs (Crustacea, Decapoda, Brachyura) from the Philippines, with description of a new species of the genus *Arcania*. Biogeography, **6**: 75-86.

KOMATSU, H. & M. TAKEDA, 2004. Two new species of the genus *Goneplax* (Decapoda, Brachyura, Goneplacidae) from East Asia. Crustaceana, **76**(10): 1243-1256.

MARUMURA, M. & M. TAKEDA, 2004. Taxonomic studies on the crabs of the Nagai Collection preserved in the Wakayama Prefectural Museum of Natural History. I. Species new to Japan (1). Nankiseibutu, **46**(2): 93-99. [In Japanese with English abstract.]

OKUNO, J., M. TAKEDA & T. KASE, 2004. Additional specimens and range extension of *Brachycarpus crosnieri* Bruce, 1998 (Decapoda, Caridea, Palaemonidae). Crustaceana, **76**(9): 1143-1146.

OSAWA, M., M. MITSUHASHI & M. TAKEDA, 2004. Discovery of the type specimen of *Lysiosquilla crassispinosa* Fukuda, 1909 (Crustacea: Stomatopoda) and some Fukuda's stomatopod specimens. Benthos Research, **59**(1): 1-10.

OSAWA, M. & M. TAKEDA, 2004. Hermit crabs (Crustacea: Decapoda: Anomura: Paguroidea) from submarine caves in the Ryukyu Islands, south-western Japan. Journal of Natural History, **38**(9): 1097-1132.

TAKEDA, M. & S. NAGAI, 2004. Four species of giant crustaceans from the Indonesian depths, with description of a new species of the family Lithodidae. Bulletin of the National Science Museum, Tokyo, (A) **30**(1): 9-21.

TAKEDA, M. & H. WATABE, 2004. Deepwater carcinoplacine crabs of the genus *Mathildella* (Crustacea, Decapoda, Brachyura), with description of a new species from the Kyushu-Palau submarine ridge, southwestern Japan. Bulletin of the National Science Museum, Tokyo, (A) **30**(4): 181-189.

2005

ABE, W. & M. TAKEDA, 2005. Semiterrestrial tardigrades from the Tokiwamatsu Imperial Villa, Tokyo, Japan. Memoirs of the National Science Museum, Tokyo, **39**: 503-510.

IMAI, H. & M. TAKEDA, 2005. A natural hybrid mud crab (Decapoda, Portunidae) from Japan. Journal of Crustacean Biology, **25**(4): 620-624.

KOMAI, T. & M. TAKEDA, 2005. First report of *Cestopagurus* Bouvier (Crustacea: Decapoda: Anomura: Paguridae) from the Pacific Ocean, and the description of a new species. Bulletin of the National Science Museum, Tokyo, (A) **31**(3): 93-104.

KOMATSU, H., M. R. MANUEL & M. TAKEDA, 2005. A small collection of leucosiid crabs (Crustacea: Decapoda: Brachyura) from Balicasag Island, Bohol, Philippines. Species Diversity, **10**(2): 105-123.

KOMATSU, H. & M. TAKEDA, 2005. Two new species of the genus *Heteronucia* (Crustacea: Decapoda: Brachyura: Leucosiidae) from Japan. Proceedings of the Biological Society of Washington, **118**(4): 731-741.

OWADA, M. & M. TAKEDA, 2005. Faunas of the Akasaka Imperial Gardens and the Tokiwamatsu Imperial Villa, Tokyo. Memoirs of the National Science Museum, Tokyo, **39**: 1-5.

SHINOHARA, G., K. HASEGAWA, A. GO, K. NAKAGUCHI, H. HORIKAWA & M. TAKEDA, 2005. An outline of "Research on Deep-sea Fauna and Pollutants in Nansei Islands, 2001-2004". In: K. HASEGAWA, G. SHINOHARA & M. TAKEDA (eds.), Deep-sea fauna and pollutants in Nansei Islands. National Science Museum Monographs, **29**: 1-15.

TAKEDA, M. & B. S. GALIL, 2005. A new crab species of the genus *Mursia* from Japanese waters (Crustacea, Decapoda, Brachyura, Calappidae). In: K. HASEGAWA, G. SHINOHARA & M. TAKEDA (eds.), Deep-sea fauna and pollutants in Nansei Islands. National Science Museum Monographs, **29**: 289-295.

TAKEDA, M. & M. KOIZUMI, 2005. Occurrence of the Chinese mitten crab, *Eriocheir sinensis* H. Milne Edwards, in Tokyo Bay, Japan. Bulletin of the National Science Museum, Tokyo, (A) **31** (1): 21-24.

TAKEDA, M. & H. KOMATSU, 2005. Collections of crabs dredged off Amami-Oshima Island, the northern Ryukyu Islands. In: K. HASEGAWA, G. SHINOHARA & M. TAKEDA (eds.), Deep-sea fauna and pollutants in Nansei Islands. National Science Museum Monographs, **29**: 271-288.

TAKEDA, M. & M. OGINO, 2005. Record of a whale louse, *Cyamus scammoni* Dall (Crustacea: Amphipoda: Cyamidae), from the gray whale strayed into Tokyo Bay, the Pacific coast of Japan. Bulletin of the National Science Museum, Tokyo, (A) **31**(4): 151-156.

TAKEDA, M., H. WATABE & S. OHTA, 2005. Deep-sea crabs collected by the R. V. *Hakuho Maru* during KH-05-01 cruise off the Ryukyu Islands. Bulletin of the National Science Museum, Tokyo, (A) **31**(3): 105-114.

TANAKA, M., M. TAKEDA & M. NAGANO, 2005a. Limnological faunae and florae of the ponds in the Akasaka Imperial Gardens, Tokyo, Japan. Memoirs of the National Science Museum, Tokyo, **39**: 515-529, pls. 1-14. [In Japanese with English abstract.]

TANAKA, M., M. TAKEDA & M. NAGANO, 2005b. Limnological fauna and flora of the garden pond of Prince Akishinonomiya Residence in the Akasaka Imperial Gardens, Tokyo, Japan. Memoirs of the National Science Museum, Tokyo, **39**: 531-540, pls. 1-6. [In Japanese with English abstract.]

TANAKA, M., M. TAKEDA & M. NAGANO, 2005c. Limnological fauna and flora of the Tokiwamatsu Imperial Villa, Tokyo, Japan. Memoirs of the National Science Museum, Tokyo, **39**: 541-550, pls. 1-7. [In Japanese with English abstract.]

WEBBER, R. & M. TAKEDA, 2005. A new spider crab of the genus *Achaeus* (Crustacea, Decapoda, Brachyura) from the Kermadec Islands. Bulletin of the National Science Museum, Tokyo, (A) **31**(2): 45-50.

YUMOTO, K. & M. TAKEDA, 2005. Taxonomical and biogeographical notes on terrestrial isopod crustaceans from Ibaraki Prefecture, central Japan, with keys to the suborders, families, genera, and species. Bulletin of Ibaraki Nature Museum, **8**: 83-113. [In Japanese with English abstract.]

2006

AOKI, M. & M. TAKEDA, 2006. Caprellid amphipods from the coastal waters of Shimoda, Izu Peninsula, central Japan. Memoirs of the National Science Museum, Tokyo, **41**: 65-70. [In Japanese with English abstract.]

IKEDA, H. & M. TAKEDA, 2006. Brachyuran fauna of Hayama at 20-30 m deep, Sagami Bay, central Japan. Memoirs of the National Science Museum, Tokyo, **41**: 173-182. [In Japanese with English abstract.]

KOMAI, T. & M. TAKEDA, 2006. A review of the pagurid hermit crab (Decapoda: Anomura: Paguroidea) fauna of the Sagami Sea, central Japan. Memoirs of the National Science Museum, Tokyo, **41**: 71-144.

MITSUHASHI, M. & M. TAKEDA, 2006. A list of mantis shrimps (Crustacea: Stomatopoda) in the collections of the Department of Zoology, the University Museum, the University of Tokyo. The University Museum, The University of Tokyo, Material Reports, **62**: 107-113.

OKUNO, J. & M. TAKEDA, 2006. Validity of *Palapedia truncatifrons* (Sakai, 1974) (Crustacea, Decapoda, Brachyura, Xanthidae). Bulletin of the National Science Museum, Tokyo, (A) **32**(1): 15-22.

OKUNO, J., M. TAKEDA & M. YOKOTA, 2006. Inshore hermit crabs (Crustacea: Decapoda: Anomura) collected from the Izu Oceanic Park, eastern coast of Izu Peninsula, Japan. Memoirs of the National Science Museum, Tokyo, **41**: 145-171. [In Japanese with English abstract.]

OWADA, M. & M. TAKEDA, 2006. Monitoring survey of fauna of the Imperial Palace, Tokyo. Memoirs of the National Science Museum, Tokyo, **43**: 1-4. [In Japanese with English abstract.]

TAKEDA, M., M. HIGUCHI, S. MATSUMOTO, H. ONO, T. KURAMOCHI & H. NAMIKAWA, 2006. Over view of the survey project "Study on Environmental Changes in the Sagami Sea and Adjacent Coastal Area with Time Serial Comparison of Fauna and Flora" III. Living organism and soil of coastal areas. Memoirs of the National Science Museum, Tokyo, **42**: 1-5. [In Japanese with English abstract.]

TAKEDA, M., T. KOMAI, H. KOMATSU & H. IKEDA, 2006. Crab fauna of the Sagami Sea, Pacific coast of central Japan. Memoirs of the National Science Museum, Tokyo, **41**: 183-208. [In Japanese with English abstract.]

TAKEDA, M. & M. R. MANUEL-SANTOS, 2006. Crabs from Balicasag Island, Bohol, the Philippines: Dromiidae, Dynomenidae, Homolidae, Raninidae, Dorippidae, and Calappidae. Memoirs of the National Science Museum, Tokyo, **44**: 83-104.

TAKEDA, M., H. NAMIKAWA, T. KURAMOCHI, H. ONO, M. HIGUCHI & S. MATSUMOTO, 2006a. Over view of the survey project "Study on Environmental Changes in the Sagami Sea and Adjacent Coastal Area with Time Serial Comparison of Fauna and Flora" I. Marine organism (brown algae and animals (sponges–annelids)). Memoirs of the National Science Museum, Tokyo, **40**: 1-6. [In Japanese with English abstract.]

TAKEDA, M., H. NAMIKAWA, T. KURAMOCHI, H. ONO, M. HIGUCHI & S. MATSUMOTO, 2006b. Over view of the survey project "Study on Environmental Changes in the Sagami Sea and Adjacent Coastal Area with Time Serial Comparison of Fauna and Flora" II. Marine organism (animals (arthropods–chordates)). Memoirs of the National Science Museum, Tokyo, **41**: 1-6. [In Japanese with English abstract.]

TAKEDA, M. & R. UESHIMA, 2006. Crabs preserved in the Department of Zoology, the University Museum, the University of Tokyo. The University Museum, The University of Tokyo, Material Reports, **62**: 61-105.

TAKEDA, M., S. UOCHI & M. KOBAYASHI, 2006. Swarms of a majid crab, *Phalangipus hystrix* (Miers), appeared in Tokyo Bay, central Japan. Journal of Teikyo Heisei University, **18**: 11-19. [In Japanese with English abstract.]

TAKEDA, M. & R. WEBBER, 2006. Crabs from the Kermadec Islands in the South Pacific. In: Y. TOMIDA, T. KUBODERA, S. AKIYAMA & T. KITATAMA (eds.), Proceedings of the Seventh and Eighth Symposia on Collection Building and Natural History Studies in Asia and the Pacific Rim. National Science Museum Monographs, **34**: 191-237.

TANAKA, M., M. TAKEDA & M. NAGANO, 2006. Protozoans, rotifers, cladocerans and copepods from the inside moats of the Imperial Palace, Tokyo. II. Memoirs of the National Science Museum, Tokyo, **43**: 419-439. [In Japanese with English abstract.]

2007

KOMATSU, H. & M. TAKEDA, 2007. First record of leucosiid crabs (Crustacea: Decapoda: Brachyura) from the Kermadec Islands, with description of a new species. Bulletin of the National Museum of Nature and Science, Tokyo, (A) **33**(2): 61-66.

OSAWA, M. & M. TAKEDA, 2007. Deep-sea Galatheidae (Crustacea, Decapoda, Anomura) from Tosa Bay and Okinawa Trough, southern Japan. Bulletin of the National Museum of Nature and Science, Tokyo, (A) **33**(4): 133-146.

TAKEDA, M., 2007. A small collection of crabs from shallow water off Amami-Oshima Island, northern Ryukyu Islands, collected by the R/V Tansei Maru (KT-02-24 cruise). Journal of Teikyo Heisei University, **19**: 1-13.

TAKEDA, M. & S. BUSSARAWIT, 2007. A new species of the genus *Paralomis* White, 1856 (Crustacea, Decapoda, Anomura, Lithodidae) from the Andamann Sea. Bulletin of the National Museum of Nature and Science, Tokyo, (A) **33**(2): 51-59.

TAKEDA, M. & R. WEBBER, 2007. Two species of the family Parthenopidae (Crustacea, Decapoda, Brachyura) from the Kermadec Islands in the South Pacific. Bulletin of the National Museum of Nature and Science, Tokyo, (A) **33**(4): 147-152.

2008

MITSUHASHI, M. & M. TAKEDA, 2008. Identity of the coral-associated pontoniine shrimp species, *Coralliocaris nudirostris* (Heller, 1861) and *C. venusta* Kemp, 1922 (Crustacea: Decapoda: Palaemonidae), with descriptions of two new species. Zootaxa, **1703**: 1-24.

2009

KOMATSU, H. & M. TAKEDA, 2009a. A new species of the genus *Pinnixa* (Decapoda: Brachyura: Pinnotheridae) from Otsuchi Bay, northeastern Japan. Bulletin of the National Museum of Nature and Science, Tokyo, (A) (Supplement) **3**: 199-204.

KOMATSU, H. & M. TAKEDA, 2009b. Rare crabs (Crustacea, Decapoda, Brachyura) from Okinawa Island, with description of a new species of the family Leucosiidae. Bulletin of the National Museum of Nature and Science, Tokyo, (A) **35**(2): 125-136.

MARUMURA, M. & M. TAKEDA, 2009. Taxonomic studies on the crabs of the Nagai Collection preserved in the Wakayama Prefectural Museum of Natural History. II. Species new to Japan (2). Nankiseibutu, **51**(2): 75-80. [In Japanese with English abstract.]

TAKEDA, M., 2009. A new spider crab of the genus *Rochinia* (Decapoda: Brachyura: Epialtidae) from the Izu Islands, central Japan. Bulletin of the National Museum of Nature and Science, Tokyo, (A) (Supplement) **3**: 167-173.

TAKEDA, M. & S. NAGAI, 2009. New material of two rare species of the leucosiid crab genus *Ixa* Leach (Crustacea: Decapoda: Brachyura) from the West Pacific. Journal of Teikyo Heisei University, **20**(1): 203-207.

TAKEDA, M., H. SUGIYAMA & B.-Z. QIAN, 2009. Records of three species of freshwater crabs from China. Journal of Teikyo Heisei University, **20**(1): 193-202.

2010

NARUSE, T. & M. TAKEDA, 2010. *Notonyx angulatus*, a new species of goneplacid crab (Crustacea, Decapoda, Brachyura) from Amami-Oshima, the Ryukyu Islands, Japan. Bulletin of the National Museum of Nature and Science, Tokyo, (A) **36**(3): 71-74.

NG, P. K. L. & M. TAKEDA, 2010. *Zehntneriana*, a replacement name for *Zehntneria* Takeda, 1972 (Crustacea, Brachyura, Pilumnidae), preoccupied by *Zehntneria* Brunner von Wattenwyl, 1907 (Insecta, Orthoptera, Phasmidae). Bulletin of the National Museum of Nature and Science, Tokyo, (A) **36**(2): 49-50.

TAKEDA, M., 2010. A new swimming crab (Crustacea, Decapoda, Brachyura, Portunidae) from a submarine cave in the Philippines. Bulletin of the National Museum of Nature and Science, Tokyo, (A) **36**(4): 107-113.

TAKEDA, M. & H. KOMATSU, 2010a. Two new species of the family Goneplacidae (Decapoda, Brachyura) from the West Pacific. In: P. CASTRO, P. J. F. DAVIE, P. K. L. NG & B. RICHER DE FORGES (eds.), Studies on Brachyrua: a Homage to Danièle Guinot. Crustaceana Monographs, **13**: 319-328. (Brill, Leiden).

TAKEDA, M. & H. KOMATSU, 2010b. A new xanthid crab (Decapoda, Brachyura) from a submarine cave in the Philippines. In: C. H. J. M. FRANSEN, S. DE GRAVE & P. K. L. NG (eds.), Studies on Malacostraca: Lipke Bijdeley Holthuis Memorial Volume. Crustaceana Monographs, **14**: 677-683. (Brill, Leiden).

TAKEDA, M. & M. MARUMURA, 2010. Spider crabs of the genus *Huenia* de Haan, 1837 (Crustacea, Decapoda, Brachyura) from Japan, with descriptions of two new species. Bulletin of the National Museum of Nature and Science, Tokyo, (A) **36**(2): 39-48.

TAKEDA, M., H. SUGIYAMA & A. RANGSIRUJI, 2010. Freshwater crabs from Surat Thani, Peninsula Thailand, as intermediate host of lung flukes. Journal of Teikyo Heisei University, **21**(1): 149-158.

2011

KOMATSU, H. & M. TAKEDA, 2011. A new species of the genus *Meractaea* (Crustacea, Decapoda, Xanthidae) from Japan. Bulletin of the National Museum of Nature and Science, Tokyo, (A) **37**(2): 79-85.

TAKEDA, M., S. FURUTA, T. MIYANAGA, A. TAMURA & T. WADA, 2011. Crabs from the southwestern Sea of Japan along Tottori Prefecture and its vicinity, Japan. Bulletin of the Tottori Prefectural Museum, **48**: 29-94. [In Japanese with English abstract.]

TAKEDA, M., S. HABE & M. KUBOTA, 2011. Records of some freshwater crabs from Laos and Cambodia. Journal of Teikyo Heisei University, **22**(1): 205-226.

TAKEDA, M. & H. KOMATSU, 2011. A new genus and species of euxanthine crab (Crustacea: Decapoda: Xanthidae) from submarine banks off the Izu Islands, central Japan. Bulletin of the National Museum of Nature and Science, Tokyo, (A) (Supplement) **5**: 119-126.

FIRST *VIRIDOTHERES* MANNING, 1996, FROM JAPAN, WITH A KEY TO THE SPECIES (DECAPODA, BRACHYURA, PINNOTHERIDAE)

BY

SHANE T. AHYONG[1,4]), TOMOYUKI KOMAI[2]) and TETSUYA WATANABE[3])

[1]) Australian Museum, 6 College St., Sydney, NSW 2010, Australia

[2]) Natural History Museum and Institute, Chiba, Chiba 260-8682, Japan

[3]) Osaka Museum of Natural History, Osaka, Osaka 546-0034, Japan

ABSTRACT

Viridotheres takedai sp. nov., the seventh species of the genus, is described from Japan. It is the fourth species known from the Indo-West Pacific and the first to be recorded from Japanese waters. *Viridotheres takedai* is also the first pinnotherid to be recorded from *Nipponoclava gigantea* (Sowerby), a tube dwelling bivalve. An unusual feature of adult male *V. takedai* is the well-developed exopod on gonopod 2, recognised here for the first time from a brachyuran as a normal feature (rather than the result of a developmental abnormality). The gonopod 2 exopod is absent in the juvenile male, so its presence in adult *V. takedai* may be derived feature. We also report the occurrence of well-developed gonopod 2 exopods in male *Nepinnotheres affinis* (Bürger, 1895), *N. androgynus* Manning, 1993, and the hexapodid *Thaumastoplax anomalipes* Miers, 1881. Most pinnotherid species are known at present only from females and when males of these species are discovered, the phenomenon may prove more widespread than currently known. A key to the species of *Viridotheres* Manning, 1996 is provided.

INTRODUCTION

Viridotheres was established by Manning (1996) for three West African species (two previously assigned to *Nepinnotheres* Manning, 1993) sharing a maxilliped 3 palp in which the dactylus articulates near the midlength of the flexor margin of the propodus, the walking legs are left-right symmetrical, and in which the second walking leg (pereiopod 3) of females not only had the longest dactylus, but was longest leg overall. Six species of *Viridotheres* are presently recognised: three from the eastern Atlantic: *V. marionae* Manning,

[4]) Corresponding author; e-mail: Shane.Ahyong@austmus.gov.au

1996 (type species; Cape Verde Islands) and *V. lillyae* (Manning, 1993) (Ivory Coast), and *V. viridis* (Manning, 1993) (Cape Verde Islands); and three from the Indo-West Pacific: *V. gracilis* (Bürger, 1895) (Philippines), and *V. burgeri* (Rathbun, 1909) (Thailand) and *V. otto* Ahyong & Ng, 2007 (Philippines). Recently, an undescribed species of *Viridotheres* was discovered associated with the tube dwelling bivalve, *Nipponoclava gigantea* (Sowerby) from the Ariake and Yatsushiro seas in Kyushu, Japan. The new species, representing the seventh known species of *Viridotheres*, is described herein.

Morphological terminology generally follows Ahyong & Ng (2007). Measurements are in millimetres (mm). Carapace length (cl) is measured along the dorsal midline. Carapace width (cw) is the greatest width. Specimens studied are deposited in the Australian Museum, Sydney (AM), Natural History Museum and Institute, Chiba, Japan (CBM) and Senckenberg Museum, Frankfurt-am-Main, Germany (SMF).

TAXONOMY

Family PINNOTHERIDAE De Haan, 1833
Genus *Viridotheres* Manning, 1996

Viridotheres takedai sp. nov.
[New Japanese name: Deguchi-tsutsugaki-pinno]
(figs. 1-3)

Type material. — Holotype: CBM-ZC 10487, female (cl 12.8 mm, cw 14.3 mm), 2.2 km north-north-west of Yushima, Shimabara Bay, Ariake Sea, Kyushu, 100 m, commercial gill net, associated with *Nipponoclava gigantea*, coll. N. Deguchi, 16 December 2007.

Paratypes: CBM-ZC 10488, 1 male (cl 8.5 mm, cw 9.4 mm), collected with holotype; AM P87160, 1 male (cl 6.5 mm, cw 7.2 mm), 1 female (cl 10.4 mm, cw 11.7 mm), 2.2 km north-north-west of Yushima, Shimabara Bay, Ariake Sea, Kyushu, 100 m, commercial gill net, associated with *Nipponoclava gigantea*, coll. N. Deguchi, 2 February 2008; CBM-ZC 10535, 1 male (cl 6.6 mm, cw 7.6 mm), 1 ovigerous female (cl 10.4 mm, cw 12.0 mm), 2.2 km north-north-west of Yushima, Shimabara Bay, Ariake Sea, Kyushu, 100 m, commercial gill net, associated with *Nipponoclava gigantea*, coll. N. Deguchi, 26 May 2008; CBM-ZC 10489, 1 female (cl 10.1 mm, cw 12.0 mm), off Goshoura, Yatsushiro Sea, Kyushu, coll. H. Shimizu, 15 February 2007; CBM-ZC 10490, 1 juvenile male (cl 2.7 mm, cw 2.7 mm), 1 ovigerous female (cl 11.3 mm, cw 12.5 mm), off Goshoura, Yatsushiro Sea, Kyushu, coll. H. Shimizu, 27 February 2007.

Diagnosis. — Carapace subcircular, dorsal surface glabrous; front slightly produced, anterior margin transverse. Maxilliped 3 with ischiomerus length less than twice width; inner proximal margin straight; inner distal margin with prominent angle; dactylus apex falling short of propodal apex. Cheliped palm dorsally glabrous. Pereiopods 2-5 dactyli about 0.5 propodi length or less; propodi and carpi sparsely setose or glabrous. Pereiopod 5 merus 0.5 length of pereiopod 3 merus.

Description of adult female. — Carapace subcircular, slightly wider than long. Front slightly produced, anterior margin transverse in dorsal view. Anterolateral margins poorly defined. Dorsum smooth, lateral margins sparsely setose, surface otherwise glabrous; dorsal midline convex in profile.

Antennular sinus larger than orbit; antennules folded slightly obliquely. Antennal articles 1 and 2 fused to epistome. Eyes not visible in dorsal view, filling orbit, cornea pigmented.

Maxilliped 3 with ischiomerus length less than twice width; inner proximal margin straight; inner distal margin with prominent angle, setose; outer margin strongly convex. Carpus shorter than propodus. Propodus spatulate, longer than wide, gently tapering to blunt, setose apex. Dactylus digitiform, distally setose, inserted slightly proximal to propodal midlength, apex falling short of propodal apex. Exopod with straight inner margin and convex outer margin; flagellum 2-articulate, distal article setose.

Chelipeds symmetrical from left to right. Dactylus and pollex relatively straight, crossing distally. Dactylus occlusal margin with blunt proximal tooth, otherwise smooth. Pollex occlusal margin with low crenulate prominence near midlength, otherwise smooth; with fringe of short setae on inner ventral margin. Propodus dorsal margin 2.2 times height, about twice length of dactylus; ventral margin sinuous. Carpus and merus unarmed, inner proximal margins setose.

Pereopods 2-5 (walking legs 1-4) similar and symmetrical from left to right, with few scattered setae; relative lengths in decreasing order pereiopod 3 > pereiopod 2 > pereiopod 4 > pereiopod 5. Dactyli distally spiniform; flexor margins with scattered long setae and rows of minute, closely spaced setae; relative dactylus lengths: pereiopod 3 > pereiopod 2 > pereiopod 4 = pereiopod 5; pereiopod 2 and pereiopod 4 dactyli 0.4 propodus length; pereiopod 3 and pereiopod 5 dactylus 0.5 propodus length. Propodi with long scattered setae along flexor margin; with minute, closely spaced setae on disto-flexor margin adjacent to dactyl articulation. Meri distinctly longer and slightly deeper than other articles; pereiopod 5 merus half-length of pereiopod 3 merus; with row of long setae along proximal extensor margin, extending onto ischium.

Abdomen broader than long, covering thorax; comprising 6 free somites and telson, widest at somite 5.

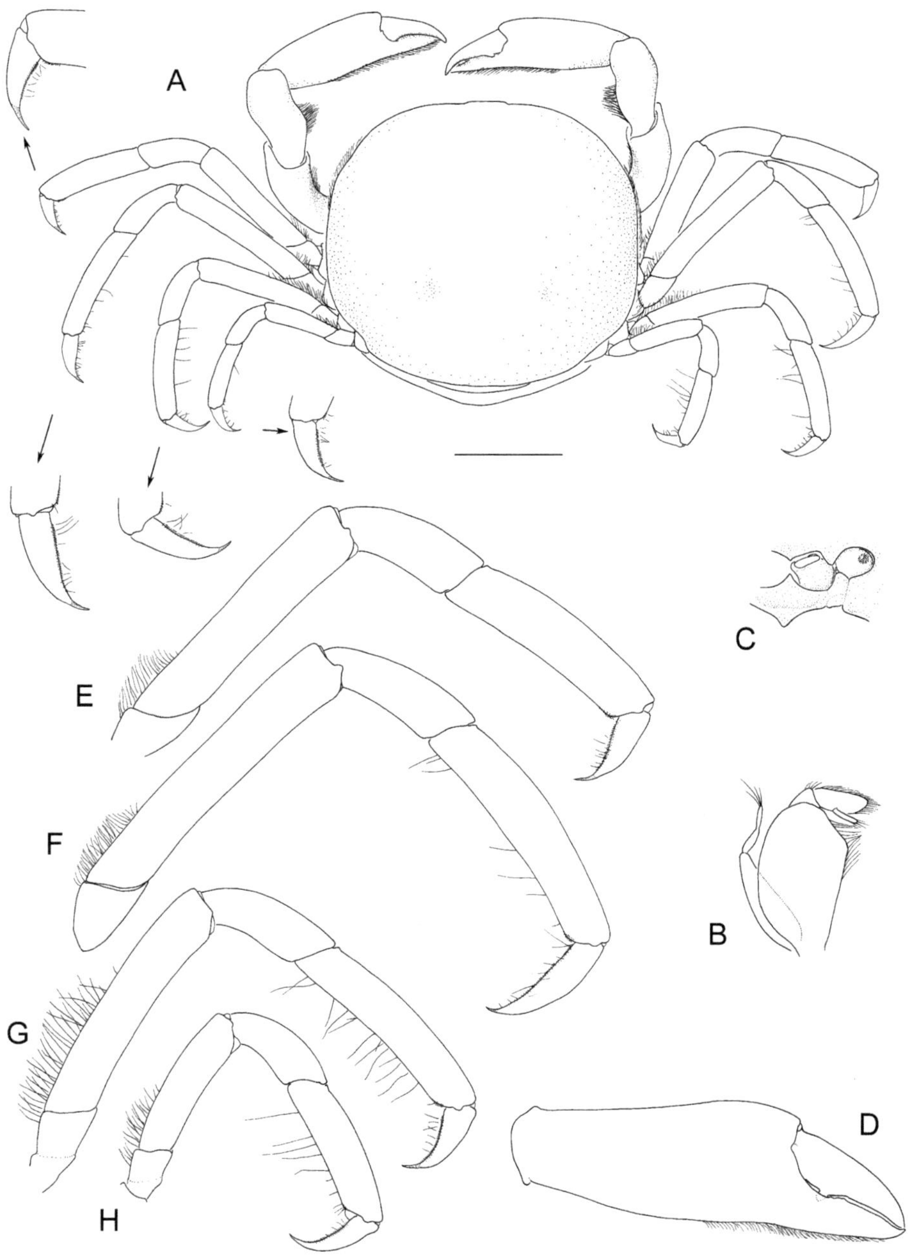

Fig. 1. *Viridotheres takedai* sp. nov., female holotype, cl 12.8 mm, cw 14.3 mm (CBM-ZC 10487). A, dorsal habitus; B, right maxilliped 3; C, left cephalothorax, anterior view; D, right chela; E–H, right pereiopod 2-5. Scale: A = 5.0 mm, B–H = 2.5 mm.

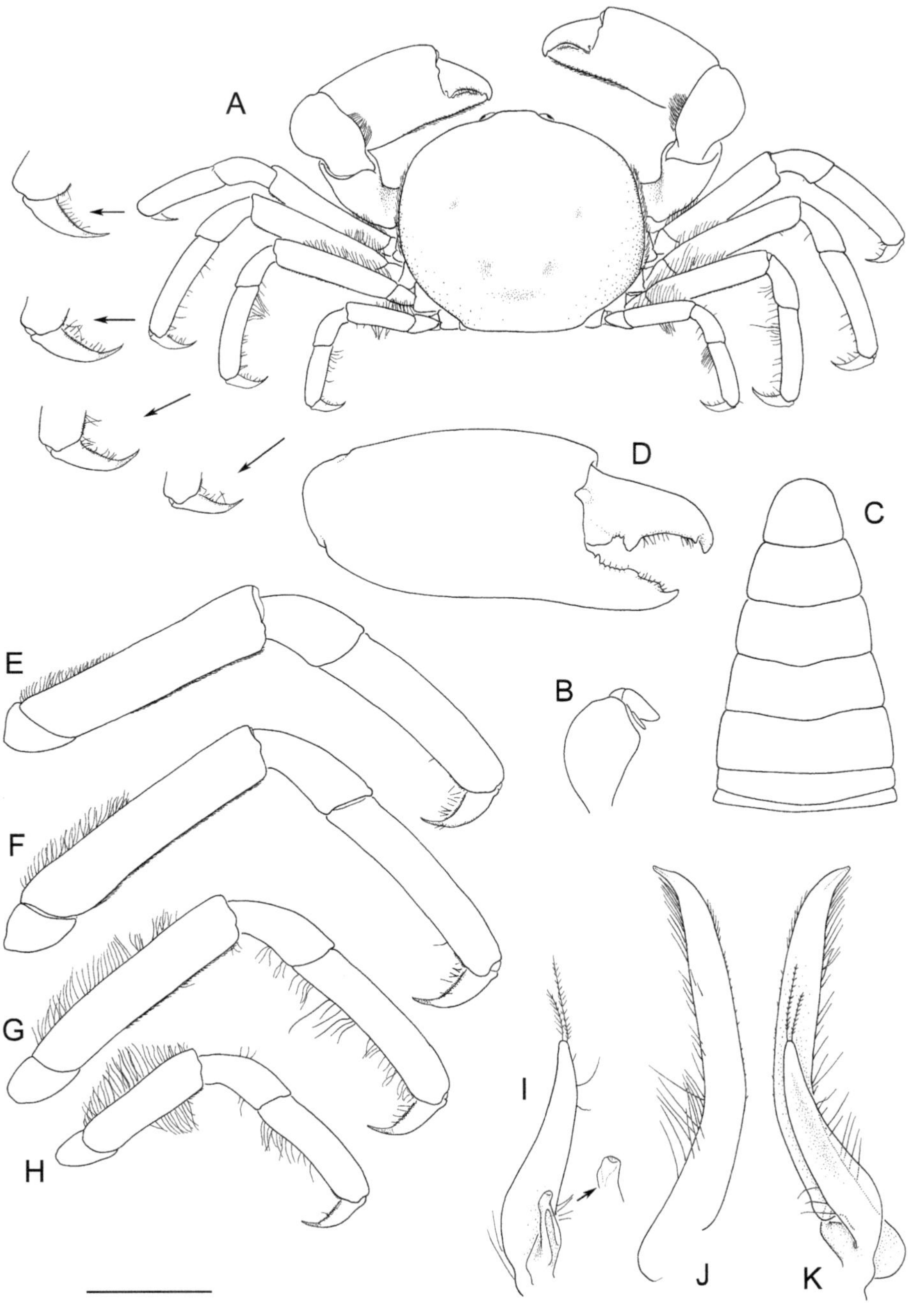

Fig. 2. *Viridotheres takedai* sp. nov., male paratype, cl 8.5 mm, cw 9.4 mm (CBM-ZC 10488). A, dorsal habitus; B, right maxilliped 3; C, abdomen; D, right chela; E–H, right pereiopod 2-5; I, left gonopod 2, sternal view; J, left gonopod 1, sternal view; K, left gonopods 1 and 2, abdominal view. Scale: A = 5.0 mm, B–H = 2.5 mm, I–K = 1.25 mm.

Description of adult male. — Carapace subcircular, slightly wider than long. Front produced, anterior margin transverse in dorsal view. Anterolateral margins poorly defined. Dorsum smooth, lateral margins sparsely setose, surface otherwise glabrous; dorsal midline convex in profile. Antennular sinus larger than orbit; antennules folded slightly obliquely. Antennal articles 1 and 2 fused to epistome. Eyes visible in dorsal view, filling orbit, cornea pigmented.

Maxilliped 3 with ischiomerus length less than twice width; inner proximal margin straight; inner distal margin with prominent angle, setose; outer margin strongly convex. Carpus shorter than propodus. Propodus spatulate, longer than wide, gently tapering to blunt, setose apex. Dactylus digitiform, distally setose, inserted slightly proximal to propodal midlength, apex falling short of propodal apex. Exopod with straight inner margin, outer margin convex; flagellum with 2 articles, distal article setose.

Chelipeds symmetrical from left to right, robust, inflated. Dactylus and pollex relatively straight, crossing distally. Dactylus occlusal margin weakly crenulate, with blunt tooth proximal to midlength and smaller distal tooth proximal to apex. Pollex occlusal margin irregularly crenulated, with low blunt prominence near midlength, a small upright tooth proximally and low angular prominence distally near apex; with fringe of short setae on inner ventral margin. Propodus dorsal margin 1.5 times height, 1.7 times length of dactylus; ventral margin sinuous. Carpus unarmed, inner proximal margin setose.

Pereiopods 2-5 (walking legs 1-4) similar and symmetrical from left to right, with few scattered setae; relative lengths in decreasing order pereiopod 3 > pereiopod 2 > pereiopod 4 > pereiopod 5. Dactyli distally spiniform; flexor margins with scattered long setae and rows of minute, closely spaced setae; relative dactylus lengths: pereiopod 2 = pereiopod 3 = pereiopod 4 > pereiopod 5; pereiopod 2-4 dactyli 0.4 propodus length; pereiopod 5 dactylus 0.5 propodus length. Propodi with minute, closely spaced setae on disto-flexor margin adjacent to dactyl articulation; pereiopod 2-3 propodi with few scattered setae along flexor margins; pereiopod 4-5 propodi with long scattered setae along flexor margin. Meri distinctly longer and slightly deeper than other articles; pereiopod 5 merus half-length of pereiopod 3 merus; with row of long setae along proximal extensor margin on pereiopods 2 and 3, extending full length on pereiopods 4 and 5; flexor margins with fine setae, and additional long setae on pereiopod 5.

Abdomen slender, margins evenly converging posteriorly; comprising 6 free somites and telson, widest at somites 1-3; telson slightly wider than long, apex rounded.

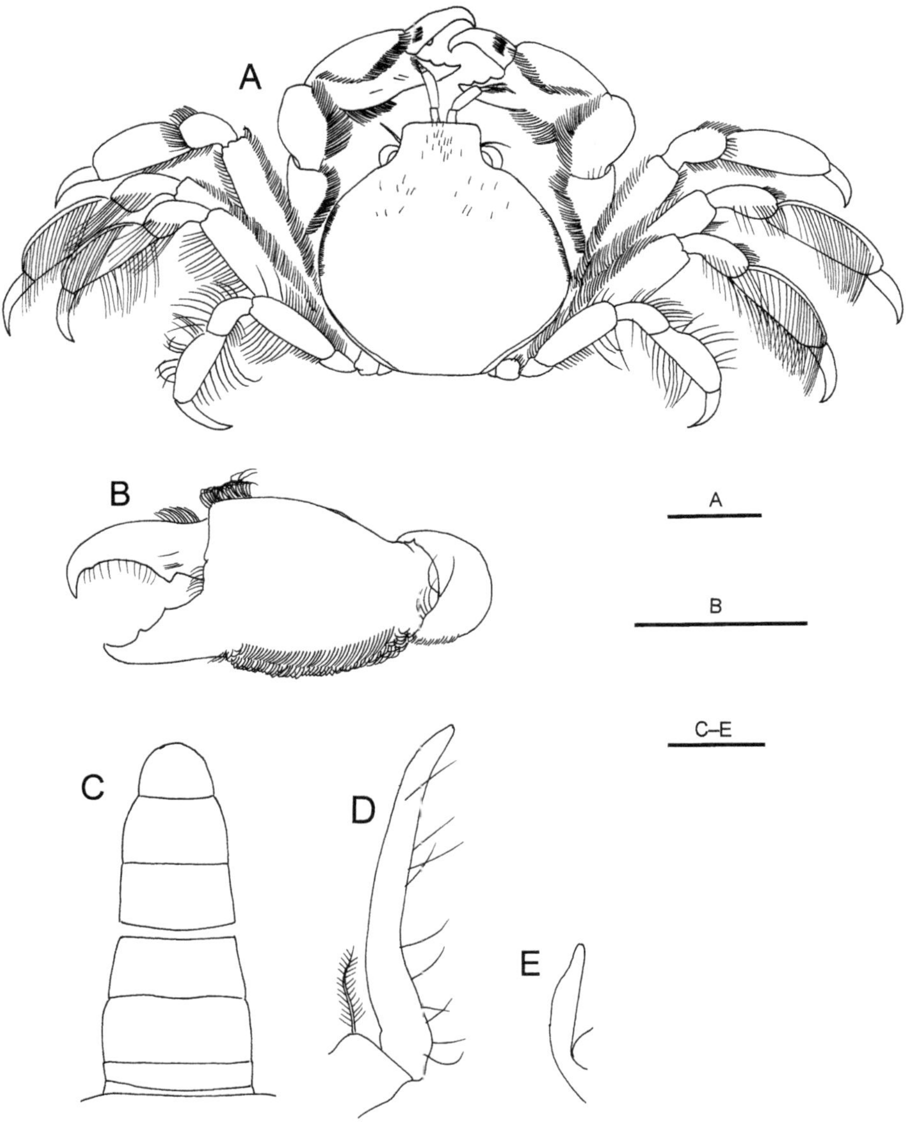

Fig. 3. *Viridotheres takedai* sp. nov., juvenile male paratype, cl 2.7 mm, cw 2.7 mm (CBM-ZC 10490). A, dorsal habitus; B, left chela, outer view; C, abdomen, ventral view; D, left gonopod 1, abdominal view; E, left gonopod 2. Scales: A = 1 mm, B–C = 0.5 mm, D–E = 0.25 mm.

Gonopod 1 elongate, slender, arcuate, dorsoventrally compressed, slightly longer than half length of abdomen, tapering to simple apex; outer margin lined with long setae. Gonopod 2 endopod simple, about 1/4 length of gonopod 1; exopod well-developed, slender, flattened, evenly tapering to rounded apex, almost triple length of endopod; about half length of gonopod 1. Pleopods 3-5 absent.

Description of juvenile male. — Carapace subcircular, as wide as long. Front strongly produced, anterior margin transverse in dorsal view; dorsal surface faintly grooved medially. Anterolateral margins poorly defined. Anteroventral margin (near base of cheliped) with prominent fringe of dense setae directed dorsally. Dorsum smooth, gastric region sparsely setose, surface otherwise glabrous; dorsal midline convex in profile.

Chelipeds symmetrical from left to right, robust, not particularly inflated. Dactylus and pollex with curved tips, crossing distally. Dactylus strongly curved, proximally setose; occlusal margin nearly smooth, with low subdistal swelling and subacute tooth proximal to midlength. Pollex nearly straight except for curved tip, occlusal margin with 2 small triangular teeth. Propodus dorsal margin 1.1 times height and 1.3 times length of dactylus; ventral margin slightly sinuous, with covering of dense setae; inner dorsal margin setose; inner surface with patch of setae near ventral margin proximally. Carpus unarmed, inner proximal and ventral margins with dense setae. Merus with dense setae on dorsal and ventral margins.

Pereiopods 2-5 (walking legs 1-4) similar and symmetrical from left to right; propodi each with dorsal and ventral row of long natatory setae on outer (posterior) surfaces; carpi of same pereiopods each with oblique medial row of natatory setae on outer surfaces. Otherwise similar to those of adult males.

Abdomen similar to that of adult males, but slightly more slender proportionally.

Gonopods not fully differentiated. Gonopod 1 about half as long as abdomen; gently curved outward, gradually tapering to subacute apex; outer margin sparsely setose. Gonopod 2 flattened, tapering to blunt apex, shorter than one-third as long as gonopod 1.

Colouration in life. — Females overall whitish or pale yellowish transparent; brown gonad visible through integument; cornea reddish brown.

Males generally whitish, carapace with symmetrical yellowish brown markings on dorsal surface.

Etymology. — Named in honour of Prof. Masatsune Takeda, in recognition of his major contributions to the taxonomy of Decapoda, especially that of the crabs of Japan.

Distribution. — Presently known only from the Ariake Sea and Yatsushiro Sea off Kyushu, Japan.

Host. — The present seven specimens were all associated with the tube dwelling bivalve, *Nipponoclava gigantea* (Sowerby), representing the first pinnotherids known from this host. Two females, one ovigerous, were found in a narrow space between the Inner demibranch and pallial retractor muscle of the host. Six of seven specimens were found in heterosexual pairs, suggesting that this species lives in pairs.

Remarks. — *Viridotheres* is very similar to *Nepinnotheres*, the two genera being differentiated only by the latter having the pereiopod 3 dactylus as long as, instead of longer than the pereiopod 2 dactylus. On the basis of the longer pereiopod 2 dactylus, the new species is assigned to *Viridotheres*. As observed by Ahyong & Ng (2007), however, the similarities between *Viridotheres* and *Nepinnotheres* are substantial and the length difference between the pereiopod 3 and 4 dactyli may be slight, as in *V. otto* Ahyong & Ng, 2007. Future study may reveal that the two genera are synonymous.

Viridotheres takedai is the seventh known species of the genus and the first to be recorded from Japan. Unfortunately, only *V. lillyae* and *V. takedai* are known from both sexes, so comparisons between congeners are usually based only of females. *Viridotheres takedai* differs from *V. marionae* in the maxilliped 3 ischiomerus having a straight versus concave inner proximal margin and differing proportions (length less than twice width versus length slightly exceeding twice width). From *V. lillyae*, *V. takedai* differs in having proportionally longer walking legs in which the dactyli do not exceed half the propodus length versus two-thirds or more. Additionally, the length of the pereiopod 5 merus of *V. takedai* is half that of pereiopod 3 versus 0.6 or more in *V. lillyae*. From *V. viridis*, *V. takedai* is readily distinguished by the absence of a prominent upright tooth near the midlength of the occlusal margin on the cheliped pollex. *Viridothereres takedai* is readily distinguished from *V. otto* and *V. burgeri* by the setose instead of glabrous walking legs and proportionally longer pereiopod 3 merus, which is twice the length of the pereiopod 5 merus (versus less than twice), and the straight rather than concave inner proximal margin of the maxilliped 3 ischiomerus. *Viridotheres gracilis* differs from *V. takedai*, at least in females, in having densely versus very sparsely setose meri, carpi and propodi of the walking legs, and in having a more elongate ischiomerus of maxilliped 3 (length about twice width versus distinctly less than twice width) in which the inner proximal margin is concave rather than straight.

The juvenile male is similar to the adult males in most major respects, and differs in expected ways given its developmental stage, such as the more pronounced front, incipient rather than distinct subdistal tooth on the occlusal margin of the cheliped dactylus, well-developed natatory setae on the walking legs and incompletely differentiated pleopods 1 and 2. The juvenile male, however, also differs in having setose dorsal margins on the cheliped palm and proximal margin of the cheliped dactyli; these setae are completely absent in adult males. It is noteworthy that the juvenile male appears to be in ‘hard stage’, a stage at which males of other pinnotherids, such as *Juxtafabia muliniarum* (Rathbun, 1918) and *Fabia subquadrata* (Dana, 1851), are sexually mature (Pearce, 1966; Campos, 1993).

Of the local Japanese species, *V. takedai* superficially resembles *Nepinnotheres cardii* (Bürger, 1895) and *Nepinnotheres pholadis* (de Haan, 1835) in the slender, sparsely setose walking legs. In both *N. cardii* and *N. pholadis*, however, the length of the pereiopod 3 dactylus is equal to, rather than longer than that of the pereiopod 2 dactylus. The maxilliped 3 of *V. takedai* differs from that of *N. cardii* in having a straight rather than concave inner proximal margin on the ischiomerus, and differs from that of *N. pholadis* in having a dactylus that does not exceed the propodus, rather than reaching or extending beyond the apex. Male *V. takedai* have two teeth on the occlusal margin of the dactylus and in this respect also resemble males of two East Asian species, *Pinnotheres bidentatus* Sakai, 1939 (Japan) and *Pinnotheres taichungae* Sakai, 2000 (Taiwan). Male *V. takedai* are readily separated from *P. bidentatus* and *P. taichungae* by the shorter dactylus of maxilliped 3, which does not reach the apex of the propodus (versus overreaching the propodus).

An unusual feature of adult male *V. takedai* is the well-developed exopod of gonopod 2, as evident in all adult males examined (fig. 2I). In-situ, with the gonopod 2 endopod inserted into gonopod 1, the gonopod 2 exopod lies flat against the abdominal surface of gonopod 1 (fig. 2K). In early juvenile male brachyurans, the rami of pleopod 2 are well-developed prior to differentiation of pleopods 1 and 2 into gonopods, in which the exopods typically become vestigial or are lost (Shen, 1935; Hebling & Regier, 2003; Guerau & Rottlant, 2009). The gonopod 2 in the juvenile male *V. takedai*, however, is uniramous, suggesting that the well-developed gonopod 2 exopod of adults is not paedomorphic and may be a derived feature.

Unfortunately, few species of Pinnotheridae are known from males, so we take this opportunity to report a similar occurrence in the gonopod 2 of adult male *Nepinnotheres affinis* (Bürger, 1895) (fig. 4). In *N. affinis*, however, the gonopod 2 endopod is about one-quarter the length of gonopod 1 as in *V. takedai*, but the exopod is as long as or slightly shorter than the endopod, rather than distinctly longer. In addition, males of the gynandromorphic *N. androgynus* Manning, 1993, and the hexapodid *Thaumastoplax anomalipes*

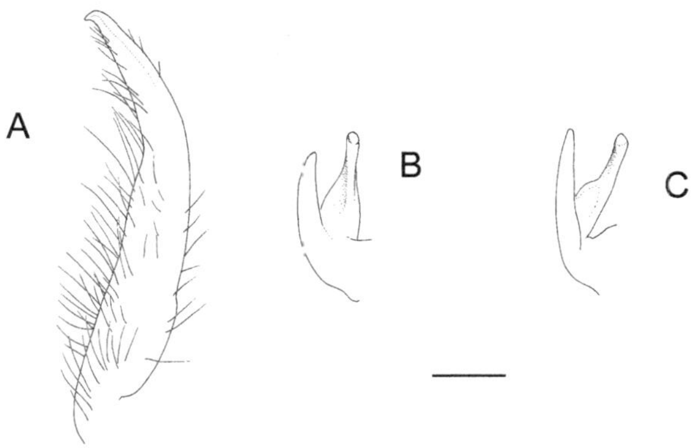

Fig. 4. *Nepinnotheres affinis* (Bürger, 1895). A–B, male paralectotype, Bohol, Philippines, cl 6.3 mm, cw 6.7 mm (SMF-ZMG 171). C, male, Ubay, Philippines, cl 7.2 mm, cw 7.2 mm (SMF-ZMG 175). A, left gonopod 1, sternal view; B–C, left gonopod 2, sternal view. Scale = 0.5 mm.

Miers, 1881, may also have well-developed gonopod 2 exopods if Monod's (1956: figs. 477, 523) and Manning's (1993: fig. 21h) partial figures are correct. Otherwise, only in the case of deformity or developmental abnormality have gonopod 2 exopods been previously recognised, as in the case of a unusual specimen of *Pleistacantha pungens* (Wood-Mason & Alcock, 1891), which, in addition, also bore pleopods 3-5, of which the latter two pairs were left-right asymmetrical (Gordon, 1963). Thus, as far as we are aware, the presence of well-developed gonopod 2 exopods reported here are the first records for normal adult male Brachyura. The gonopod 2, however, is often overlooked in pinnotheroid taxonomy, warranting a more focussed survey of their gonopods (as well as those of many other Brachyura) to determine the extent of exopod development. Most pinnotherid species are known only from females; when males of these species are discovered, the phenomenon may prove more widespread than currently known and the degree of development of gonopod 2 exopod may prove to be taxonomically and phylogenetically useful.

KEY TO SPECIES OF *VIRIDOTHERES* (ADULT FEMALES)

1. Cheliped pollex with prominent, upright tooth near midlength of occlusal margin .... *V. viridis*
– Cheliped pollex with at most a low swelling or obtusely angular projection near midlength of occlusal margin 2
2. Maxilliped 3 ischiomerus broad, length distinctly less than twice width 3
– Maxilliped 3 ischiomerus comparatively narrow, length twice width 4

3. Pereiopod 2-5 dactyli about two-thirds propodi length. Pereiopod 5 merus 0.6 length of pereiopod 3 merus . . . . *V. lillyae*
– Pereiopods 2-5 dactyli 0.5 propodi length or less. Pereiopod 5 merus 0.5 length of pereiopod 3 merus . . . . *V. takedai* sp. nov.
4. Pereiopods 2-5 strongly setose. Pereiopods 3 and 4 dactyli 0.4 propodus length 5
– Pereiopods 2-5 glabrous. Pereiopods 3 and 4 dactyli 0.5-0.6 propodus length 6
5. Dorsal margin of cheliped palm with fine setal patch adjacent to dactyl articulation . . . . *V. gracilis*
– Dorsal margin of cheliped palm glabrous . . . . *V. marionae*
6. Pereiopods 2-5 dactyli not exceeding 0.6 propodus length. Maxilliped 3 dactylus not reaching propodal apex . . . . *V. otto*
– Pereiopods 2-5 dactyli exceeding 0.7 propodus length. Maxilliped 3 dactylus reaching propodal apex . . . . *V. burgeri*

ACKNOWLEDGMENTS

We sincerely thank Mr. Norihiko Deguchi and Dr. Hiroaki Ugai (Goshoura Cretaceous Museum) for collecting and making available specimens of this interesting new species for study. We are grateful to Ernesto Campos and an anonymous reviewer for constructive comments on the manuscript, Michael Türkay and Andreas Allspach for access to the Bürger pinnotherid material, and the editors of this volume for inviting our contribution.

REFERENCES

AHYONG, S. T. & P. K. L. NG, 2007. The pinnotherid type material of Semper (1880), Nauck (1880) and Bürger (1895) (Crustacea: Decapoda: Brachyura). Raffles Bulletin of Zoology, (Supplement) **16**: 191-226.

BÜRGER, O., 1895. Ein Beitrag zur kenntniss der Pinnotherinen. Zoologische Jahrbücher, Abtheilung für Systematik, Geographie und Biologie der Thiere, **8**: 361-390, pls. 9, 10.

CAMPOS, E., 1993. Systematic and taxonomic remarks on *Pinnotheres muliniarum* Rathbun, 1918 (Crustacea: Brachyura: Pinnotheridae). Proceedings of the Biological Society of Washington, **106**(1): 92-101.

DANA, J. D., 1851. On the classification of the Crustacea Grapsoidea. American Journal of Science and Arts, **2**(12): 283-291.

GORDON, I., 1963. An anomalous male spider crab with five pairs of pleopods. Crustaceana, **5**(2): 151-154.

GUERAO, G. & G. ROTLLANT, 2009. Post-larval development and sexual dimorphism of the spider crab *Maja brachydactyla* (Brachyura: Majidae). Scientia Marina, **73**: 797-808.

HAAN, W. DE, 1833-1850. Crustacea. In: P. F. VON SIEBOLD (ed.), Fauna Japonica, sive Descriptio Animalium, quae in Itinerne per Japoniam, Jussu et Auspiciis Superiorum, qui summum in India Batavia Imperium Tenent, Suscepto, Annis 1823-1830 collegit, notis, obervationibus et adumbrationibus illustravit: i–xvii, i–xxxi, ix–xvii, 1-243, pls. A–J, L–Q, 1-55, circ. table 2. (J. Müller et Co., Lugduni Batavorum [= Leiden]).

HEBLING, N. J. & P. J. RIEGER, 2003. Desenvolvimento juvenil de *Hepatus pudibundus* (Herbst) (Crustacea, Decapoda, Calappidae), em laboratório. Revista Brasilieira de Zoologia, **20**: 531-539.

MANNING, R. B., 1993. West African pinnotherid crabs subfamily Pinnotherinae (Crustacea, Decapoda, Brachyura). Bulletin du Muséum national d'Histoire naturelle, Paris (4) **15A**(1-4): 125-177.

— —, 1996. *Viridotheres marionae*, a new genus and species of pinnotherid crab from West Africa (Crustacea: Decapoda: Brachyura). Zoologische Mededelingen, **70**(19): 271-273.

MIERS, E. J., 1881. On a collection of Crustacea made by Baron Hermann Maltzam at Goree Island, Senegambia. Annals and Magazine of Natural History, (5) **8**: 432-434.

MONOD, T., 1956. Hippidea et Brachyura ouest-africains. Mémoires de l'Institut Français d'Afrique noire, **45**: 1-674.

PEARCE, J. B., 1966. The biology of the mussel crab, *Fabia subquadrata*, from the waters of the San Juan Archipelago, Washington. Pacific Science, **20**: 3-35.

RATHBUN, M. J., 1909. New crabs from the Gulf of Siam. Proceedings of the Biological Society of Washington, **22**: 107-114.

— —, 1918. The grapsoid crabs of America. United States National Museum Bulletin, **97**: 1-461.

SAKAI, K., 2000. On the occurrence of three species of crabs on Shikoku Island, Japan, and a new species, *Pinnotheres taichungae* nov. spec., from Taiwan (Decapoda, Brachyura). Crustaceana, **73**(9): 1155-1162.

SAKAI, T., 1939. Studies on the crabs of Japan. IV. Brachygnatha, Brachyrhyncha, **3**: 365-741, pls. 42-111. (Yokendo Co. Ltd., Tokyo).

SHEN, C. J., 1935. An investigation of the post-larval development of the shore-crab *Carcinus maenas*, with special reference to the external secondary sexual characters. Proceedings of the Zoological Society of London, **1935**(1): 1-33.

WOOD-MASON, J. & A. ALCOCK, 1891. Natural history notes from H.M. Marine Survey Steamer 'Investigator,' Commander R.F. Hoskyn, R.N., late commanding. No. 21. Note on the results of the last season's deep-sea dredging. Annals and Magazine of Natural History, (6) **7**: 258-272.

First received 24 August 2011.
Final version accepted 16 January 2012.

THYMOPS TAKEDAI, A NEW SPECIES OF DEEPWATER LOBSTER FROM THE SOUTHWEST ATLANTIC OCEAN WITH ADDITIONAL RECORDS OF 'THYMOPINE' LOBSTERS (DECAPODA, NEPHROPIDAE)

BY

SHANE T. AHYONG[1,4]), W. RICHARD WEBBER[2,5]) and TIN-YAM CHAN[3,6])

[1]) Australian Museum, 6 College St., Sydney, NSW 2010, Australia

[2]) Museum of New Zealand Te Papa Tongarewa, P.O. Box 467, Wellington, New Zealand

[3]) Institute of Marine Biology and Center of Excellence for Marine Bioenviroment and Biotechnology, National Taiwan Ocean University, Keelung 20224, Taiwan, R.O.C.

ABSTRACT

A new species of deep-water clawed lobster, *Thymops takedai*, is described from the Scotia Sea, near South Georgia, representing the second species of the genus. The new species differs from *Thymops birsteini* (Zarenkov & Semenov, 1972) in having a ventrally spinose or serrated rostrum (versus smooth or finely granular), a non-flagellate maxilliped 3 exopod (versus exopod with flagellum), and a longer branchiocardiac groove that reaches to the posterior quarter of the carapace (versus slightly beyond midlength). We also report additional specimens of *T. birsteini* and *Thymopsis nilenta* Holthuis, 1974, and document considerable variation in the condition of the exopods on maxillipeds 2 and 3 of *Thymopsis*. This variation in the maxillipedal exopods in *Thymopsis* overlaps with that of *Thymops*, rendering the two genera otherwise almost indistinguishable apart from subtle distinctions in carapace groove patterns. Thus, the taxonomic utility of recognising both genera as separate is questionable and further study may show that the two genera should be synonymised.

INTRODUCTION

Nephropidae Dana, 1852 currently includes 54 species and 14 genera of clawed lobsters (Chan, 2010). Whereas the best known nephropids are the edible shallow water species of *Homarus* Weber, 1795 and *Nephrops* Leach,

[4]) Corresponding author; e-mail: Shane.Ahyong@austmus.gov.au

[5]) e-mail: RickW@tepapa.govt.nz

[6]) e-mail: tychan@mail.ntou.edu.tw

1814, most nephropids live in deep-water on the outer continental shelf and slope. The higher taxonomy of the clawed lobsters has undergone several important changes in the past three decades. Holthuis (1974, 1991) recognised three subfamilies within the Nephropidae: Nephropinae, Neophoberinae and Thymopinae. Recent morphological and molecular phylogenetic studies, however, have not corroborated Holthuis' subfamily divisions, and at present, subfamilies are not formally recognised within Nephropidae (Chan, 2010). Within the former Thymopinae, two morphologically similar and phylogenetically allied genera, *Thymops* Holthuis, 1974 and *Thymopsis* Holthuis, 1974, are known only from the southwestern Atlantic, each represented by a single species, *Thymops birsteini* (Zarenkov & Semenov, 1972) and *Thymopsis nilenta* Holthuis, 1974 (Ahyong, 2006). Both occur in deep, subantarctic waters around the southern part of South America including the Falkland Islands (Malvinas) and South Georgia. *Thymops birsteini* appears to be relatively common, although *Thymopsis nilenta* is seldom reported (Holthuis, 1991). Among unstudied material collected by the USS "Eltanin" in waters around the Falkland Islands in 1963 were specimens of *Thymops birsteini*, *Thymopsis nilenta* and an undescribed species of *Thymops*. The new species is described herein and additional 'thymopine' records reported.

MATERIALS AND METHODS

Descriptive terminology generally follows Holthuis (1974) and Ahyong et al. (2007). Carapace length (cl) is measured along the dorsal midline and includes the rostrum. Postorbital carapace length (pcl) is measured from the posterior margin of the orbit to the posterior margin of the carapace. Specimens are deposited in the Australian Museum, Sydney (AM), Museum of New Zealand Te Papa Tongarewa (NMNZ) and the National Museum of Natural History, Smithsonian Institution, Washington D.C. (USNM).

TAXONOMY

Family NEPHROPIDAE Dana, 1852

Genus *Thymops* Holthuis, 1974

Thymops takedai sp. nov.

(figs. 1-4)

Type material. — Holotype: USNM 1156923, female (cl 58.8 mm, pcl 38.3 mm), Burdwood Bank, Scotia Sea, 54°03.6′S 56°04.9′W, 940-950 ftm [1720-1739 m], "Eltanin" 376-29, rock dredge, 20-21 Dec. 1962. Paratypes: USNM 1156927, 1 male (pcl 47.8 mm), 1 female (pcl 51.2 mm), Scotia Sea, N of South Georgia, 53°36′S 36°51-54′W, 220-265 m, "Eltanin" 732-35, 10′ Blake trawl, 12 Sep. 1963.

Diagnosis. — Rostrum ventrally serrated to spinose along distal half. Maxilliped 3 exopod reaching slightly beyond midlength of ischium; flagellum absent. Branchiocardiac grooves reaching posteriorly to about the posterior quarter of the carapace. Sternal ridges between pereopod 4 coxae in males separated by hour-glass shaped sinus, minimum gap narrower than one-fifth maximum distance between ridges.

Description. — Rostrum slender, 0.55 pcl in holotype (rostra damaged in paratypes), cross-section triangular; dorsal surface with pair of granulose or spinose carinae extending posteriorly from posterior third of rostrum onto carapace proper, near level of base of supraorbital spines; dorsal median groove of rostrum extending posteriorly to posterior margin of carapace; lateral margins with 6 or 7 lateral spines or spinules, 2 prominent; apex acute, slightly upturned; ventral margin serrated to spinose along distal half.

Carapace granular over entire surface, finest laterally, largest anterodorsally; surface minutely setose; supraorbital spine with short granular carinae; antennal spine without carinae, but slightly swollen basally; orbital fossa rounded, concave; pterygostomial and anterolateral margin broadly convex evenly merging; ventral branchiostegal margin unarmed. Postcervical groove distinct, meeting sellar groove (indistinct) dorsally, ventrally continuous with cervical, hepatic and antennal grooves. Urogastric groove short. Branchiocardiac groove faintly indicated, extending onto posterior quarter of carapace, slightly oblique to midline; parabranchial groove faintly indicated; cervical groove extending dorsally slightly above α point; gastro-orbital groove faintly indicated; marginal carina flanked by distinct groove, extending anteriorly to incisura clavicularis.

Abdomen subcylindrical; somites with smooth articular surfaces, demarcated from non-articular surfaces by setose, transverse groove. Somite 1 surface with short median tubercle, and shallow, setose pits, posterior surface of somite smooth; pleuron subquadrate, sparsely pitted, anteriorly blunt, overlapping posterior carapace margin. Somites 2-5 with low blunt median carina, extending almost to posterior somite margins, smooth dorsally, margins with irregular setose pits; with broad, shallow, transverse setose groove on either side of median carina, extending onto pleura; midlateral surfaces almost glabrous, rugose and pitted, more pronounced in paratypes than in holotype; posterior margin smooth, glabrous; pleura broad, with blunt, angular terminations, sur-

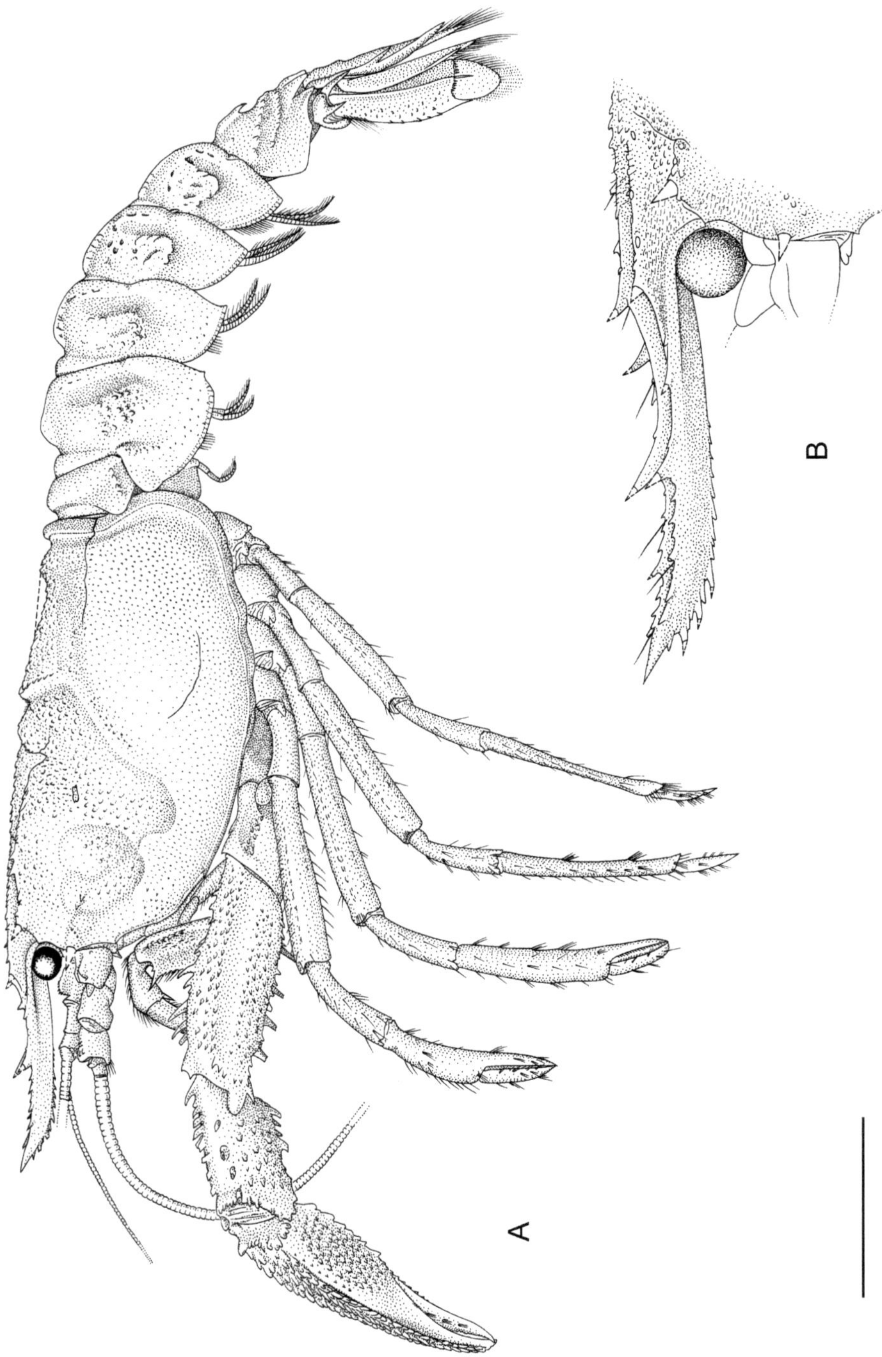

Fig. 1. *Thymops takedai* sp. nov., holotype female, cl 58.8 mm, pcl 37.3 mm (USNM 1156923): A, habitus, lateral view; B, rostrum, lateral view. Scale: A = 16 mm, B = 8 mm.

faces finely granular, sparsely setose. Somite 6 surface uneven, sparsely pitted; median carina more slender than that of preceding somite, extending to midlength of somite, margins irregular, sparsely setose, with posterior spine; with 2 arcuate, oblique rows of sparsely setose tubercles on either side of midline; posterior margin slightly concave, with median and posterolateral spines; pleuron broad, triangular.

Telson longer than wide, widest posteriorly; surface sparsely granular and sparsely setose; posterior margin evenly convex; with prominent posterolateral and midlateral spines. Uropodal protopod with 2 prominent spines, 1 each above endopod and exopod articulations, respectively. Endopod with prominent posterolateral spine; surface with single longitudinal carina; posterior margin evenly convex. Exopod with diaeresis; proximal segment with longitudinal carina; lateral margin with 2 or 3 low spines; posterior margin with 18-20 spinules and strong posterolateral spine; distal segment half as long as wide, broadly rounded.

Eyes small, movable, cornea subglobular, not wider than stalk.

Antennular peduncle extending anteriorly to midlength of rostrum. Articles unarmed.

Antennal peduncle extending anteriorly to midlength of rostrum. Basal article with minute distoventral spinule on margin anterior to antennal pore. Article 3 with small distoventral spinule. Scaphocerite reduced to blunt sclerite.

Epistome surface smooth, unarmed; posterior margin with blunt clavicular ridge; anterior margin adjacent to antennular articulations forming rounded, thickened ridge.

Maxilliped 2 exopod extending to distal quarter of the merus; flagellum absent.

Maxilliped 3 endopod with unarmed dactylus and propodus; carpus with 1 or 2 distal flexor spines; merus distal margin with small extensor and large flexor spine flanked on either side by small spine; ischium flexor margin dentate. Exopod slender, reaching slightly beyond midlength of ischium of endopod; flagellum absent.

Sternites of pereopods 1-3 each with paired, posteriorly divergent ridges between coxae, increasing in size posteriorly. Sternite of pereopod 5 unarmed.

Sternite of pereopod 4 in male with paired ridges, fused anteriorly, divergent posteriorly, forming hood-like structure; each ridge posteriorly with rounded, mesially inclined lobe, together forming hour-glass shaped median sinus; minimum gap between posterior lobes less than one-fifth maximum width of

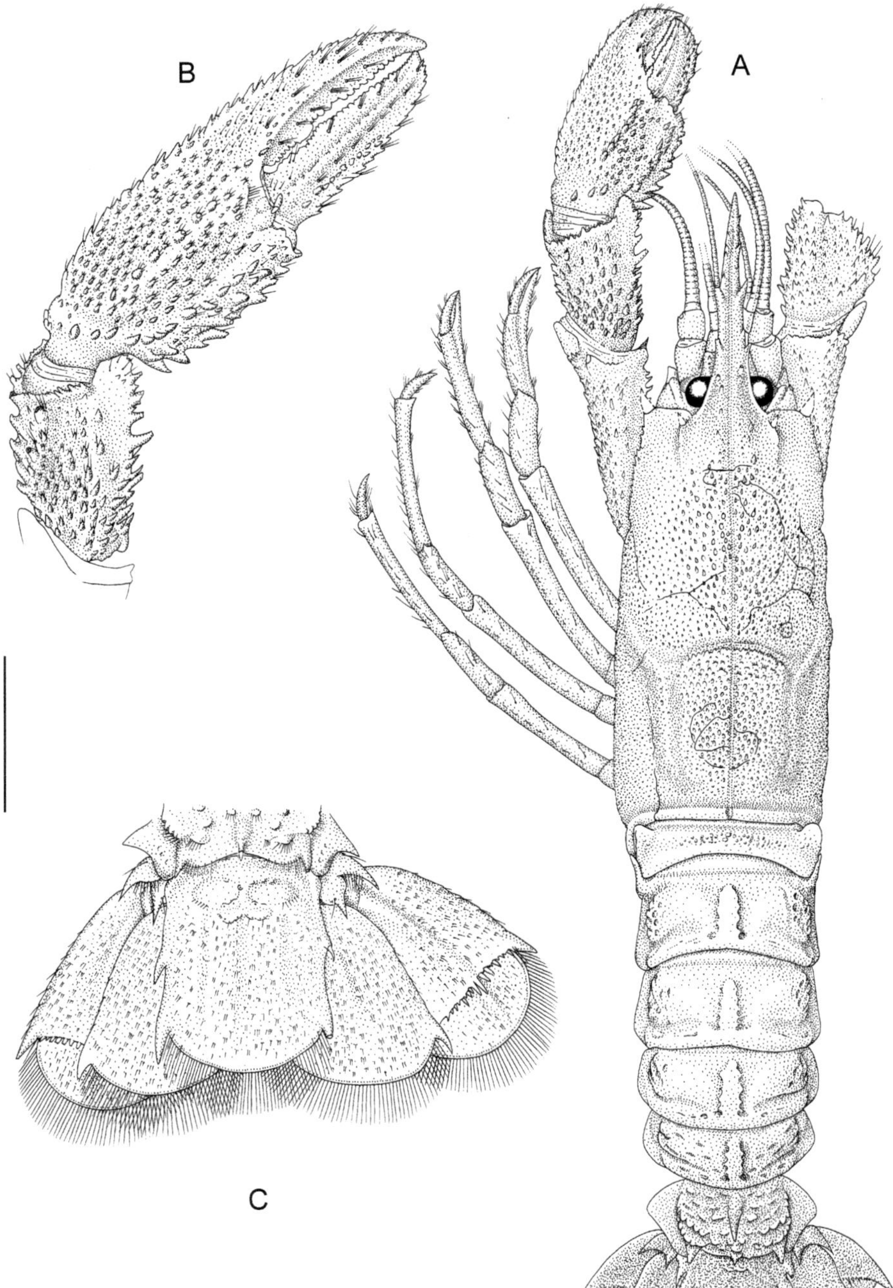

Fig. 2. *Thymops takedai* sp. nov., holotype female, cl 58.8 mm, pcl 37.3 mm (USNM 1156923): A, habitus, dorsal view; B, left cheliped 1, dorsal view; C, telson, dorsal view. Scale: A = 15 mm, B = 12.5 mm, C = 10 mm.

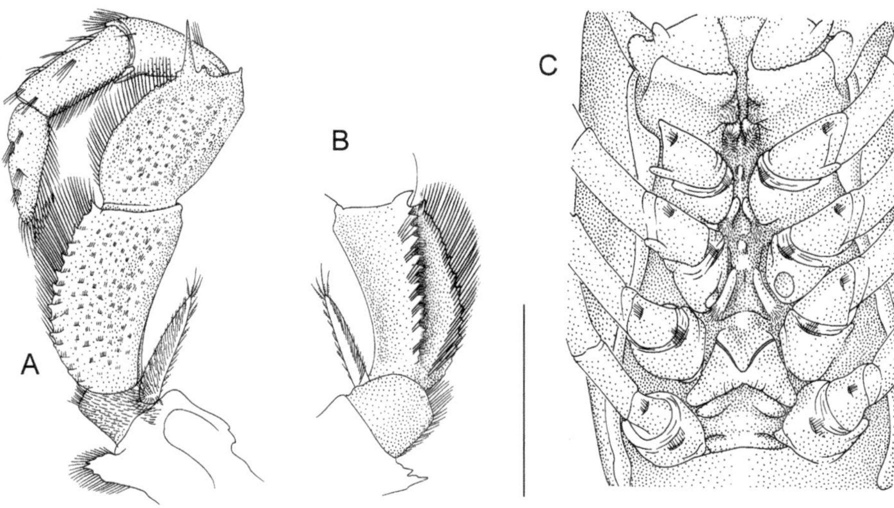

Fig. 3. *Thymops takedai* sp. nov., holotype female, cl 58.8 mm, pcl 37.3 mm (USNM 1156923): A, left maxilliped 3, posterior (external) view; B, left maxilliped 3 basis and ischium, anterior (internal) view; C, sternum. Scale: A–B = 8 mm, C = 12.5 mm.

'hood'; 'hood' partially overhanging pereopod 5 coxae and anterior half of first pleopods.

Sternite of pereopod 4 in females with swollen, triangular thelycum. Thelycum anteriorly with small, blunt triangular anterior median lobe abutting pair of large, swollen, posterolaterally directed triangular lobes, separated by median fissure and forming broad V-shape posteriorly.

Pereopod 1 chelate, left–right equal and symmetrical; sparsely setose; surfaces covered with granules and acute tubercles and spines along margins; dactylus and pollex longer than palm; occlusal margins dentate; lateral and mesial surfaces with smooth, slender longitudinal carina; carpus about as long as palm, spinose and tuberculate; merus compressed proximally, distally trigonal in cross-section, spinose along distal, dorsal and ventral margins; ischium compressed, ventral margin spinose; coxa ventral margin denticulate.

Pereopods 2-3 chelate, reaching anteriorly slightly beyond rostrum; surfaces with few sparse setae; coxa with small weakly serrate or crenulated ventral margin, otherwise segments unarmed; pollex and dactylus with finely serrate occlusal margins; pereopod 2 dactylus two-thirds as long as palm; pereopod 3 dactylus slightly shorter than half palm length.

Pereopod 4 non-chelate, reaching anteriorly to rostral apex; segments unarmed.

Pereopod 5 sub-chelate in females, non-chelate in males reaching anteriorly to end of antennal peduncle; segments unarmed; propodus with short, setose pollex, about one-third as long as dactylus.

Abdominal sternites unarmed. Pleopod 1 uniramous. Pleopods 2-5 biramous.

Male pleopod 1 with distal and proximal segments fused, forming copulatory organ in form of spatulate, twisted plate with plane of distal segment approximately perpendicular to plain of proximal segment. Distal segment with blunt, subtruncate apex. Proximal segment with sparsely setose mesial surface, mesial edge with long setae.

Male pleopod 2 biramous; basis with posterior setal field extending to inner distal margin; exopod and endopod elongate-ovate with setose margins; exopod longer than endopod, appendix masculina present.

Etymology. — We are pleased to name the new species in honour of Prof. Masatsune Takeda, for his major contributions to carcinology. Although focussed primarily on Brachyura, Prof. Takeda published on a wide range of decapods, including clawed lobsters (e.g., Takeda & Hatanaka, 1984; Takeda, 1986; Takeda & Nagai, 2004).

Remarks. — *Thymops takedai* sp. nov. differs from its only congener, *T. birsteini*, in the following features: the ventral margin of the rostrum is spinose or serrated; the maxilliped 3 exopod lacks a flagellum and reaches only slightly beyond the midlength of the ischium; the branchiocardiac groove on the carapace reaches posteriorly to about the posterior quarter of the carapace, rather than not extending much beyond the midlength; and the sternal ridges between the pereopod 4 coxae in males are separated by an hour-glass shaped sinus. In contrast, the rostrum of *T. birsteini* is ventrally smooth or granular; the maxilliped 3 exopod bears a multiarticulate flagellum that reaches beyond the midlength of the merus; the branchiocardiac groove extends posteriorly scarcely beyond the postcervical groove; and the sinus separating the sternal ridges between the pereopod 4 coxae in males are separated by a wide U-shaped sinus.

Unfortunately, all specimens of *T. takedai* were damaged during collection and the paratypes, both larger than the holotype, have broken rostra. Although the original rostral lengths cannot be determined, the conformation of the intact portions indicates that the proportional rostral length decreases with increasing body size as in other nephropids.

The holotype of *T. takedai* is a subadult female as indicated by the smaller gonopores on the pereopod 3 coxae and development of the sternal thelycum, being considerably less swollen than that of the female paratype.

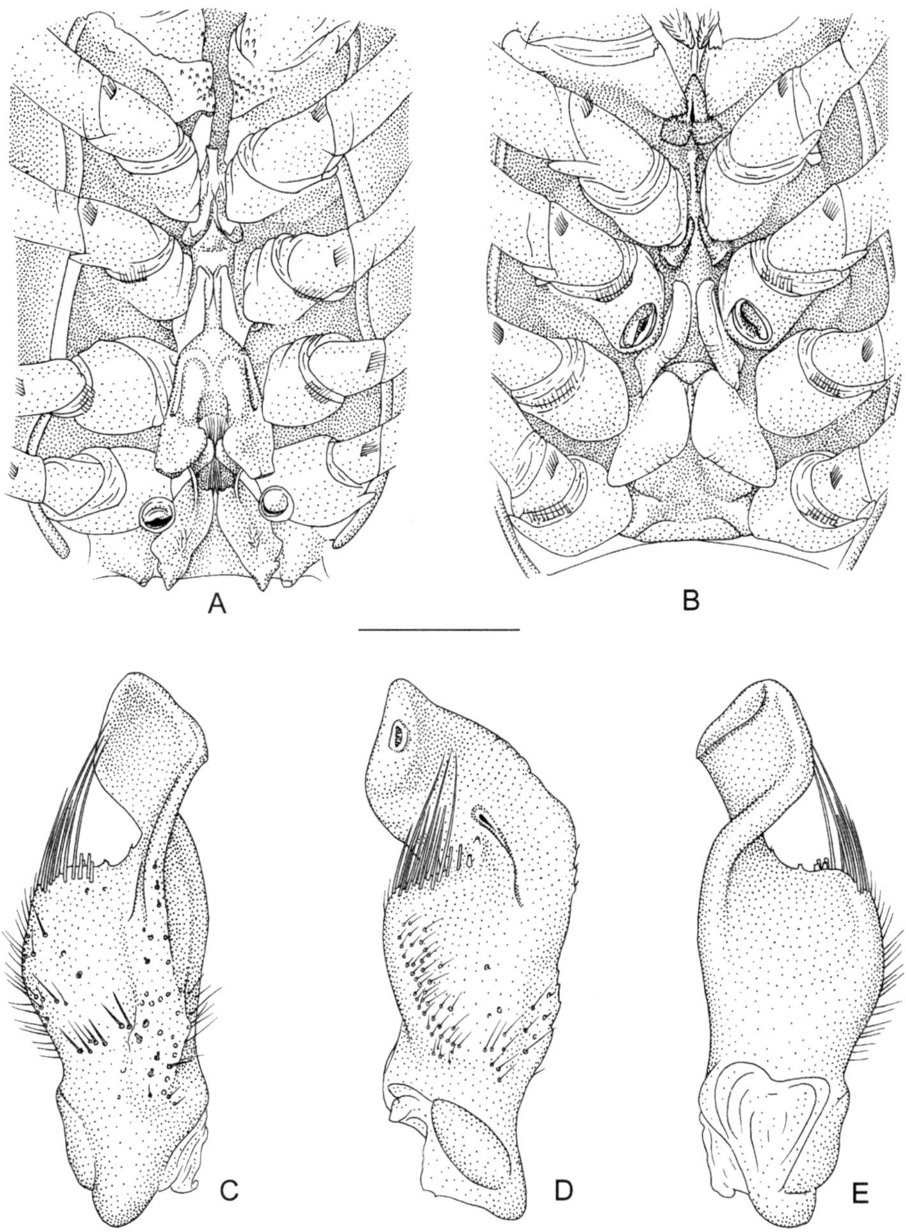

Fig. 4. *Thymops takedai* sp. nov. Paratype male, pcl 47.8 mm (USNM 1156927): A, sternum; C–E, left pleopod 1, C, posterior (external) view, D, mesial view, E, anterior (internal) view. Paratype female, pcl 51.2 mm (USNM 1156927): B, sternum. Scale: A–B = 10 mm, C–E = 2.8 mm.

It is somewhat unexpected that a new species of *Thymops* should be discovered in such close geographic proximity to the respective type localities of *Thymops birsteini* and *Thymopsis nilenta*, especially given the significant level of fishery activity in the region. *Thymopsis nilenta* and both species of *Thymops*, however, are all superficially similar and could be misidentified in the field. In particular, *T. takedai* could easily be mistaken for *Thymopsis nilenta*, especially given that the two species can be sympatric and both have a ventrally spinose rostrum. Whether *T. takedai* is naturally rare or has been overlooked requires further research.

Distribution. — Scotia Sea in the vicinity of the Falkland (Malvinas) Islands and South Georgia; 220-1739 m.

Thymops birsteini (Zarenkov & Semenov, 1972)

Nephropides birsteini Zarenkov & Semenov, 1972: 599-601, figs. 1-6.
Thymops birsteini. — Holthuis, 1974: 764-772, figs. 13, 14; Arnaud, 1981: 37-38; Takeda & Hatanaka, 1984: 11; Takeda, 1986: 328, fig. 136; Holthuis, 1991: 48, figs. 45, 86, 87; Arntz et al., 1999: 309, tabs. 1, 2; Laptikhovsky & Reyes, 2009: 35.

Material examined. — USNM 1156925, 2 males (cl 32.2 mm, pcl 23.7 mm; cl 34.6 mm, pcl 25.2 mm), Scotia Sea, Falkland Islands (Malvinas), 51°58.2′-52°01.0′S 56°38′W, 462-353 ftm [845.5-646 m], "Eltanin" 558-35, 14 Mar. 1963.

Remarks. — The specimens conform extremely well to the type description (Zarenkov & Semenov, 1972) and the full account of adults provided by Holthuis (1974).

Distribution. — Continental shelf off southern South America ranging from the Pacific coast off Chile eastwards to the vicinity of the Falkland Islands (Malvinas) and South Georgia; 122-2516 m (Holthuis, 1991; Laptikhovsky & Reyes, 2009).

Genus *Thymopsis* Holthuis, 1974
Thymopsis nilenta Holthuis, 1974
(fig. 5)

Thymopsis nilenta Holthuis, 1974: 756-763, figs. 10-12; 1991: 49-50, figs. 44, 88, 89.

Material examined. — USNM 1156929, 2 males (cl 49.2 mm, pcl 38.7 mm; cl 56.0 mm, pcl 43.9 mm), 2 females (cl 52.6 mm, pcl 43.6 mm; pcl 49.1 mm), N of South Georgia, 53°36′S 36°51-54′W, 220-265 m, 10′ Blake trawl, "Eltanin" 732-35, 12 Sep. 1963; USNM 1156924, 1 female (pcl 35.1 mm), 53°06.0-05.5′S 44°54.5-49.5′W, 1215-1136 ftm [2223.5-2078.9 m], "Eltanin" 458-35, with few arthropods and tunicates, rock dredge, 9 Feb. 1963; AM P87157, 1 male (cl 51.0 mm, pcl 41.0 mm), Scotia Sea, N of South Georgia Island, 53°22.7-23.4′S 37°11.1-20.9′W, 765-710 ftm [1400-1299 m], gravel with snails and gorgonians, 5′ Blake

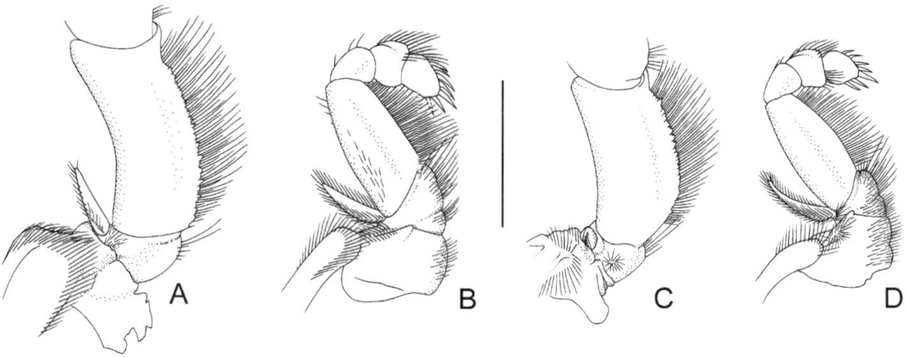

Fig. 5. *Thymopsis nilenta* Holthuis, 1974, right maxilliped 3 exopod and proximal endopod articles (coxa-basis, ischium, merus), external view (A, C) and maxilliped 2, external view (B, D). A–B, female, pcl 49.1 mm (USNM 1156929); C–D, female, pcl 45.1 mm (NMNZ CR.22921). Scale = 8 mm.

trawl, "Eltanin" 734-35, 12 Sep. 1963; NMNZ Cr.22192, 1 female (pcl 45.1 mm), South Georgia, FV "San Aspiring".

Remarks. — The specimens are in poor condition but exhibit some important variation not documented by Holthuis (1974, 1991). *Thymopsis nilenta* was originally distinguished from *Thymops* on the basis of a spinose rather than unarmed ventral rostral margin, vestigial rather than well-developed exopods on maxillipeds 2 and 3, an unarmed posterior margin of abdominal somite 6 and slightly different expression of the postcervical groove (Holthuis, 1974). The ventrally spinose rostrum of *Thymops takedai* invalidates the first mentioned generic distinction. The second 'diagnostic' feature of *Thymopsis*, vestigial maxilliped 2 and 3 exopods, is not consistent in four specimens of *T. nilenta*: two (AM P87157; NMNZ Cr.22192, fig. 5C, D) have a vestigial exopod on maxilliped 3 as expected, but on maxilliped 2, have a short, non-flagellate exopod; the third has short exopods on maxillipeds 2 and 3 (USNM 1156929, male, pcl 38.7 mm); and the fourth (USNM 1156929, female, pcl 49.1 mm, fig. 5A, B) has a small non-flagellate exopod on left and right maxilliped 2 and the right side maxilliped 3, but none on the left maxilliped 3. Likewise, the spination of the posterior margin of abdominal somite 6 overlaps between *Thymops* and *Thymopsis*, with a median and posterolateral spines variously present or absent in the latter. Thus, three of four diagnostic characters putatively separating *Thymops* and *Thymopsis* are unreliable. The two genera are otherwise almost indistinguishable apart from subtle distinctions in carapace groove patterns, the most important being slightly different arrangements of the postcervical and sellar grooves. In *Thymops*, the postcervical groove reaches the dorsal midline of the carapace where it meets the sellar groove. In *Thymopsis*, however, the postcervical groove does not reach the carapace dorsum; the postcervical and sellar grooves do not meet and only the latter crosses the dorsal midline. This single remaining distinction between *Thymops* and *Thymopsis* is slight, undermining the taxonomic utility of recognising both genera as separate, especially for a clade of only three species. Further study may show that the two genera should be synonymised.

Thymopsis nilenta was previously known only from the type material collected off the Falkland Islands and South Georgia at depths between 2886-3040 m and 1976-2068 m (Holthuis, 1974). The specimen from "Eltanin" stn 734-35 was noted as "very light salmon" in colour. Capture depths of previous records of *Thymopsis nilenta* all exceed 1900 m, so the specimens taken at 220-265 m (together with two paratypes of *Thymops takedai*) represent a significant bathymetric range extension into shallower water.

Distribution. — South-western Atlantic from the vicinity of the Falkland Islands (Malvinas) and South Georgia; between 220 and 2886-3040 m.

ACKNOWLEDGMENTS

We are grateful to Karen Reed and Linda Ward (USNM) for arranging the loan of specimens. This work was partially supported by the National Science Council and Academia Sinica, Taiwan, R.O.C. to TYC.

REFERENCES

AHYONG, S. T., 2006. Phylogeny of the clawed lobsters (Crustacea: Decapoda: Homarida). Zootaxa, **1109**: 1-14.

AHYONG, S. T., K. H. CHU & T. Y. CHAN, 2007. Description of a new species of *Thaumastochelopsis* from the Coral Sea (Decapoda: Nephropoidea). Bulletin of Marine Science, **80**(1): 201-208.

ARNAUD, P. M., 1981. *Thymops birsteini* (Zarenkov & Semenov, 1972) (Crustacea Decapoda Nephropidae) et sa prédation par *Dissostichus eleginoides* Smitt, 1898 (Pisces Nototheniidae) en Géorgie du sud. Tethys, **10**: 37-38.

ARNTZ, W. E., M. GORNY, R. SOTO, M. A. LARDIES, M. RETAMAL & I. S. WEHRTMANN, 1999. Species composition and distribution of decapod crustaceans in the waters off Patagonia and Tierra del Fuego, South America. Scientia Marina, **63**(Supplement 1): 303-314.

CHAN, T. Y., 2010. Annotated checklist of the world's marine lobsters (Crustacea: Decapoda: Astacidea, Glypheidea, Achelata, Polychelida). Raffles Bulletin of Zoology, (Supplement) **23**: 153-181.

DANA, J. D., 1852. Conspectus Crustaceorum quae in Orbis Terrarum circumnavigatione, Carolo Wilkes e Classe Reipublicae Foederatae Duce, lexit et descripsit. Proceedings of the Academy of Natural Sciences of Philadelphia, **1852**: 10-28.

HOLTHUIS, L. B., 1974. The lobsters of the superfamily Nephropoidea of the Atlantic Ocean (Crustacea: Decapoda). Bulletin of Marine Science, **24**(4): 723-884.

— —, 1991. Marine lobsters of the world. An annotated and illustrated catalogue of species of interest to fisheries known to date. FAO Fisheries Synopsis, **125**: 1-292. (FAO, Rome).

LAPTIKHOVSKY, V. & P. R. REYES, 2009. Distribution and reproductive biology of a subantarctic deep-sea lobster, the Patagonian lobsterette *Thymops birsteini* (Zarenkov and Semenov, 1972) (Decapoda, Astacidea, Nephropidae). Journal of Natural History, **43**: 35-46.

LEACH, W. E., 1814. Crustaceology. In: D. BREWSTER (ed.), The Edinburgh Encyclopaedia, **7**: 383-437, pl. 221. (William Blackwood, Edinburgh).

TAKEDA, M., 1986. Nephropidae. In: I. NAKAMURA (ed.), Important fishes trawled off Patagonia: 328-329. (JAMARC, Tokyo).

TAKEDA, M. & H. HATANAKA, 1984. Records of decapod crustaceans from the southwestern Atlantic collected by the Japanese fisheries research trawlers. Bulletin of the National Science Museum, Tokyo, (A) **10**: 7-24.

TAKEDA, M. & S. NAGAI, 2004. Four species of giant crustaceans from the Indonesian depths, with description of a new species of the family Lithodidae. Bulletin of the National Science Museum, Tokyo, (A) **30**: 9-21.

WEBER, F., 1795. Nomenclator entomologicus secundum entomologiam systematicam ill. Fabricii adjectis speciebus recens detectis et varietatibus: i–viii, 1-172. (Chilonii et Hamburgi).

ZARENKOV, N. A. & V. N. SEMENOV, 1972. A new species of the genus *Nephropides* from the South-West Atlantic. Zoologichesky Zhurnal, **51**: 599-602.

First received 11 September 2011.
Final version accepted 17 November 2011.

CAPRELLA TAKEDAI, A NEW SPECIES OF CAPRELLID AMPHIPOD FROM OFF ITO, THE EAST COAST OF IZU PENINSULA, SHIZUOKA, JAPAN

BY

MASAKAZU N. AOKI[1,3]) and ATSUSHI ITO[2])

[1]) Graduate School of Agricultural Science, Tohoku University, 1-1 Tsutsumi-dori Amamiya-machi, Aoba, Sendai, Miyagi 981-8555, Japan

[2]) Graduate School of Life and Environmental Sciences, University of Tsukuba, Tsukuba, Ibaraki 305-8572, Japan

ABSTRACT

Caprella takedai, a new caprellid amphipod is described from off Ito, east coast of the Izu Peninsula, Shizuoka, Japan. The species was found attached to green alga, *Cladophora wrightiana* Harvey, at a depth of 15 m. Morphological characteristics of the species are very close to *C. subinermis* Mayer, but it is easy to identify the species from others by the presence of a distinctive triangular projection on the palmar process of gnathopod 2 propodus.

INTRODUCTION

A series of scuba diving surveys of amphipod fauna along the coastal areas of the Izu Peninsula were conducted from September 2006 to February 2007. During these surveys, we collected a species of caprellid amphipod clinging to the green alga *Cladophora wrightiana* Harvey, 1859 at the site called 'Shirane-Minami', off Ito City, at a depth of 15 m. The species is similar to *Caprella subinermis* Mayer, 1890, but has a distinctive triangular projection on the ventral part of the palmar process of the gnathopod 2 propodus, and thus is quite easy to distinguish from other species in the genus *Caprella*, as well as *C. subinermis*. Consultation of the literature shows that it represents a new species.

The type material is deposited in the National Museum of Nature and Science, Tokyo (NSMT).

[3]) Corresponding author; e-mail: m-aoki@m.tohoku.ac.jp

TAXONOMY

Family CAPRELLIDAE Leach, 1814
Genus *Caprella* Lamarck, 1801
Caprella takedai, new species
[New Japanese name: Takeda-warekara]
(figs. 1-4)

Type material. — Holotype, male, 12.0 mm (NSMT-Cr 22090); paratypes, 2 males, 13.0 mm, 10.0 mm (NSMT-Cr 22091); on *Cladophora wrightiana*, 15 m, Shirane Minami, off Ito City, Shizuoka, 34°30′N 139°08′E, 21 September 2006, coll. M. N. Aoki & A. Ito.

Diagnosis. — Body surface smooth with no spines, head with weak frontal process, antenna 2 bearing dense long setae, gnathopod 2 arising from proximal part of pereonite 2, elongated propodus of gnathopod 2 with distally located palm slightly longer than pereonite 2, basis of gnathopod 2 shorter than 1/4 of pereonite 2, palmar projection on propodus of gnathopod 2 with a ventral triangular projection, palmar margin on propodus of pereopods 6-7 defined by long and serrated proximal grasping spines, abdomen with a pair of appendages with serrated margin.

Description. — Head with a weak process directed forward. Pereonites 2-7 smooth. No pleura on pereonites 2-4. Pereonite 2 subequal in length

Fig. 1. *Caprella takedai*, new species, holotype male, 12 mm. Photo of the live material clinging to the green alga, *Cladophora wrightiana* Harvey.

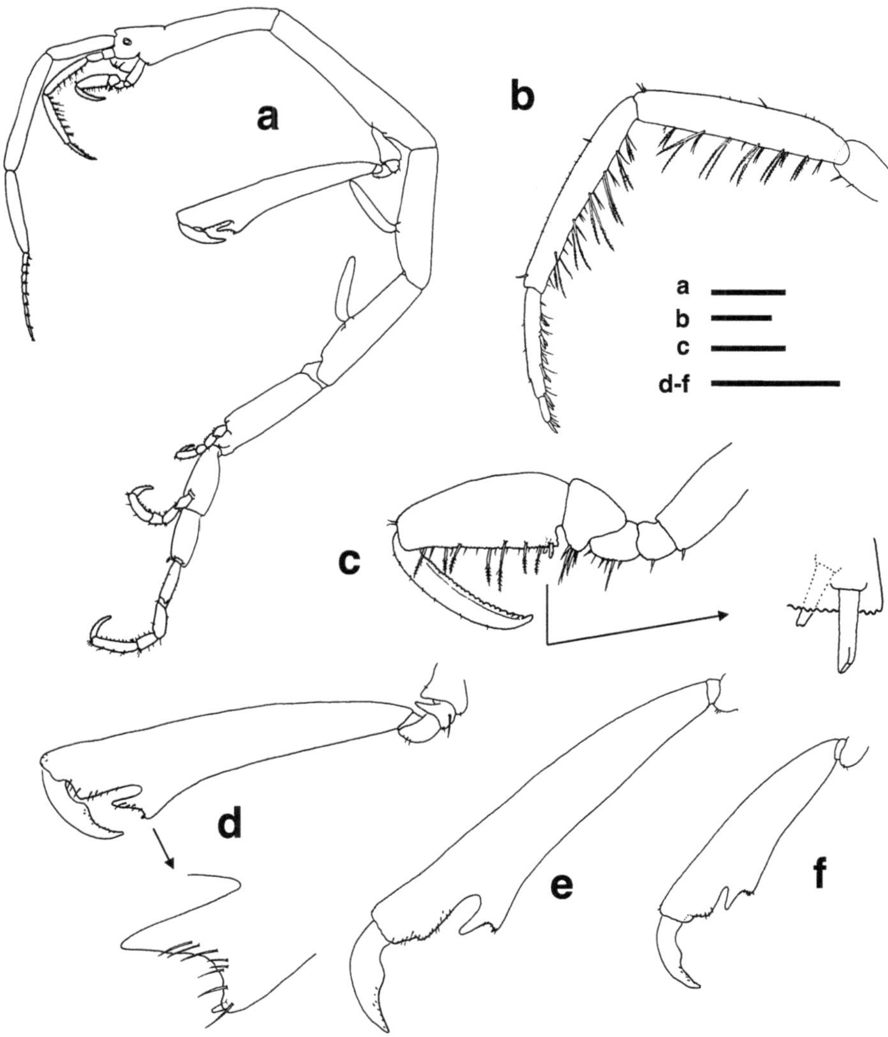

Fig. 2. *Caprella takedai*, new species, a–d, holotype male, 12 mm; e, paratype male, 13 mm; f, paratype male, 10 mm. a, body, lateral view; b, left antenna 2; c, left gnathopod 1; d–f: left gnathopod 2. Scales for a, d–f = 1 mm; b–c = 0.2 mm.

to pereonite 3 × 1.5. Pereonite 5 subequal in length to pereonites 6 and 7 combined (figs. 1, 2a). Maximum body length, 13.0 mm.

Antenna 1 subequal in length to pereonites 2 and 3 combined; flagellum as long as 1/3 of peduncle (fig. 2a). Antenna 2 shorter than 1/2 of antenna 1 (fig. 2a); peduncle densely bearing long, simple and plumose setae; flagellum with hooks and simple setae (fig. 2b).

Mouthparts typical of genus. Upper lip symmetrically bilobed, anterior margin covered with dense short setae (fig. 3a). Mandibles with no palp; left mandible with incisor five-toothed, lacinia mobilis five-toothed and three raker spines (fig. 3b); right mandible with incisor five-toothed, lacinia mobilis indistinctly five-toothed and two raker spines (fig. 3c). Lower lip with distinctive inner lobes; inner and outer lobes finely setose apically; mandibular projection pointed (fig. 3d). Maxilla 1: outer lobe with seven robust spines and a plumose seta, palp carrying eight apical spines and several long setae (fig. 3e). Maxilla 2: oval inner lobe slightly shorter than rectangular outer lobe; inner and outer lobes bearing dense long setae apically (fig. 3f). Maxilliped: inner lobe with short plumose setae apically, outer lobe with one or two spines and fringed with several short plumose or simple setae; palp four-articulate; inner margin of propodus and carpus bearing long simple setae (fig. 3g).

Gnathopod 1 having palmer margin of propodus serrate and with a pair of proximal grasping spines; grasping margin of dactylus serrate (fig. 2c). Gnathopod 2 arising at proximal part of pereonite 2; elongated propodus with distally located palm slightly longer than pereonite 2 (fig. 2a); palm of propodus barely setose, with 2 processes, distal one triangular, proximal one spiniform with ventral triangular process bearing a grasping spine; basis 1/5 as long as propodus, slightly longer than pereonite 2, basis and ischium with distolateral triangular process; dactylus apex spiniform (fig. 2d). Ventral triangular projection on the palmar process becomes more distinctive in the larger males (fig. 2d–f).

Gills elliptical (fig. 2a).

Pereopods 5, 6 and 7 increasing in length posteriorly; pereopod 7 subequal in length to pereonites 5 and 6 combined; grasping margin of propodus of pereopods 5-7 weakly concaved and smooth with several short setae and paired proximal serrate grasping spines (fig. 4a–c).

Abdomen: unsegmented appendages fringed with a row of serrate spines; pair of lobes bearing a few long plumose setae and several short plumose or simple setae; penes medial (fig. 4d).

Etymology. — The name of the species is dedicated to Dr. Masatsune Takeda for his contribution to the caprellid research in Japan. His appropriate guidance about a quarter century ago gave M. N. Aoki a good starting point for his caprellid study.

Systematics. — *Caprella takedai*, new species, is most similar to *C. subinermis*. Both species have a unique elongated propodus of gnathopod 2 and

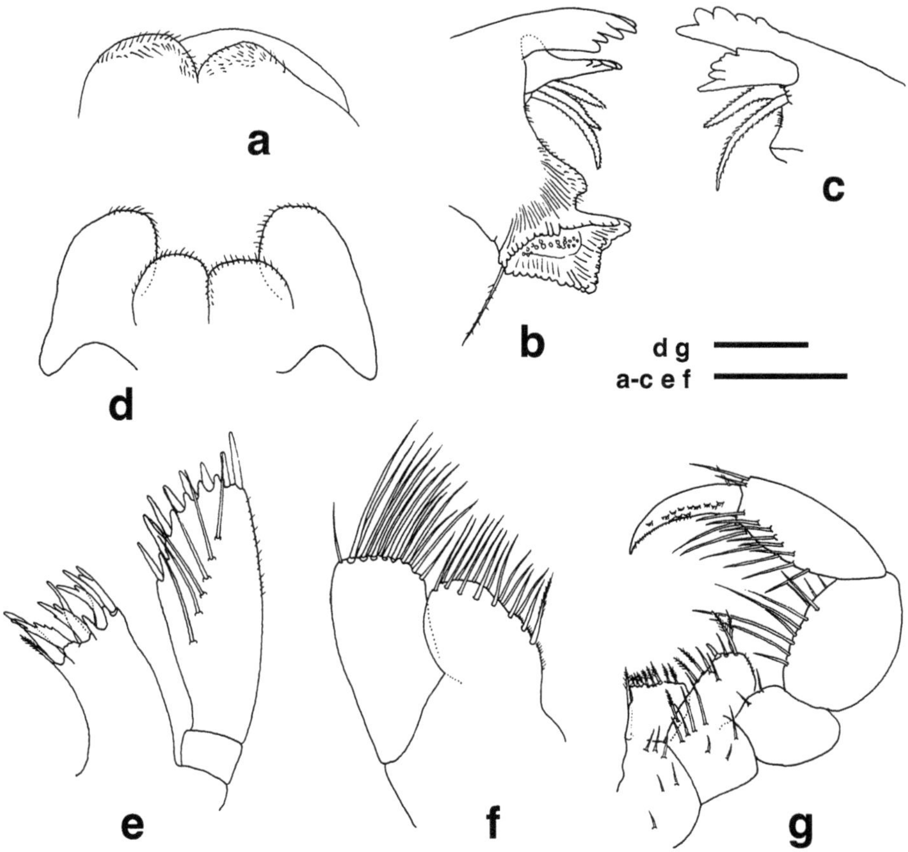

Fig. 3. *Caprella takedai*, new species, holotype male: a, anterior part of upper lip; b, left mandible; c, right mandible; d, lower lip; e, maxilla 1; f, maxilla 2; g, maxilliped. Scales for a–g = 0.1 mm.

share several morphological characteristics such as weak frontal process on the head, smooth pereopods with no spines, short basis arising from the proximal part of pereopod 2, paired serrate grasping spines on the palm of pereopods 6 and 7 and abdominal appendages with serrate margin (Mayer, 1890; Arimoto, 1976). However, the ventral triangular projection on the spiniform palmar process of gnathopod 2 differentiates *C. takedai* from *C. subnermis*. Except for the elongated propodus of gnathopod 2, *C. takedai* and *C. subinermis* also share morphological characteristics with *C. tsugarensis* Utinomi, 1947, an endemic species to Japanese waters (Utinomi, 1947; Vassilenko, 1974; Arimoto, 1976; Hosono & Munehara, 2001). It can be hypothesized that *C. takedai* was de-

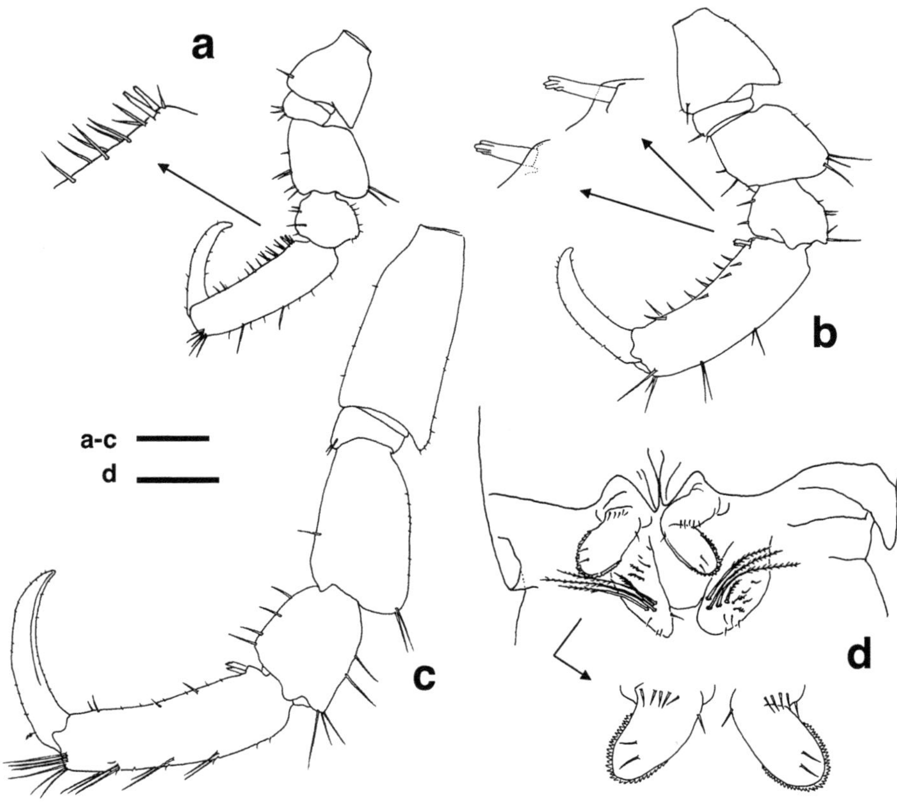

Fig. 4. *Caprella takedai*, new species, holotype male: a, left pereopod 5; b, left pereopod 6; c, left pereopod 7; d, abdomen, ventral view. Scales for a–c = 0.2 mm; d = 0.1 mm.

rived from these two species or the common ancestral species of them through local speciation.

Ecology. — *Caprella takedai* was collected from the green alga, *Cladophora wrightiana*, on rocks exposed to water current at a depth of 15 m off Ito, Shizuoka, Japan. Takeuchi et al. (1987, 1990) reported that both *Caprella subinermis* and *C. tsugarensis*, the most similar species to *C. takedai*, were collected from *Cladophora wrightiana* at a moderately exposed site in Uchiura Bay, Chiba, Japan. Their clinging behavior to the habitat substratum is apparently the "parallel" type in Takeuchi & Hirano (1995) using gnathopod 1 and three pairs of pereopods. *Caprella takedai* shows the same behavior to cling to the algal substrate (fig. 1). The species appears adapted to cling to thin substrata, withstanding even the most exposed current conditions.

ACKNOWLEDGMENTS

We thank the staff of Ito Diving Service for their help with field sampling. Thanks are also to Dr. Jiro Tanaka for the identification of the green algae. This work was supported by the FY2006 research grant of the Research Institute of Marine Invertebrates. This is contribution No. 757 from Shimoda Marine Research Center, University of Tsukuba.

REFERENCES

ARIMOTO, I., 1976. Taxonomic studies of caprellids (Crustacea, Amphipoda, Caprellidae) found in the Japanese and adjacent waters. Special Publications from the Seto Marine Biological Laboratory, **3**: 1-229.

HARVEY, W. H., 1859. Characters of new algae, chiefly from Japan and adjacent regions collected by Charles Wright in the North Pacific Exploring Expedition under Capt. John Rodgers. Proceedings of the American Academy of Arts and Sciences, **4**: 327-334.

HOSONO, T. & H. MUNEHARA, 2001. Caprellids (Crustacea, Amphipoda, Caprellidea) from Usujiri, Pacific coast of southern Hokkaido. Bulletin of Fisheries Sciences, **52**(1): 11-37.

LAMARCK, J. B., 1801. Systéme des animaux sans vertèbres, ou, tableau général des classes, des ordres et des genres de ces animaux: i–viii, 1-432. (Paris).

LEACH, W. E., 1814. Crustaceology. In: D. BREWSTER (ed.), The Edinburgh encyclopaedia, **7**(2): 385-437. (W. Blackwood, Edinburgh).

MAYER, P., 1890. Die Caprelliden des Golfes von Neapel und der angrenzenden Meeres-Abschnitte. Eine Monographie. Nachtrag zur Monographie derselben. Fauna und Flora des Golfes von Neapel, **17**: i–vii, 1-157. (R. Friedländer & Sohn, Berlin).

TAKEUCHI, I. & R. HIRANO, 1995. Clinging behavior of the epifaunal caprellids (Amphipoda) inhabiting the *Sargassum* zone on the Pacific coast of Japan, with its evolutionary implications. Journal of Crustacean Biology, **15**(3): 481-492.

TAKEUCHI, I., R. KUWABARA, R. HIRANO & H. YAMAKAWA, 1987. Species compositions of the Caprellidea (Crustacea: Amphipoda) of the *Sargassum* zone on the Pacific coast of Japan. Bulletin of Marine Science, **41**(2): 253-267.

TAKEUCHI, I., H. YAMAKAWA & M. FUJIWARA, 1990. The Caprellidea (Crustacea: Amphipoda) inhabiting the green alga *Cladophora wrightiana*: The influence of wave exposure on the species composition. La mer, **28**(2-3): 139-145.

UTINOMI, H., 1947. Caprellidea of Japan and adjacent waters. Seibutu Supplement, **1**: 68-82.

VASSILENKO, S. V., 1974. Caprellids (skeleton shrimps) of the seas of the USSR and adjacent waters. Order Amphipoda (Families Paracercopidae, Phtisicidae, Caprellidae). Opredeliteli Nauk SSSR, **107**: 1-287.

First received 7 October 2011.
Final version accepted 6 January 2012.

A NEW SPECIES OF *BOLLEGIDIA* (AMPHIPODA, BOGIDIELLIDAE SENSU LATO) FROM KUSHIMOTO COAST, CENTRAL JAPAN

BY

HIROYUKI ARIYAMA[1])
Marine Fisheries Research Center, Research Institute of Environment, Agriculture and Fisheries, Osaka Prefectural Government, Tanagawa, Misaki, Osaka 599-0311, Japan

ABSTRACT

Bollegidia takedai sp. nov. (Crustacea: Amphipoda: Bogidiellidae sensu lato) is described, based on specimens collected from Kushimoto coast in Wakayama Prefecture, central Japan. This new species is the third species in the genus *Bollegidia* Ruffo, 1974, which includes *B. sootai* (Coineau & Rao, 1972) from the Andaman and Nicobar Islands and *B. capensis* Ruffo, 1974 from South Africa. *Bollegidia takedai* differs from its congeners in an expanded palp article 2 of mandible, the presence of a setal row on the medial surface of gnathopod 2 and the shape of the pleopods, especially in male pleopod 1 with a complex membranous structure and male pleopod 2 with a 3-articulate ramus.

RESUMEN

Bollegidia takedai sp. nov. (Crustacea: Amphipoda: Bogidiellidae sensu lato) se describe sobre la base de muestras obtenidas para la costa de Kushimoto en la prefectura de Wakayama, en el centro de Japón. Esta nueva especie es la tercera especie del género *Bollegidia*, que incluye *B. sootai* (Coineau & Rao, 1972) de las islas de Andaman y Nicobar y *B. capensis* Ruffo, 1974 del África del Sur. *Bollegidia takedai* difiere de sus congéneres en (1) ampliar el artejo 2 del palpo mandibular, (2) la presencia de la fila de setas en la superficie medial de gnatópodo 2, y (3) formas de pleópodos, especialmente en pleópodo 1 del macho con complicada estructura membranosa y pleópodo 2 del macho con 3-articular rama.

INTRODUCTION

The amphipod genus *Bollegidia* was established by Ruffo (1974) with *B. capensis* Ruffo, 1974 as its type species, and he included *Bogidiella sootai*

[1]) e-mail: AriyamaH@mbox.epcc.pref.osaka.jp

Coineau & Rao, 1972 into this genus. *Bollegidia* was originally a member of the family Bogidiellidae Hertzog, 1936. However, Koenemann & Holsinger (1999) excluded the genus from this family, as the unusual combination of a uniramous uropod 1 and the extremely reduced rami of pleopods 1-3 is not diagnostic for bogidiellids according to their concept. During a survey of the amphipod fauna in Japan, a new species of the genus was obtained, herein described as *B. takedai*.

MATERIALS AND METHODS

About 2 liters of coarse sand were collected from sediment (ca. 10 cm sampling depth) in 4 m of water by snorkeling. The specimens were extracted from the sand using the decantation method, and fixed in a 10% formalin-seawater solution. Ten individuals of the amphipod were dissected and their appendages examined and illustrated under a phase-contrast microscope. Body length was measured from the apex of the rostrum along the dorsal margin to the distal end of the urosomites. The type series is deposited in the Osaka Museum of Natural History (OMNH), Osaka, Japan.

TAXONOMIC ACCOUNT

Genus *Bollegidia* Ruffo, 1974

Bollegidia takedai sp. nov.

(figs. 1-4)

Material examined. — Holotype: female (OMNH-Ar 8781), 1.4 mm, 33°26′49″N 135°47′39″E, Izumozaki coast, Kushimoto, Wakayama Prefecture, 8 August 2009, coll. H. Ariyama. Paratypes: 4 males and 5 females (OMNH-Ar 8782-8790), 1.2, 1.2, 1.2, 1.2, 1.4, 1.4, 1.4, 1.5, 1.3 mm (dissected), and a female (OMNH-Ar 8791), 1.5 mm (not dissected), same data as holotype.

Description. — Female [holotype and paratypes (OMNH-Ar 8786-8789, 8791)]. Body (fig. 1) extremely flattened laterally; rostrum indistinct, eyes lacking; pereonite 2 narrow, pereonites 5-7 wide, each pereonites with 1-2 dorsal setae; pleonites wide, without dorsal tooth, pleonites 2 and 3 bearing 1 and 2 dorsal setae, respectively; urosomites narrow, dorsal margins bare; coxae scarcely overlapping.

Antenna 1 (fig. 2A) relatively slender, with length ratio of peduncular articles 1-3 1.0 : 0.7 : 0.4, ventral margin of article 1 with 2 robust setae;

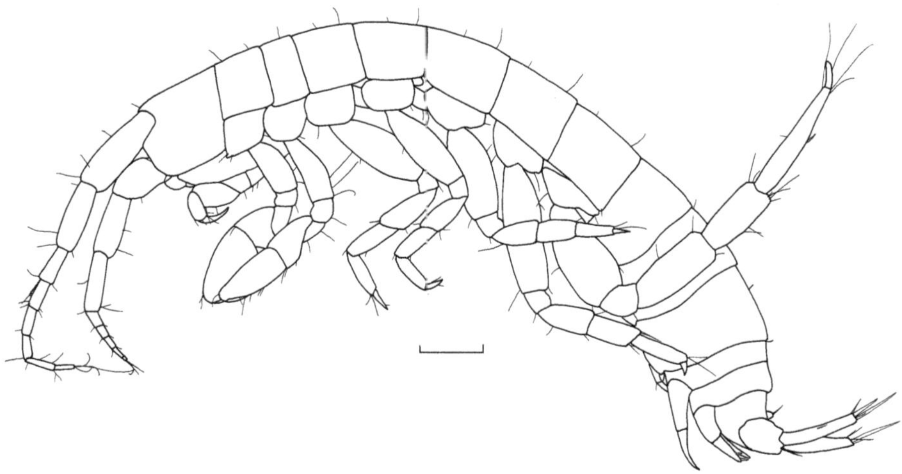

Fig. 1. *Bollegidia takedai* sp. nov., holotype, female (OMNH-Ar 8781), 1.4 mm, habitus, left lateral view. Scale: 0.1 mm.

accessory flagellum 2-articulate, with 3 setae on tip; flagellum with 5 articles, articles 3 and 4 each with aesthetasc at distal end. Antenna 2 (fig. 2B) relatively stout, shorter than antenna 1; ratio of lengths of peduncular articles 3-5 1.0 : 1.7 : 1.3, article 3 with anterior robust seta; flagellum short, with 4 articles.

Upper lip (fig. 2C) with ventral margin rounded, bare. Mandibles (fig. 2D, E) with left and right incisors bearing 5 and 4 cusps, respectively; both laciniae mobilis 2-toothed, both accessory blades 3 in number (1 wide and 2 narrow), molar small, simple; palp stout, article length ratio 1.0 : 1.3 : 1.3, article 2 expanded, articles 1-3 with 0, 1 and 3 setae, respectively. Lower lip (fig. 2F) with short mandibular process, apical margin bare. Maxilla 1 (fig. 2G) with round inner plate without setae; outer plate bearing 7 dentate robust setae apically; palp uniarticulate, with 2 setae on tip. Maxilla 2 (fig. 2H) with inner plate bearing 5 apical setae, outer plate with 4 apical setae. Maxilliped (fig. 2I) large; inner and outer plates small, each bearing robust seta on tip; palp stout, consisting of 4 articles, article 2 with long seta distally, distal part of article 3 and medial part of article 4 bearing dorsal setal row, article 4 with apical nail.

Gnathopod 1 (fig. 3A, A1) small, subchelate; coxa produced anterodistally, bearing 2 setae; basis stout, with 3 posterior long setae; ischium with seta at posterodistal corner; merus with 2 posterodistal setae, posteromedial surface covered with fine setae; carpus with 4 thick setae at posterodistal corner and many fine setae on posteromedial surface; propodus longer than carpus, palm oblique, with 2 large and 2 small robust setae; dactylus slender. Gnathopod 2 (fig. 3B, B1) small, a little longer than gnathopod 1, subchelate; coxa with 2

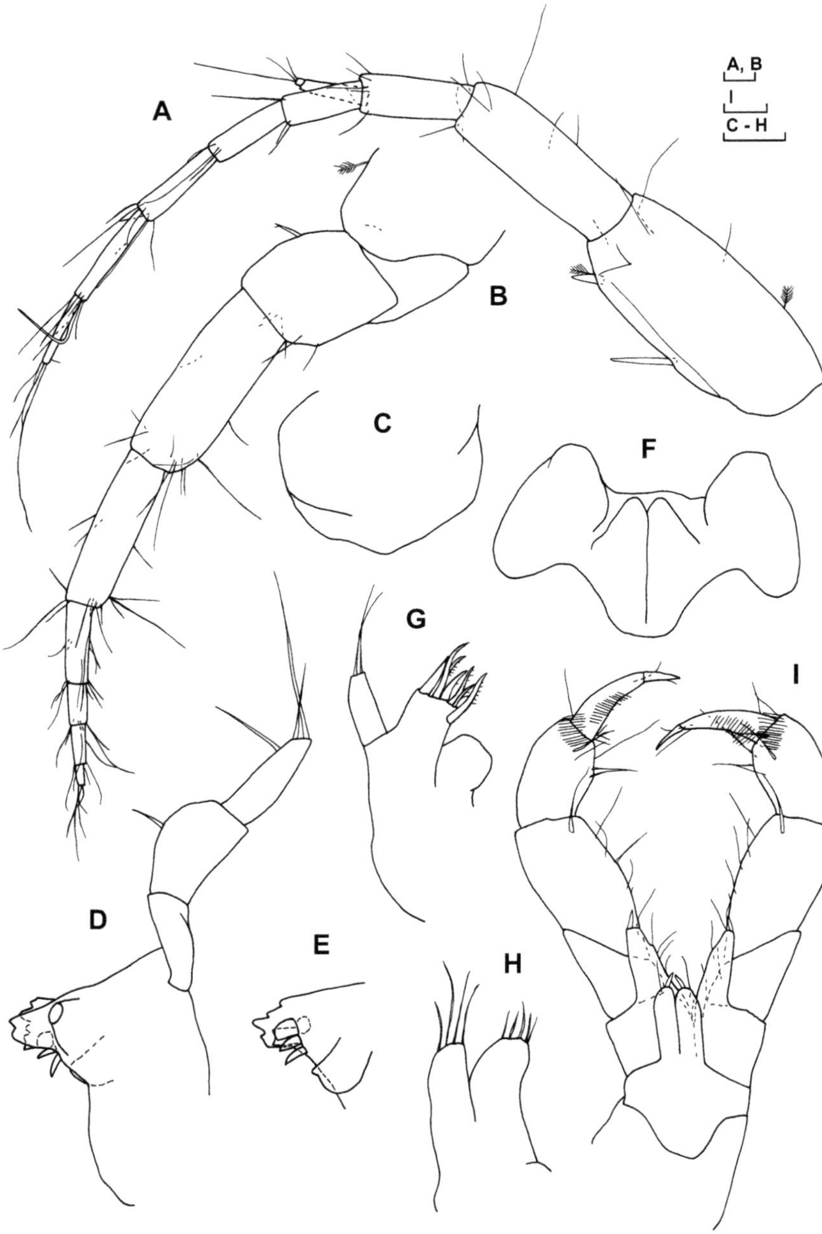

Fig. 2. *Bollegidia takedai* sp. nov. A, B, holotype, female (OMNH-Ar 8781), 1.4 mm; C, F, paratype, female (OMNH-Ar 8787), 1.4 mm; D, E, G–I, paratype, female (OMNH-Ar 8786), 1.4 mm. A, left antenna 1, lateral view; B, right antenna 2, medial view; C, upper lip, anterior view; D, left mandible, ventral view; E, right mandible (part), dorsal view; F, lower lip, ventral view; G, right maxilla 1, ventral view; H, right maxilla 2, ventral view; I, maxillipeds, dorsal view. Scales: 0.02 mm.

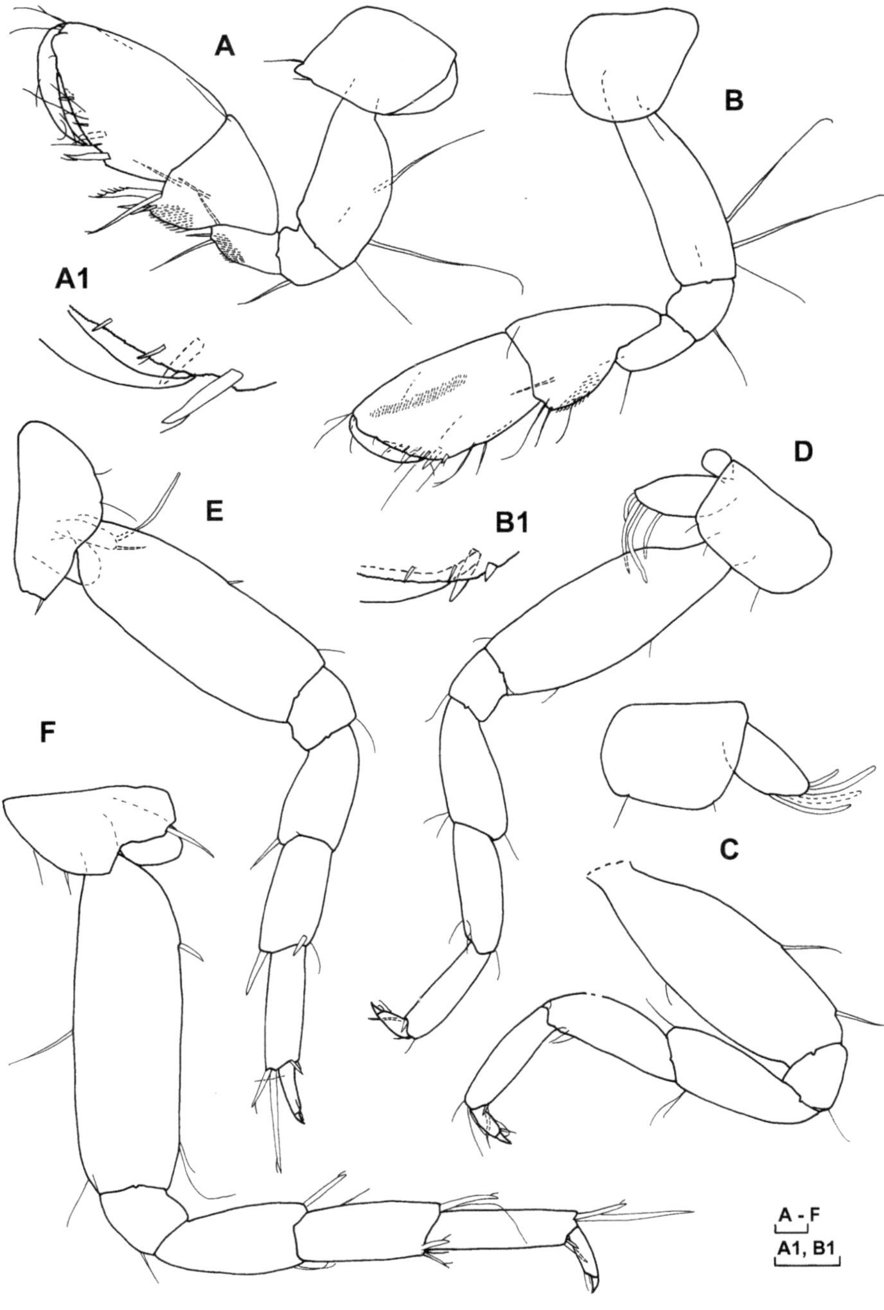

Fig. 3. *Bollegidia takedai* sp. nov., holotype, female (OMNH-Ar 8781), 1.4 mm. A, B, left gnathopods 1, 2, lateral views; A1, B1, palms of left gnathopods 1, 2, lateral views; C, left pereopod 3, lateral view; D, E, right pereopods 4, 5, lateral views; F, left pereopod 6, lateral view. Scales: 0.02 mm.

ventral setae, gill and oostegite lacking; basis slender, with 3 posterior long setae; ischium bearing posterior seta; merus with posterodistal seta; carpus with 3 posterodistal setae, posteromedial surface covered with fine setae; propodus a little longer than carpus, palm oblique, with 2 large and 2 small robust setae, anteromedial surface with row of fine setae; dactylus slender.

Pereopod 3 (fig. 3C) slender, longer than gnathopod 2; coxa short, with 2 ventral setae, gill absent, oostegite bearing 4 setae (1 lost); basis long, with 1 anterior and 2 posterior setae; merus, carpus and propodus slender, posterodistal corner of propodus with 1 long and 1 short robust setae; dactylus short, with nail. Pereopod 4 (fig. 3D) similar to pereopod 3, but coxa with 3 ventral setae, gill present, and basis with 2 anterior and 1 posterior setae. Pereopod 5 (fig. 3E) a little longer than pereopods 3 and 4; coxa triangular, bearing 2 ventral setae and posterodistal robust seta, gill present, oostegite with 2 setae; basis long, with 2 anterior setae; merus with posterodistal robust seta; carpus bearing 1 anterodistal and 1 posterodistal robust setae; propodus with 2 anterodistal robust setae and 2 posterodistal long thick setae; dactylus short, with nail. Pereopod 6 (fig. 3F) 1.2 times as long as pereopod 5; coxa bilobed, bearing 2 ventral setae and posterodistal robust seta, gill present; basis long, with 2 anterior setae, posteroproximal margin with robust seta; merus with posterodistal robust seta; carpus bearing 2 anterodistal and 1 posterodistal robust setae; propodus with 2 anterodistal robust setae and 2 posterodistal long thick setae; dactylus short, with nail. Pereopod 7 (fig. 4A) extremely long, 1.5 times as long as pereopod 6; coxa slightly bilobed, bearing ventral seta and posterodistal robust seta, gill absent; basis stout, anterior and posterior margins each with 2 robust setae; merus stout, with 1 anterodistal, 1 posterior and 1 posterodistal robust setae; carpus stout, with 2 anterior, 2 anterodistal and 3 posterodistal robust setae; propodus long, with 3 anterior, 3 anterodistal and 2 posterior long thick setae; dactylus slender, with nail.

Pleonal epimera 1-3 (fig. 4B) each with posteroventral tooth and short seta. Pleopods 1 and 2 (fig. 4C, D) similar; peduncles flattened dorsoventrally, each with small single coupling hook and mediodistal minute seta; rami lacking. Pleopod 3 (fig. 4E) much longer than pleopods 1 and 2; peduncle flattened dorsoventrally, with small single coupling hook and laterodistal seta; single

---

Fig. 4. *Bollegidia takedai* sp. nov. A, F–I, holotype, female (OMNH-Ar 8781), 1.4 mm; B, paratype, female (OMNH-Ar 8791), 1.5 mm; C–E, paratype, female (OMNH-Ar 8789), 1.4 mm; J, paratype, female (OMNH-Ar 8788), 1.4 mm; K–O, paratype, male (OMNH-Ar 8782), 1.2 mm. A, right pereopod 7, lateral view; B, left pleonal epimera 1–3, lateral view; C, D, left pleopods 1, 2, dorsal views; E, right pleopod 3, ventral view; F, left uropod 1, lateral view;

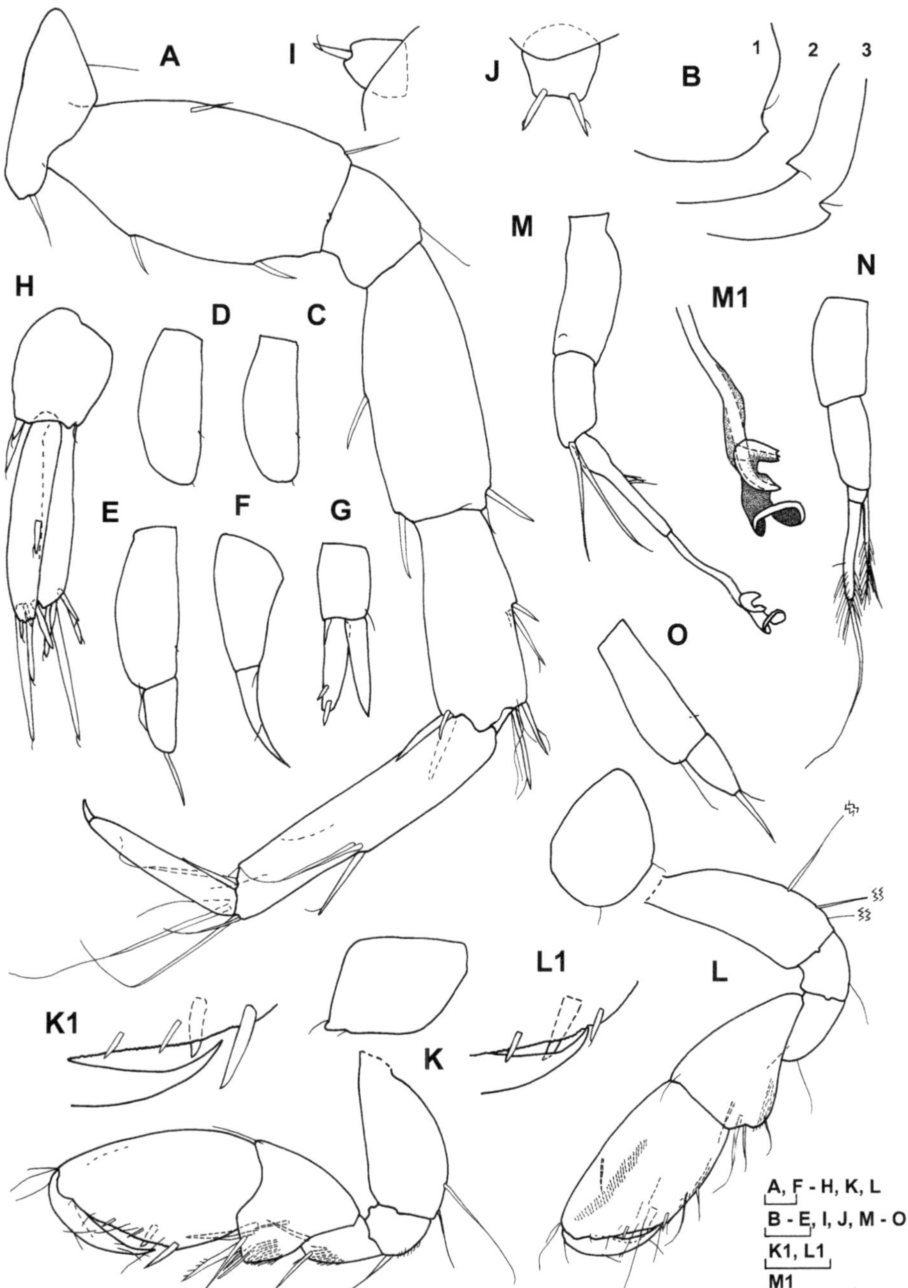

G, left uropod 2, dorsal view; H, left uropod 3, lateral view; I, telson, right lateral view; J, telson, dorsal view; K, L, left gnathopods 1, 2, lateral views; K1, L1, palms of left gnathopods 1, 2, lateral views; M, N, left pleopods 1, 2, ventral views; M1, tip of left pleopod 1, ventral view; O, right pleopod 3, ventral view. Scales: 0.02 mm.

ramus uniarticulate, bearing distal robust seta. Uropod 1 (fig. 4F) uniramous; peduncle with laterodistal seta; ramus pointed distally, with seta. Uropod 2 (fig. 4G) biramous, shorter than uropod 1; peduncle bearing mediodistal seta; both rami longer than peduncle, outer ramus with 2 distal robust setae, inner ramus bare. Uropod 3 (fig. 4H) biramous, long, 1.9 times as long as uropod 2; peduncle short, with 2 laterodistal robust setae and mediodistal seta; outer and inner rami 1.8 and 1.6 times as long as peduncle, respectively, outer ramus with 1 dorsal and 5 distal robust setae, inner ramus bearing 5 distal robust setae. Telson (fig. 4I, J) small, a little broader than long, fleshy, with 2 robust setae distally.

Male [paratype (OMNH-Ar 8782)]. Almost same as females except for oostegites and pleopods. Gnathopods 1, 2 (fig. 4K, K1, L, L1) similar to those of female, but basis of gnathopod 1 with 2 posterior long setae, and palm of gnathopod 2 with 1 large and 2 small robust setae. Pleopod 1 (fig. 4M, M1) long, not flattened; peduncle slender, with small single coupling hook; single ramus 2-articulate, article 1 with 2 long distal setae; article 2 with 4 short lateral setae and strange-shaped long thick setae bearing complicated membranous structure. Pleopod 2 (fig. 4N) long, not flattened; peduncle short; single ramus 3-aticulate, article 1 relatively long, with plumose seta, article 2 very short, with simple seta, article 3 setose distally, bearing long thick seta on tip. Pleopod 3 (fig. 4O) similar to that of female, but single ramus bearing 1 robust and 1 normal setae distally.

Etymology. — The species name is dedicated to Dr. Masatsune Takeda for his great contribution to crustacean taxonomy.

DISCUSSION

Bollegidia takedai sp. nov. resembles *B. sootai* and *B. capensis* in most of the morphological characters. However, this new species differs from its congeners in the expanded palp article 2 of mandible, the presence of a setal row on the medial surface of gnathopod 2 and the shape of the pleopods, especially in male pleopod 1 with a complex membranous structure and male pleopod 2 with a 3-articulate ramus. In contrast, the palp article 2 of the mandible is straight and the setal row on the medial surface of gnathopod 2 is absent in both *B. sootai* and *B. capensis* (Coineau & Rao, 1972; Ruffo, 1974). In the female pleopods of both species, the peduncle of pleopod 2 bears relatively long seta, and the ramus of pleopod 3 is rather short, without seta. Moreover in the male pleopods of the both species, ramus of pleopod 1 does

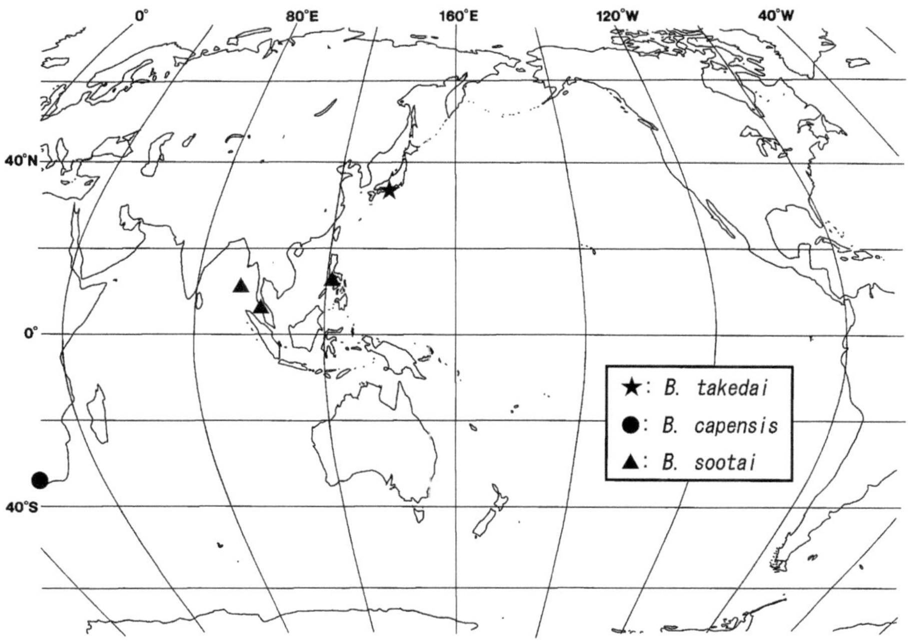

Fig. 5. Global distribution of *Bollegidia* species.

not possess membranous structure, ramus of pleopod 2 is 2-articulate in *B. sootai* and uniarticulate in *B. capensis*, and pleopod 3 is more slender (Coineau & Rao, 1972; Ruffo, 1994). In addition, *B. capensis* has a short dorsal tooth on each of pleonites 1-3; which is absent in *B. sootai* and *B. takedai*.

Figure 5 shows the global distribution of *Bollegidia* species. In the present study, *Bollegidia takedai* constitutes the first record of the genus for Japan, *Bollegidia sootai* was collected from the Andaman and Nicobar Islands, Malaysia and the Philippine Islands (Coineau & Rao, 1972; Ruffo, 1985, 1994), whilst *B. capensis* is only known from South Africa (Ruffo, 1974). Considering the very small body sizes of the species (1.2, 0.9 and 1.5 mm in *B. sootai*, *B. capensis* and *B. takedai*, respectively) and the relatively few studies so far on marine interstitial fauna, *Bollegidia* species may be more widespread in the Indo-West Pacific area.

ACKNOWLEDGEMENTS

I wish to thank the late Prof. Sandro Ruffo and Dr. Traudl Krapp-Schickel (Forschungsinstitut und Museum Alexander Koenig) for sending literature, and Ikuko Ariyama for translation of the abstract into Spanish. I am also grateful to Dr. Kouki Fukuoka of Seikai National Fisheries Research Institute and two anonymous reviewers for a critical reading of the manuscript.

REFERENCES

COINEAU, N. & G. C. RAO, 1972. Isopodes et Amphipodes des sables intertidaux des Iles Andaman et Nicobar (Golfe du Bengale). Vie et Milieu, **23**(1A): 65-100.

HERTZOG, L., 1936. Crustacés de biotopes hypogées de la vallée du Rhin d'Alsace. Bulletin de la Société Zoologique de France, **61**: 356-372.

KOENEMANN, S. & J. R. HOLSINGER, 1999. Phylogenetic analysis of the amphipod family Bogidiellidae s. lat., and revision of taxa above the species level. Crustaceana, **72**(8): 781-816.

RUFFO, S., 1974. Nouvi Anfipodi interstiziali delle coste del Sud Africa. Atti dell'Instituto Veneto di Scienze, Lettere ed Arti, **132**: 399-419.

— —, 1985. Nuovi Anfipodi mesopsammici delle Isole Andamane (Crust. Amphipoda). Bollettino del Museo Civico di Storia Naturale, Verona, **10**: 485-509.

— —, 1994. New stygobiont amphipods (Crustacea Amphipoda) from the Philippine Islands. Tropical Zoology, **7**: 355-366.

First received 29 August 2011.
Final version accepted 19 December 2011.

MITHRACITES TAKEDAI, A NEW HOMOLOID CRAB (DECAPODA, BRACHYURA) FROM THE LOWER CRETACEOUS (APTIAN) OF COLOMBIA

BY

BARRY W. M. VAN BAKEL[1,2,5]), DANIÈLE GUINOT[3,6]), JOHN W. M. JAGT[4,7]) and RENÉ H. B. FRAAIJE[1,8])

[1]) Oertijdmuseum De Groene Poort, Bosscheweg 80, NL-5283 WB Boxtel, The Netherlands
[2]) Nederlands Centrum voor Biodiversiteit (Naturalis), P.O. Box 9517, NL-2300 RA Leiden, The Netherlands
[3]) Département Milieux et peuplements aquatiques, Muséum national d'Histoire naturelle, 61 rue Buffon, CP 53, F-75231 Paris Cedex 5, France
[4]) Natuurhistorisch Museum Maastricht, de Bosquetplein 6-7, NL-6211 KJ Maastricht, The Netherlands

ABSTRACT

A new homoloid crab, *Mithracites takedai*, is described on the basis of two specimens from Colombia, preserved in a limestone nodule of early Aptian (Early Cretaceous) age, as dated by co-occurring ammonite species. The new species differs from the type species, *M. vectensis*, by a distinctly sulcate front, a narrower anterior mesogastric process, and a differently shaped cardiac region. The status of the family Mithracitidae Števčić, 2005, is revised and an emended diagnosis is provided; the appearance of the linea homolica in extinct homoloids is discussed.

INTRODUCTION

During a recent survey of fossil decapod crustaceans housed in the collections of the Nederlands Centrum voor Biodiversiteit (Naturalis, Leiden), a new species of homoloid crab was discovered in limestone nodules from the Paja Formation at Santander, Colombia. The associated ammonite fauna (Kakabadze & Hoedemaeker, 1997; Hoedemaeker, 2004; Kakabadze et al., 2004)

[5]) Corresponding author; e-mail: barryvanbakel@gmail.com
[6]) e-mail: guinot@mnhn.fr
[7]) e-mail: john.jagt@maastricht.nl
[8]) e-mail: info@oertijdmuseum.nl

dates this material as Aptian (Early Cretaceous). One subspecies in particular, *Cheloniceras crassum crassum* Spath, 1930, found in the same piece of matrix as the crab material, is known from the lower Aptian (c. 125-120 Ma) of southern England, Georgia, Bulgaria and Colombia (Kakabadze et al., 2004: 528).

Previously, only a single species of crab has been recorded from the Paja Formation. *Dakoticancer olssoni* Rathbun, 1937, originally described from the Barremian of Santander (Colombia), was transferred to *Necrocarcinus* by Feldmann et al. (1999: 91) and synonymised with *Orithopsis tricarinata* by Vega et al. (2010: 275). The latter authors pointed out that the type specimen was from the Paja Formation, which they inconsistently noted to be of late Aptian or late Albian age (Vega et al., 2010: 276). Examination of the original illustration (Rathbun, 1937: pl. 5 fig. 6) clearly shows that this specimen cannot be attributed either to *Orithopsis tricarinata* or to the Orithopsidae as already stated by Schweitzer & Feldmann (2011a: 5); we prefer placing it in *Paranecrocarcinus* Van Straelen, 1936.

Mithracites takedai n. sp. constitutes the second formally named species of a genus which hitherto was known solely from one species from southern England. The taxonomic status of *Mithracites* has recently been the subject of discussion. A historical account was presented by Guinot & Tavares (2001), and Števčić (2005) subsequently erected a new family, Mithracitidae, for it. Karasawa et al. (2011) and Schweitzer & Feldmann (2011b) provided new diagnoses, but neither of these contains a good synapomorphy for the alleged family. The possible absence of a linea homolica could be an important feature of the family, and to verify the appearance of this in the fossil state, extant material should be screened for new insights.

SYSTEMATIC PART

Abbreviations used in the text are as follows: P1–P5: pereiopods 1 to 5; mxp3: third maxilliped; MAB: Oertijdmuseum De Groene Poort, Boxtel, The Netherlands; MNHN: Muséum national d'Histoire naturelle, Paris, France; RGM: Nederlands Centrum voor Biodiversiteit (Naturalis; formerly Rijksmuseum van Geologie en Mineralogie), Leiden, the Netherlands.

Section PODOTREMATA Guinot, 1977
Superfamily HOMOLOIDEA De Haan, 1839
Family ?MITHRACITIDAE Števčić, 2005

Included genera. — *Mithracites* Gould, 1859; *Enodicarcinus* Schweitzer & Feldmann, 2011.

Diagnosis. — Carapace suboval, width equalling length or slightly longer than wide, widest in posterior third. Dorsal surface areolated; cervical and branchial grooves well marked. Rostrum single, deflexed, connecting epistome through narrow proepistome. Anterolateral margin short, straight, distinct; posterolateral margin long, convex, indistinct, blunt with few tubercles. Posterior margin wide, convex to weakly concave. Linea homolica absent (see Remarks). False orbits large, in contact with cervical groove; extraorbital tooth robust. Buccal cavern widened anteriorly; mxp3 operculiform, longitudinally grooved. Chelipeds slightly unequal, fingers short; P2–P4 elongated; P5 reduced, subdorsal. Abdomen with six somites plus telson; male abdomen large, completely covering thoracic sternum; female abdomen slightly larger, fixed anteriorly between P1–P3 coxae, telson entering between mxp3 coxae; somites 1 and 2 narrow, exposed dorsally. Thoracic sternum with sterno-coxal depressions. Thoracic sternite 4 with homoloid press-button; uropods of abdominal somite 6 transformed into sockets.

Remarks. — Guinot and Tavares (2001), who discussed at length the status of *Mithracites*, suggested affiliation with Homoloidea, mainly based on a typical suite of characters associated with the homoloid abdominal locking system (Bouchard, 2000: fig. 27B). Števčić (2005: 23) subsequently erected the Mithracitidae for the genus, but his diagnosis does not provide any synapomorphy or unique character set for this new family. An assumed absence of a linea homolica in *Mithracites* is not mentioned; instead, the author stated, 'lateral margins distinct'. Schweitzer & Feldmann (2011a: 2) did not include the absence of a linea homolica in their diagnosis either, but mentioned it in their discussion, and stated that Števčić (2005) referred *Mithracites* to its own family 'based upon its unique features', while nearly all characters used are those of the Homolidae De Haan, 1839. In contrast, Karasawa et al. (2011: 547) did include the absence of a linea homolica in their characterisation of the family. Here we refine previous diagnoses (Števčić, 2005; Karasawa et al., 2011; Schweitzer & Feldmann, 2011a), based on our observations of the type species, *Mithracites vectensis* Gould, 1859, and the present new Colombian species.

The absence of a linea homolica in the Mithracitidae requires discussion. In homoloids in which the posterior carapace does not expand ventrally and envelop the cephalothorax, as in the Poupiniidae Guinot, 1991, and Latreilliidae Stimpson, 1858, a linea homolica is absent (a lateral linea is partially present in Latreilliidae, being only at the level of the long 'neck'). Among homoloids, only homolids possess a (paired) dorsal linea homolica; and, this would be a significant difference between the Mithracitidae and Homolidae. No linea homolica is observed in the two specimens of the new Colombian species. It has not been reported or illustrated in all previously described individuals of *M. vectensis* (cf. Withers, 1951: pl. 17 figs. 1-5; Wright & Collins, 1972: pl. 5 figs. 1-7; Guinot & Tavares, 2001: fig. 19; Schweitzer & Feldmann, 2011a: figs. 1.2, 1.3), of *M. vectensis*? sensu Schweitzer & Feldmann (2011a: fig. 1.1) and is also absent in a newly collected specimen of *M. vectensis* (MAB k.2376) examined for the present study.

A survey of the linea homolica in extant Homolidae housed in the MNHN collections has shown that there is quite some variation in the appearance of this moulting line, ranging from wide to attenuated, even obsolete. The linea homolica is sometimes undifferentiated in the anterior carapace, but visible in the surface of the false orbit. In most cases the linea homolica is well defined particularly along the large branchial area as a thin line, semi-transparent, along which cuticle granulation is interrupted. However, this condition was found to be inconsistent; the linea homolica varies markedly, being occasionally very thin, and is sometimes virtually imperceptible (Guinot & Richer de Forges, 1995: 299).

To test what the linea homolica would look like in the fossil state (in those cases where cuticle is preserved), a piece of modelling clay was used to produce imprints of the branchial area of several homolids. In *Homolochunia valdiviae* Doflein, 1904, a linea homolica is practically invisible; only a weakly transparent line is discernible at carapace mid-length (Guinot & Richer de Forges, 1995: fig. 50d). In *Lamoha inflata* (Guinot & Richer de Forges, 1981), the linea homolica is recognisable as a faintly transparent line, while in *Homologenus levii* Guinot & Richer de Forges, 1995, it can be seen as a thin, transparent line but as it is lined with numerous small granules (Guinot & Richer de Forges, 1995: figs. 60b, 67f), the linea would not be visible in the fossil state. Thus, there may be considerable difference in the appearance of the linea homolica in fossil members of the family Homolidae. Some have a clearly differentiated linea homolica, while in others it may appear obsolete. In addition, preservational issues need to be considered as well. The outer layer of the cuticle may either be preserved intact, or became peeled, and as such, the cuticle may be modified. Extinct homolid carapaces are occasionally also found to be disarticulated at the level of the lineae (see Crawford, 2008), and as such, the extralineal portions of the carapace are not preserved in some material (see e.g., Beschin et al., 2009: fig. 5). These preliminary observations on the poorly studied linea homolica indicate that caution needs to be exercised before conclusions can be drawn and more studies are needed.

Schweitzer & Feldmann (2011a: 2) stated that the Mithracitidae also differed from other homoloids in probably having a subdorsal P4, while Karasawa et al. (2011: 547) affirmed that a subdorsal P4 was a characteristic feature of the family. After a detailed comparison with numerous different members of the Homoloidea, we have found the disposition of the pereiopods in *Mithracites* to be similar to that of extant homoloids. The P1–P5 coxae are not in the same plane, in accordance with the posterior curvature of the thoracic sternites (see illustrations in Garassino, 2009). On account of the width and curvature of the thoracic sternum, the P2–P4 coxae may be visible in dorsal view, but the legs are not 'subdorsal'; these function as normal walking legs (in contrast to P5, used for carrying behaviour).

For the time being, we maintain the Mithracitidae, but with a query. In view of the fact that none of the specimens of the two known *Mithracites* species known to date display any trace of a moulting line, a linea homolica appears to be absent in the Mithracitidae. In addition, the carapace is very short for a homoloid (although some homolids do have an equally short carapace, e.g., *Paromolopsis boasi* Wood-Mason & Alcock, 1891). It is hoped that additional material of *Mithracites* will become available for study, in order to determine the status of this genus.

Several taxa that are known in particular from carapace material have been assigned to the Glaessneropsoidea Patrulius, 1959, and Homolodromioidea Alcock, 1900, based solely on the absence of a linea homolica. These may turn out to be members of the Homoloidea. Examples include *Palehomola* Rathbun, 1926 (see Schweitzer et al., 2004: fig. 7) and several species assigned to *Glaessnerella* Wright & Collins, 1975 (see e.g., Schweitzer & Feldmann, 2011b). We must await the discovery of specimens retaining ventral characters to identify their taxonomic position in more detail.

Genus **Mithracites** Gould, 1859

Type species. — *Mithracites vectensis* Gould, 1859, by monotypy, from the Aptian of the Isle of Wight, southern England.

Mithracites takedai n. sp.
(fig. 1)

Types. — The holotype, a carapace of a presumably adult specimen with partial left and right chelipeds preserved, is RGM M988/C-1 (ex P. H. Creutzberg Collection); a single paratype (RGM M988/C-2) is available; this probably is a juvenile. Both specimens are preserved in the same piece of matrix and associated with an unidentified gastropod and the ammonite

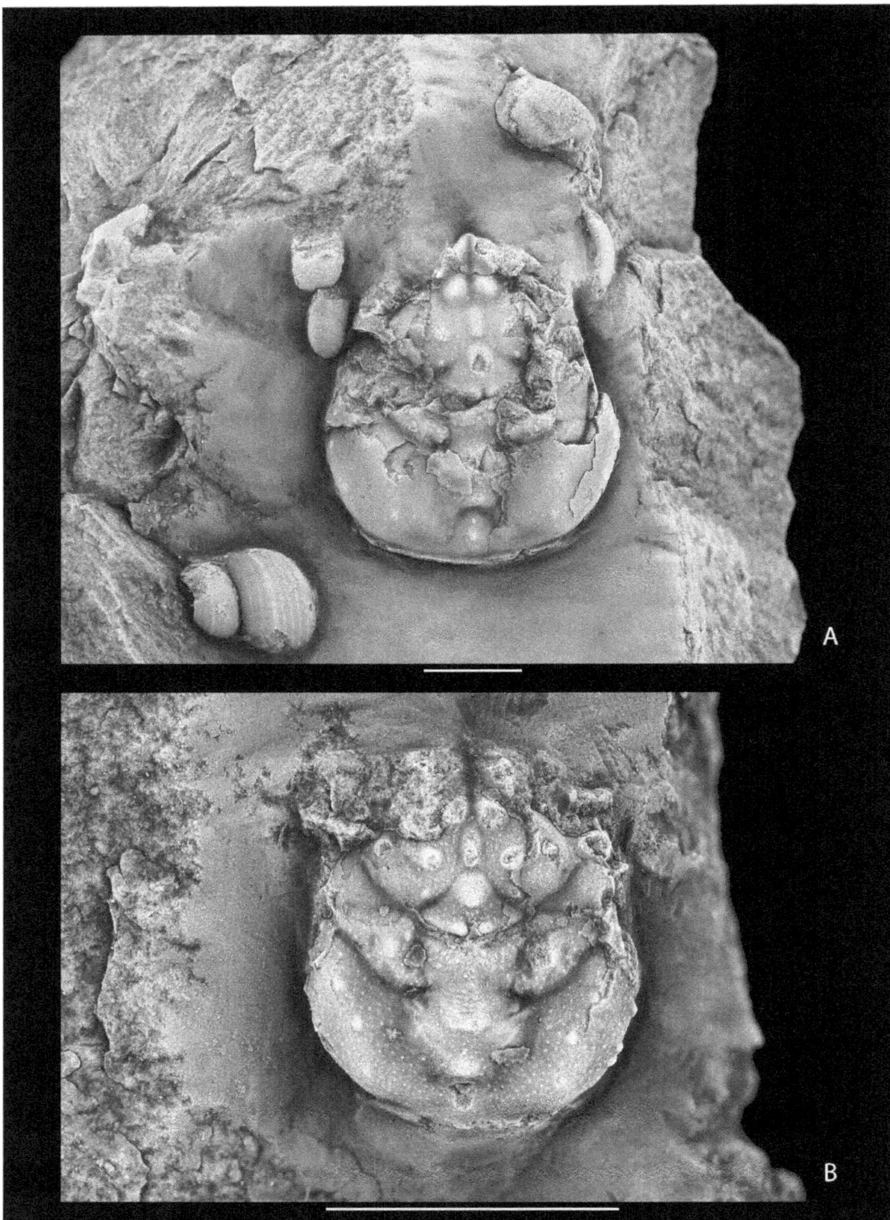

Fig. 1. *Mithracites takedai* n. sp. A, Holotype, RGM M988/C-1, presumably adult; B, Paratype, RGM M988/C-2, a probable juvenile. Scale bars equal 5 mm.

Cheloniceras crassum crassum (RGM 282 770 M988/B). The type locality is 'along the road near house of fam. Rivera', near Guane (Santander, northwest Colombia), and the stratigraphic level is the Paja Formation of Aptian age.

Etymology. — The species name honours Masatsune Takeda in recognition of his valuable contributions to our knowledge of Brachyura.

Diagnosis. — *Mithracites* with distinctly sulcate front, narrow anterior mesogastric process, and triangular cardiac region.

Description. — Carapace small, suboval, longer than wide, widest in posterior half, convex in transverse cross section, gently arched in longitudinal cross section. Carapace regions inflated, well defined by distinct grooves, tuberculate. Orbitofrontal margin wide. Front triangular, long, projected, slightly downturned, distinctly sulcate up to tip. Orbits large, anteriorly directed, with blunt, sinuous supraorbital margin. Exorbital tooth strong. Anterolateral margin short, straight. Posterolateral margin long, continuous, convex, with rounded edge and three small tubercles. Posterior carapace margin very wide, convex, with subtle, acute rim. Epigastric regions small, well defined, inflated, oval. Anterior mesogastric process narrow, anteriorly restricted by epigastric swellings, posteriorly with tubercle. Posterior mesogastric region wide, triangular, anteriorly with central tubercle, posteriorly with short axial sulcus, posterior margin arched. Protogastric region large, with outwardly-directed groove, anteriorly with two large tubercles, posteriorly with single, small, acute tubercle. Hepatic region small, triangular, without tubercle. Urogastric lobe short, weakly inflated. Cardiac region subtriangular with apex downwards, anteriorly and posteriorly with tubercle. Intestinal region small, with central tubercle. Branchial region large, posteriorly with three tubercles arranged in an outwardly-directed arc. Cervical groove distinct, continuous, arched, reaching exorbital teeth. Branchiocardiac grooves shallow, short. Branchial grooves distinct, deep, parallel to cervical groove, notching carapace margins. Linea homolica apparently not present. Carapace surface covered entirely with small, evenly spaced granules (juvenile), or only granular on posterior branchial region (adult). P1 merus elongate, curved; carpus short, stout; right cheliped short, outer surface tumid, with subtle groove, fixed finger slender.

Remarks. — *Mithracites takedai* n. sp. is similar to the type species, *M. vectensis*, in carapace shape and curvature, shape of orbits, and division of regions and tuberculation. The new species differs in having a front that is distinctly sulcate up to the apex; in having a narrower anterior mesogastric process, and a more triangular cardiac region that is narrow posteriorly.

The branchial groove is continuous in *M. takedai* whereas it is sinuous in *M. vectensis*. In addition, in the adult specimen of *M. takedai*, the dorsal granulation is nearly obsolete, while adult specimens of *M. vectensis* have a clearly granular dorsal surface.

Mithracites takedai differs from *M. vectensis*? sensu Schweitzer & Feldmann (2011a: 2) in having a convex, rather than concave, posterior carapace margin, and in having tubercular axial and branchial regions.

ACKNOWLEDGEMENTS

We thank Frank P. Wesselingh, Stephen K. Donovan and Wouter Wildenberg (all NCB-RGM, Leiden) for access to collections in their care and Bart Fraaije for donating new material of *Mithracites vectensis* to the Oertijdmuseum De Groene Poort (Boxtel) collections. We acknowledge the research support from the SYNTHESYS Project http://www.synthesys.info/ which is financed by European Community Research Infrastructure Action under the FP7 "Capacities" Program, which allowed the senior author to visit the MNHN in July 2011 to examine extant homoloids. We are grateful to Francisco J. Vega (UNAM, Mexico) and an anonymous reviewer for their comments on an earlier manuscript.

REFERENCES

ALCOCK, A., 1900. Materials for a carcinological fauna of India, 5: The Brachyura Primigenia or Dromiacea. Journal of the Asiatic Society of Bengal, (II) **68**(3): 123-169.

BESCHIN, C., A. DE ANGELI & R. ZORZIN, 2009. Crostacei fossili del Veneto: una inedita fauna eocenica dei Lessini oriental (Monte Serea di San Giovanni Ilarione, Verona), con descrizione di tre nuove specie. Bollettino del Museo civico di Storia naturale di Verona, **33**: 59-83.

BOUCHARD, J.-M., 2000. Morphologie fonctionnelle des systèmes de rétention de l'abdomen chez les Brachyoures (Crustacea Decapoda). Microstructures, implications phylogénétiques et systématiques: 1-694. (Unpublished Ph.D. thesis, Muséum national d'Histoire naturelle, Paris).

CRAWFORD, R. S., 2008. A new species of fossil homolid crab (Decapoda: Brachyura) from the Río Foyel Formation (Paleogene), Río Negro Province, Argentina. Journal of Paleontology, **82**: 835-841.

DE HAAN, F., 1833-1850. Crustacea. In: P. F. VON SIEBOLD (ed.), Fauna Japonica, sive descriptio animalium, quae in itinere per Japoniam, jussu et auspiciis superiorum, qui summum in India Batava Imperium tenent, suscepto, annis 1823-1830 collegit, notis, observationibus e adumbrationibus illustravit: i–lxxi, i–vii, i–xvii, i–ixxvi, 1-243, pls. 1-55, A–Q, pl. 2. A. (Arnz, Lugduni Batavorum (Leiden)). [see Holthuis & Sakai, 1970 for dates of publication].

DOFLEIN, F., 1904. Brachyura. Wissenschaftliche Ergebnisse der Deutschen Tiefsee-Expedition auf dem Dampfer 'Valdivia' 1898-1899, **6**: i–xiv, 1-314, pls. 1-7.

FELDMANN, R. M., T. VILLAMIL & E. G. KAUFFMAN, 1999. Decapod and stomatopod crustaceans from mass mortality Lagerstatten: Turonian (Cretaceous) of Colombia. Journal of Paleontology, **73**: 91-101.

GARASSINO, A., 2009. The thoracic sternum and spermatheca in the extant genera of the family Homolidae De Haan, 1839 (Crustacea, Decapoda, Brachyura). Memorie della Società Italiana di Scienze Naturali e del Museo Civico di Storia Naturale di Milano, **36**: 1-80.

GOULD, C., 1859. Description of a new fossil crustacean from the Lower Greensand. Quarterly Journal of the Geological Society London, **15**: 237-238.

GUINOT, D., 1977. Propositions pour une nouvelle classification des Crustacés Décapodes Brachyoures. Comptes Rendus des Séances hebdomadaires de l'Académie des Sciences, Paris, (D) **285**: 1049-1052.

— —, 1991. Établissement de la famille des Poupiniidae pour *Poupinia hirsuta* gen. nov., sp. nov. de Polynésie (Crustacea Decapoda Brachyura Homoloidea). Bulletin du Muséum national d'Histoire naturelle, Paris, (4) (A) **12**: 577-605.

GUINOT, D. & B. RICHER DE FORGES, 1981. Homolidae, rares ou nouveaux, de l'Indo-Pacifique (Crustacea, Decapoda, Brachyura). Bulletin du Muséum national d'Histoire naturelle, Paris, (4) (A) **3**: 523-581.

— — & — —, 1995. Crustacea Decapoda Brachyura: révision de la famille des Homolidae De Haan, 1839. In: A. CROSNIER (ed.), Résultats des campagnes MUSORSTOM, vol. 13. Mémoires du Muséum national d'Histoire naturelle, Paris, **163**: 283-517.

GUINOT, D. & M. TAVARES, 2001. Une nouvelle famille de Crabes du Crétacé et la notion de Podotremata Guinot, 1977 (Crustacea, Decapoda, Brachyura). Zoosystema, **23**: 507-546.

HOEDEMAEKER, P. J., 2004. On the Barremian-lower Albian stratigraphy of Colombia. Scripta Geologica, **128**: 3-15.

HOLTHUIS, L. B. & T. SAKAI, 1970. Ph.F. Von Siebold and Fauna Japonica. A history of early Japanese Zoology: 1-323, pls. I–XXXII. (Academic Press of Japan, Tokyo).

KAKABADZE, M. V. & P. J. HOEDEMAEKER, 1997. New and less known Barremian-Albian ammonites from Colombia. Scripta Geologica, **114**: 57-117.

KAKABADZE, M. V., P. J. HOEDEMAEKER, T. N. BOGDANOVA & M. Z. SHARIKADZE, 2004. On the Barremian-Early Albian biogeography (by ammonites) of Colombia. Scripta Geologica, **128**: 515-558.

KARASAWA, H., C. E. SCHWEITZER & R. M. FELDMANN, 2011. Phylogenetic analysis and revised classification of podotrematous Brachyura (Decapoda) including extinct and extant families. Journal of Crustacean Biology, **31**: 523-565.

PATRULIUS, D., 1959. Contributions à la systématique des décapodes néojurassiques. Revue de Géologie et Géographie, **3**: 249-257.

RATHBUN, M. J., 1926. The fossil stalk-eyed Crustacea of the Pacific Slope of North America. United States National Museum Bulletin, **138**: 1-155.

— —, 1937. Cretaceous and Tertiary crabs from Panama and Colombia. Journal of Paleontology, **11**: 26-28.

SCHWEITZER, C. E. & R. M. FELDMANN, 2011a. New fossil Brachyura (Decapoda: Homoloidea, Dorippoidea, Carpilioidea) from the United Kingdom. Bulletin of the Mizunami Fossil Museum, **37**: 1-11.

— — & — —, 2011b. Revision of some fossil podotrematous Brachyura (Homolodromiidae; Longodromitidae; Torynommidae). Neues Jahrbuch für Geologie und Paläontologie Abhandlungen, **260**: 237-256.

SCHWEITZER, C. E., T. G. NYBORG, R. M. FELDMANN & R. L. M. ROSS, 2004. Homolidae De Haan, 1839 and Homolodromiidae Alcock, 1900 (Crustacea: Decapoda: Brachyura) from the Pacific Northwest of North America and a reassessment of their fossil records. Journal of Paleontology, **78**: 133-149.

SPATH, L. F., 1930. On some Ammonoidea from the Lower Greensand. Annals and Magazine of Natural History, (10) **5**: 417-464.

ŠTEVČIĆ, Z., 2005. The reclassification of brachyuran crabs (Crustacea: Decapoda: Brachyura). Natura Croatica, **14**: 1-159.

STIMPSON, W., 1858. Crustacea Anomura, Pars VII. Prodromus descriptionis animalium evertebratorum, quae in Expeditione ad Oceanum Pacificum Septentrionalem, a Republica Federata missa, Cadwaladaro Ringgold et Johanne Rodgers Ducibus, observavit et descripsit. Proceedings of the Academy of Natural Sciences of Philadelphia, **10**: 225-252.

VAN STRAELEN, V., 1936. Crustacés décapodes nouveaux ou peu connus de l'époque crétacique. Bulletin du Musée royal d'Histoire naturelle de Belgique, **12**(45): 1-50.

VEGA, F. J., T. NYBORG, G. KOVALCHUK, F. ETAYO, J. LUQUE, A. ROJAS-BRICEÑO, P. PATARROYO, H. PORRAS-MÚZQUIZ, A. ARMSTRONG, H. BERMÚDEZ & L. GARIBAY, 2010. On some Panamerican Cretaceous crabs (Decapoda: Raninoida). Boletín de la Sociedad Geológica Mexicana, **62**: 263-279.

WITHERS, T. H., 1951. Some Jurassic and Cretaceous crabs (Prosoponidae). Bulletin of the British Museum (Natural History), **6**: 171-186.

WOOD-MASON, J. & A. ALCOCK, 1891. A note on the result of the last season's deep-sea dredging: natural history notes from H. M. Indian Marine Survey Steamer "Investigator": Commander R. F. Hoskyn, R. N. Commanding, no. 21. Annals and Magazine of Natural History, (6) **7**: 258-272.

WRIGHT, C. W. & J. S. H. COLLINS, 1972. British Cretaceous crabs. Monograph of the Palaeontographical Society London, **126**(533): 1-114.

—— & ——, 1975. *Glaessnerella* (Crustacea, Decapoda, Cymonomidae), a replacement name for *Glaessneria* Wright & Collins, 1972 *non* Takeda & Miyake, 1969. Palaeontology, **18**: 441.

First received 3 October 2011.
Final version accepted 25 January 2012.

GONEPLACID CRABS (DECAPODA, BRACHYURA, GONEPLACIDAE) OF THE MAINBAZA AND MIRIKI EXPEDITIONS TO THE MOZAMBIQUE CHANNEL, WITH THE DESCRIPTION OF A NEW SPECIES OF *PYCNOPLAX* CASTRO, 2007

BY

PETER CASTRO[1])
Biological Sciences Department, California State Polytechnic University, Pomona, CA 91768-4032, U.S.A.

ABSTRACT

A study of the goneplacid crabs (Goneplacidae) collected in the Mozambique Channel, southwestern Indian Ocean by the MAINBAZA expedition along the Mozambique coast and the MIRIKI expedition off northwestern Madagascar revealed the presence of four species, one of which is described as a new species of *Pycnoplax* Castro, 2007. The new species differs from its six congeners by the morphology of the carapace front, G1, and vulva, and by having a complete thoracic sternal suture 7/8. It is the first species of *Pycnoplax* recorded from the Indian Ocean, its congeners being restricted to the Pacific. The discovery of a new species has necessitated a review of the diagnosis of *Pycnoplax* and the key to its species.

RÉSUMÉ

L'étude des crabes Goneplacidae récoltés dans l'océan Indien occidental lors la campagne MAINBAZA au large de la côte de Mozambique et la campagne MIRIKI au large de la côte nord-ouest de Madagascar a identifié quatre espèces, dont une espèce appartenant a *Pycnoplax* Castro, 2007, est décrite comme nouvelle. La nouvelle espèce se distingue de ses six congénères par la forme du front, du premier gonopode mâle, et de la vulve, et par la présence d'une suture sternale thoracique 7/8 complète. Elle est la première espèce de *Pycnoplax* trouvée dans l'océan Indien, ses congénères ne se trouvent que dans l'océan Pacifique. La découverte de la nouvelle espèce mérite mettre au point la diagnose de *Pycnoplax* et de la clé de ses espèces.

---

[1]) e-mail: pcastro@csupomona.edu

INTRODUCTION

Four species of goneplacid crabs were collected in the southwestern Indian Ocean by B/O *Vizconde de Eza* during the MAINBAZA expedition to the Mozambique coast in April 2009 and by the trawler *Miriky* off northwestern Madagascar during the MIRIKY expedition in June/July 2009. One of these species proved to be an undescribed species of *Pycnoplax* Castro, 2007, and it is described herein.

MATERIALS AND METHODS

The material examined was collected using a shrimp trawl (CC stations), beam trawl (CP stations), and Warén dredge (DW stations). All specimens are deposited in the Muséum national d'Histoire naturelle, Paris (MNHN). The morphological terminology and measurements used follow Castro (2007). The ambulatory legs are referred to by the abbreviations P2 to P5, the first and second pairs of male pleopods by G1 and G2, respectively; carapace length and carapace width by cl and cw, respectively. These and other paired appendages are referred to in the singular for simplicity. Only the most important references are listed in the synonymies.

TAXONOMY

Superfamily Goneplacoidea MacLeay, 1838
Family Goneplacidae MacLeay, 1838
Carcinoplax ischurodous (Stebbing, 1923)

Geryon ischurodous Stebbing, 1923: 2, pl. 11 [type locality: South Africa].
Carcinoplax eurysternum Guinot & Richer de Forges, 1981a: 1113, pl. 6 fig. 5, 5a, 5b [nomen nudum]; 1981b: 249, figs. 9E, 10A–D, pl. 6 fig. 5, 5a, 5b; Guinot, 1989: 314, fig. 44, pl. 13 fig. F.
Carcinoplax ischurodous — Manning & Holthuis, 1988: 78, fig. 1; Castro, 2007: 635.

Material examined. — Mozambique Channel, MAINBAZA: 2 males, stn CP3138, 25°12.8′S 35°21.04′E, 707 m, 10 April 2009 (MNHN-IU-2010-1527); 1 pre-adult female, photograph CP3138-40 (MNHN-IU-2010-1530); 1 male, photograph CP3141-27, stn CP3141, 23°31.95′S 35°55.74′E, 698 m, 11 April 2009 (MNHN-IU-2010-1529); 3 males, 6 females, stn CC3165, 24°15.7′S 25°42.12′E, 612 m, 15 April 2009 (MNHN-IU-2010-1525); 1 male, photograph CC3165-27 (MNHN-IU-2010-1528); 1 male, stn CC3166, 24°21.22′S 35°42.35′E, 715 m, 15 April 2009 (MNHN-IU-2010-1526); 4 males, 2 pre-adult females, 6 females, stn CC3172, 25°58.86′S 34°36.51′E, 638 m, 16 April 2009 (MNHN-IU-2010-1524).

Colouration. — Light pink-orange carapace and pereopods.

Geographical distribution. — Western Indian Ocean and southwestern Pacific as far east as Tonga; depth: 464-800 m (Castro, 2007).

Remarks. — The largest male in the collection (cl 35.2 × cw 51.8 mm; MNHN-B32594) had chelipeds that were proportionally larger than in smaller individuals.

Carcinoplax longimana (De Haan, 1833)

Cancer (*Curtonotus*) *longimanus* De Haan, 1833: 50, pl. 6 fig. 1 [type locality: Japan].
Carcinoplax longimanus — Guinot, 1969: 524, fig. 61; 1989: 273, figs. 2, 4, 6, pl. 1 figs. A–D, pl. 2 figs. A–D.
Carcinoplax longimana — Castro, 2007: 627, fig. 1.

Material examined. — Mozambique Channel off Mozambique, MAINBAZA: 1 pre-adult male, cl 6.3 × cw 9.3 mm, photograph 3132-64, stn CP3132, 25°11.24′S 35°01.51′E, 102 m, 10 April 2009 (MNHN-IU-2010-1532); 1 pre-adult female, cl 8.0 × cw 10.7 mm, photograph 3133-30, stn CP3133, 25°10.56′S 35°10.56′E, 201 m, 10 April 2009 (MNHN-IU-2010-1531).

Mozambique Channel off northwestern Madagascar between Nosy Be and Leven Bank, MIRIKI: 1 pre-adult female, stn DW3183, 12°38′S 48°14′E, 420-436 m, 26 June 2009 (MNHN-IU-2010-947); 1 male, stn CP3241, 14°30′S 47°27′E, 274-325 m, 6 July 2009 (MNHN-IU-2010-935); 2 males, stn CP3262, 15°34′S 45°44′E, 227-283 m, 10 July 2009.

(MNHN-IU-2010-936), 4 males, 2 females (MNHN-IU-2010-938); 2 males, stn CP3264, 15°34′S 45°44′E, 227-283 m, 10 July 2009 (MNHN-IU-2010-934); 1 male, stn CP3289, 14°29′S 47°26′E, 332-379 m, 14 July 2009 (MNHN-IU-2010-933); 1 male, 3 females, stn CP3294, 14°29′S 47°29′E, 263-331 m, 14 July 2009 (MNHN-IU-2010-937).

Distribution. — Indian Ocean to western Pacific as far south as Indonesia; depth: 66-436 m.

Remarks. — The first anterolateral teeth in the two pre-adult specimens examined, particularly the male (MNHN-IU-2010-1532), are not as acute as that shown by Guinot (1989: fig. 4A) for a slightly larger Madagascar specimen. The carapaces of both pre-adults have the short tubercles diagnostic of the species.

Coloration. — Color photographs of the freshly collected pre-adults (MNHN-IU-2010-1531, MNHN-IU-2010-1532) allow, apparently for the first time, the description of the color pattern of pre-adults. The dorsal surface of the carapace and chelipeds shows orange-red irregular spots and reticulations. Three broad bands of the same color are present on the P2–P4 meri and carpi.

Psopheticus crosnieri Guinot, 1990

Psopheticus crosnieri Guinot, 1990: 351, figs. 29-34, 50, 51 [type locality: Madagascar]; Castro, 2007: 742.

Material examined. — Mozambique Channel off Mozambique, MAINBAZA: 1 female (with rhizocephalan parasite), photograph 1533-33, stn CC3151, 19°32.8′S 36°45.96′E, 355 m, 13 April 2009 (MNHN-IU-2010-1533).

Mozambique Channel off northwestern Madagascar between Nosy Be and Leven Bank, MIRIKI: 1 male, photograph DW3181-78, stn DW3181, 12°36′S 48°17′E, 316-324 m, 26 June 2009 (MNHN-IU-2010-1566); 1 female, photograph DW3189-08, stn DW3189, 12°30′S 48°18′E, 346-376 m, 27 June 2009 (MNHN-IU-2010-1567); 1 female, stn DW3197, 12°07′S 48°58′E, 362-431 m, 28 June 2009 (MNHN-IU-2010-950); 1 female, photograph DW3223-03, stn DW3223, 12°46′S 48°11′E, 430-488 m, 2 July 2009 (MNHN-IU-2010-1568); 1 female, stn DW3234, 13°27′S 47°55′E, 187-247 m, 3 July 2009 (MNHN-IU-2010-944); 2 females, stn DW3241, 14°30′S 47°27′E, 274-325 m, 6 July 2009 (MNHN-IU-2010-942); 1 male, 1 female, stn DW3248, 14°50′S 46°57′E, 340-446 m, 7 July 2009 (MNHN-IU-2010-941); 1 male, stn DW3250, 15°22′S 46°00′E, 493-750 m, 8 July 2009 (MNHN-IU-2010-943); 1 male, 1 female, stn DW3262, 15°34′S 45°44′E, 227-283 m, 10 July 2009 (MNHN-IU-2010-948); 4 males, stn DW3263, 15°34′S 45°44′E, 287-450 m, 10 July 2009 (MNHN-IU-2010-939); 2 males, 1 female (MNHN-IU-2010-949); 4 male, 2 females, stn DW3264, 15°33′S 45°44′E, 325-560 m, 10 July 2009 (MNHN-IU-2010-940); 1 pre-adult female, stn DW3265, 15°32′S 45°45′E, 327-730 m, 10 July 2009 (MNHN-IU-2010-945); 1 pre-adult male, stn DW3275, 15°26.66′S 46°3.3′E, 53-59 m, 12 July 2009 (MNHN-IU-2010-946); 1 pre-adult female, stn DW3282, 14°52′S 46°58′E, 215-261 m, 13 July 2009 (MNHN-IU-2010-952); 2 males, 1 female, stn DW3285, 14°47′S 46°58′E, 512-680 m, 13 July 2009 (MNHN-IU-2010-951).

Distribution. — Southwestern Indian Ocean: Madagascar, Comoro Is., Réunion, and now off Mozambique; depth: 53-730 m.

Colouration. — Orange carapace and pereopods; thin, white band across posterior third of carapace and lighter color posterior to the band; bright orange eye peduncles.

Pycnoplax Castro, 2007

Carcinoplax — Guinot, 1969: 524; 1989: 273 (part; not *Carcinoplax* H. Milne Edwards, 1852).
Pycnoplax Castro, 2007: 661.

Species included: *Pycnoplax aspera* Castro, 2009, *P. bispinosa* (Rathbun, 1914), *P. coryphaea* n. sp., *P. latifolia* Castro, 2007, *P. meridionalis* (Rathbun, 1923), *P. victoriensis* (Rathbun, 1923), and *P. surugensis* (Rathbun, 1932) (type species).

Remarks. — The description of a new species herein as well as of a sixth species (*P. aspera*) described after the original description of *Pycnoplax* (Castro, 2007: 661) necessitates updating the diagnosis and the key to its species (Castro, 2007: 662). The description of the genus stated "median sulcus on thoracic sternite 4 absent" in error. It is indeed present and although it is not clearly seen in a photograph of *P. surugensis* (Castro, 2007, fig. 14), it is conspicuous in *P. aspera* (Castro, 2009, fig. 2E), *P. bispinosa* (Castro, 2007, fig. 15), *P. coryphaea* n. sp. (fig. 2C), and *P. victoriensis* (Castro, 2007, fig. 18).

Diagnosis. — Carapace transversely quadrate, slightly wider than long, widest at junction of anterolateral, posterolateral borders; front lamellar, straight (slightly concave in *P. coryphaea* n. sp.), not marked by median notch or projection. Notch between front, inner edge of supraorbital border (absent or barely discernible in *P. bispinosa*, *P. coryphaea* n. sp.); orbits narrow, not expanded distally; supraorbital borders slightly sinuous; suborbital borders rounded, with short, blunt inner tooth not visible dorsally (except in *P. aspera*); anterolateral borders convex; posterolateral borders long, arched. Dorsal surface of carapace smooth (2 slight horizontal ridges across carapace in *P. victoriensis*; granular in *P. aspera*) slightly to strongly convex, without clear indication of regions. Outer orbital angle with short, triangular tooth (absent in *P. bispinosa*); 2 anterolateral teeth on each side of carapace. Basal antennal article short, thick, distalmost (third) article reaching front. Eye peduncles short, much shorter than front (0.2 front width; 0.1 in *P. aspera*); cornea spherical, only slightly expanded distally. Anterior border of endostome well demarcated from buccal cavern, ridges faint but clearly defined. Third maxillipeds almost completely or completely closing buccal cavern. No obvious stridulating mechanism other than possible rubbing of proximal portion of cheliped (P1) merus against pterygostomian ridge. Thoracic sternum wide; median sulcus on thoracic sternite 4. Thoracic sternite sutures 4/5, 5/6, 7/8 interrupted medially, 6/7 complete (7/8 complete in *P. coryphaea*). Anterior end of sterno-abdominal cavity anterior to thoracic sternite 4. Cheliped fingers moderately slender, shorter than propodus (longer in *P. aspera*, *P.latifolia*), tips darker in color (completely pale in *P. victoriensis*). Carpus with tooth on inner margin. Dorsal margins of meri, carpi, propodi of ambulatory legs (P2–P5) unarmed; P5 dactylus slender, smooth (small teeth in *P. aspera*; slender spines in *P. coryphaea*), setose. Male abdomen with 6 freely-movable somites plus telson, relatively wide, somites 4-6 gradually decreasing in length from somite 3 (widest somite); telson wider than long (slightly wider than long in *P. aspera*; nearly as wide as long in *P. coryphaea*); somite 3 covering space between P5 coxae; somite 2 slightly narrower than or as wide as somite 3 so that thoracic sternite 8 not visible or only leaving small portion visible (*P. aspera*, *P. bispinosa*, *P. meridionalis*) when abdomen is closed. G1 stout, thick, proximally broad, curved or straight, small spinules variously distributed along tip. G2 slender, almost as long, slightly longer or noticeably longer than G1, dark brown in large males of some species, flagellum shorter or almost as long as proximal part (peduncle); tip acuminate (*P. coryphaea*, *P. surugensis*), pointed with several spinules (*P. meridionalis*), slightly-expanded with 2 terminal spinules (*P. bispinosa*, *P. latifolia*, *P. victoriensis*), or leaf-like (*P. aspera*). Penis arising from P5 coxa, moderate size; broad, soft proximal portion. Female abdomen with 6 freely-movable somites, wide; telson wider than long; somite

3 covering space between P5 coxae, somite 2 narrower than somite 3 so that it leaves small portion of thoracic sternite 8 visible, or somite 2 as wide as somite 3 so that thoracic sternite 8 not visible. Vulva of mature females proportionally small to large but not greatly enlarged as in *Carcinoplax*, extending from suture 5/6 to suture 6/7, spherical, typically thick, salient margins, sternal vulvar cover on outer margin (thin and placed on outer-posterior margin in *P. coryphaea*; no evidence of a vulvar cover in *P. surugensis*).

KEY TO SPECIES OF *PYCNOPLAX*

1. Two thin, spine-like anterolateral teeth (Castro, 2007, fig. 13B) .... *P. bispinosa*
– Two triangular or rounded anterolateral teeth 2
2. Ambulatory legs (P2–P5) long, proportionally slender, distal end of merus of folded P5 extending beyond tip of second anterolateral tooth (Castro, 2007, fig. 13A) .... *P. surugensis*
– Ambulatory legs (P2–P5) relatively short, distal end of merus of folded P5 only reaching or barely reaching tip of second anterolateral tooth 3
3. Outer surface of cheliped dactylus, propodus with small, short granules (Castro, 2009, fig. 2F); ventral surface, anterior margin of merus with large granules .... *P. aspera*
– Outer surface of cheliped dactylus, propodus, ventral surface of merus smooth 4
4. Truncated, granular, shallow tooth on each outer orbital angle (Castro, 2007, figs. 16A, 17) .... *P. latifolia*
– Triangular to rounded, smooth tooth on each outer orbital angle 5
5. First anterolateral tooth broad, with rounded outline (fig. 1A). Short suborbital tooth ... . *P. coryphaea* n. sp.
– First anterolateral tooth triangular, pointed. Conspicuous suborbital tooth 6
6. Second anterolateral tooth curved, hook-like, acute tip (Ahyong, 2008, fig.15A). Cheliped fingers dark brown .... *P. meridionalis*
– Second anterolateral tooth triangular, obtuse tip (Ahyong, 2008, fig.15B). Cheliped fingers unpigmented .... *P. victoriensis*

Pycnoplax coryphaea n. sp.
(figs. 1A, B; 2A-C)

Material examined. — Holotype: Mozambique Channel off Mozambique, MAINBAZA, stn CP3141, 23°31.95′S 35°55.74′E, 698 m, 11 April 2009, male, cl 4.5 × cw 5.7 mm (MNHN-IU-2010-1534).

Paratypes: Mozambique Channel, MAINBAZA, stn CP3143, 23°30.94′S 35°46.3′E, 277 m, female, cl 7.1 × cw 9.4 mm, female, cl 6.1 × cw 8.1 mm, ovigerous female cl 5.7 × cw 7.4 mm, female cl 4.7 × cw 6.1 mm (MNHN-IU-2010-1535).

Type locality. — Southeast Africa, Mozambique, off the Mozambique Channel coast, 23°31.95′S 35°55.74′E, 698 m.

Etymology. — Named in honour of Professor M. Takeda for his contribution to carcinology. The epithet derives from *koryphaios*, Greek for "leader" or "chief" in reference to Prof. Takeda's work.

Description. — Carapace (fig. 1A) transversely rectangular, slightly wider than long (1.3 as wide in all specimens), widest at second anterolateral teeth at junction of anterolateral, posterolateral borders; anterolateral borders arched, with two anterolateral teeth on each border, first largest, short, with rounded margin, second minute, only as slight rise on posterior of anterolateral carapace borders; dorsal surface slightly convex, without clear indication of regions. Front lamellar, slightly concave, marked by slight median notch in some specimens. Distinct notch between front, inner edge of supraorbital border. Orbits narrow, shorter than front. Supraorbital borders sinuous, slight median notch, orbits low along anterior border. Suborbital borders sinuous, microscopically dentate, each with short, wide inner tooth not visible dorsally. Outer orbital angle with broad, rounded prominence continuing as a carina-like crest to just anterior of second anterolateral tooth. Posterolateral borders long, arched.

Basal antennal article short, wide, distalmost (third) article reaches front. Eye peduncles short (0.2 front width), smooth; dorsal margin bilobed, with median notch; cornea slightly expanded distally. Anterior border of endostome well demarcated from buccal cavern, ridges faint but clearly defined. Third maxillipeds almost completely closing buccal cavern; merus quadrate, anteroexternal margin slightly auriculiform.

Thoracic sternum wide; conspicuous median sulcus on thoracic sternite 4 (fig. 1B). Anterior end of sterno-abdominal cavity at about same level as thoracic sternite 4. Thoracic sternal sutures 4/5, 5/6 interrupted, 6/7, 7/8 complete; median line along sternites 7, 8, small posterior portion of 6 (fig. 2C).

Chelipeds (P1) (fig. 1A, B) nearly equal in both sexes; fingers slender, as long as propodus, with blunt teeth in both sexes, about three-quarters to nearly all of distal portion of fingers dark brown. Coarse tubercles along anterior, inner margins of carpus; broad, blunt tooth on inner margin. Ambulatory legs (P2–P5) moderately short, stout; length of P5 merus 0.5 cl; margins smooth but microscopically dentate; long, simple setae, few plumose; P2–P4 dactyli slender, with simple setae. P5 propodus conspicuously wide in relation to dactylus; dactylus long, slender, with slender spines along both margins, distally increasing in relative thickness.

Male abdomen (fig. 1B) broadly triangular, with 6 freely-movable somites plus telson; telson nearly as wide as long; thoracic sternite 8 not visible ven-

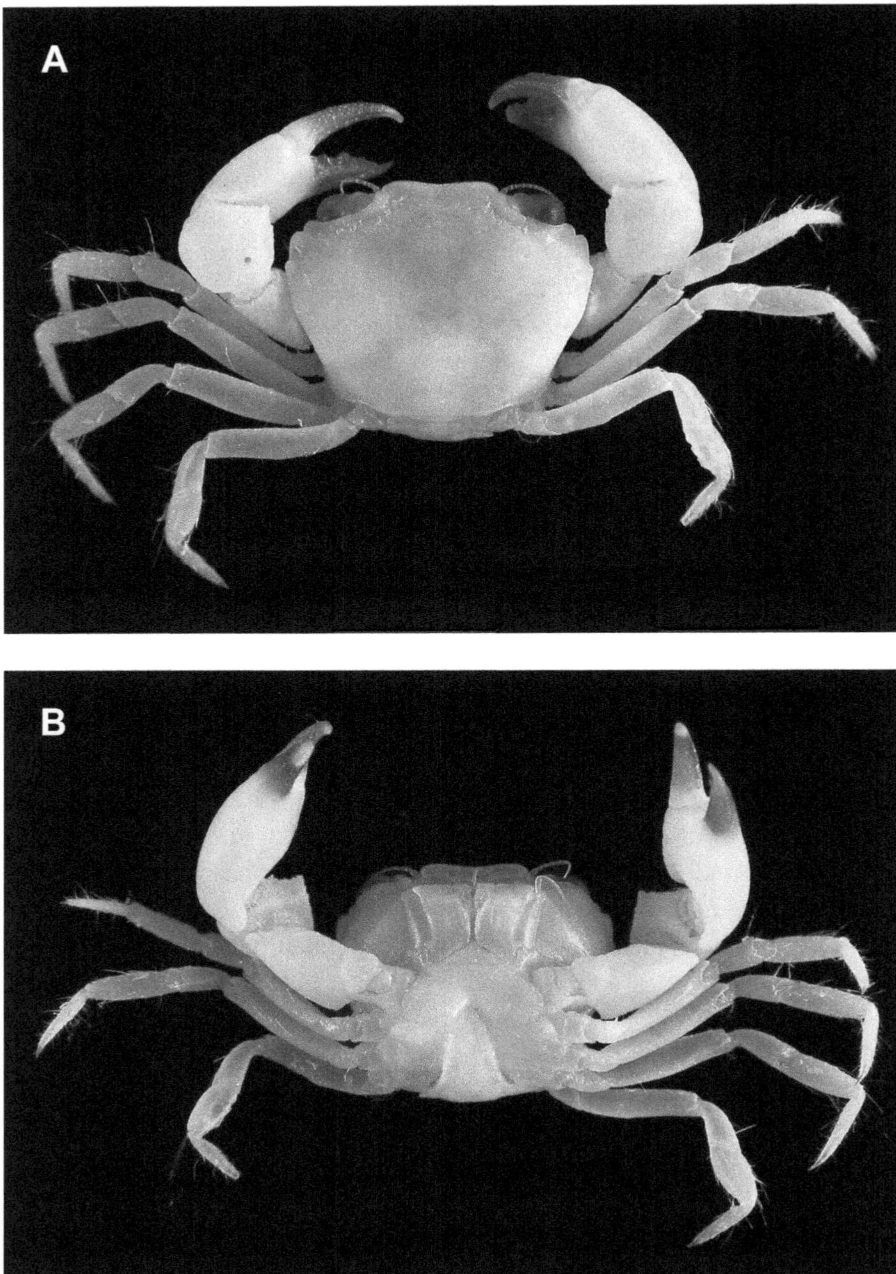

Fig. 1. *Pycnoplax coryphaea* n. sp. A, B, male holotype (cl 4.5 × cw 5.7 mm), Mozambique Channel (MNHN-IU-2010-1534). A, dorsal view; B, ventral view.

trally when abdomen closed. Somite 1 arched, partially covered by carapace; somite 2 about as broad as somite 3; somite 3 covers space between P5 coxae, fits under episternite 7. Penis opening close to coxo-sternal condyle of P5 coxa, long, slender; protected by somite 3 of closed abdomen, expanded episternite 7; concavity under episternite 7 continuing along posterior margin of thoracic somite 7 overhanging thoracic somite 8 but penis enters G1 directly without being aligned along concavity. G1 (fig. 2A) stout, straight, proximally broad; short, thick spinules mostly along outer margin, small, salient, rectangular process on inner margin of base armed with similar spinules plus few longer spines; slender, straight, slightly rounded apex. G2 (fig. 2B) slender, slightly longer than G1, curved flagellum slightly shorter than proximal part (peduncle); acuminate apex.

Female abdomen wide, with 6 freely movable somites; thoracic sternite 8 not visible ventrally when abdomen closed; somite 1 partially covered by carapace, somite 2 about as broad as somite 3, somite 3 nearly covers entire space between P5 coxae. Telson triangular. Vulva of mature females (fig. 2C) round, slightly elevated margin, extending from edge of anteriorly deflected thoracic sternal suture 5/6 to close to suture 6/7; mostly covered by thin but rigid, flat sternal vulvar cover continuous with sternite on outer-posterior margin, leaving narrow aperture along curved anterior, inner margins.

Distribution. — Only known from the Mozambique Channel at depths of 277 and 698 m.

Colour. — All specimens were off-white and the tip of the fingers dark brown without any indications of a particular colour pattern on the carapace after preservation in alcohol for a few months.

Remarks. — The new species is assigned to *Pycnoplax* Castro, 2007, sharing with its six known congeners the general morphology of the carapace, G1, and vulvae, the main characters used by Castro (2007) in the revision of the Goneplacidae that resulted in the description of *Pycnoplax* and nine other genera. These similarities include a transversely rectangular carapace with two well-developed anterolateral teeth, orbits that are much shorter than the front, short eye peduncles, stout G1 with a proportionally broad proximal part, long G2 that is as long or longer than the G1, a vulva that extends from sternal suture 5/6 to suture 6/7 without being greatly expanded, and the presence of a sternal vulvar cover.

The carapace of *P. coryphaea* n. sp. bears two well-developed anterolateral teeth (fig. 1A), a diagnostic character of *Pycnoplax*. The shape of the carapace and of the anterolateral teeth are very similar to those of *P. latifolia* Castro,

2007 (Castro, 2007, figs. 16A, 17), and *P. aspera* Castro, 2009 (Castro, 2009, fig. 2A).

The carapace of the new species is similar to that of *Carcinoplax* H. Milne Edwards, 1852. Diagnostic to the 19 species of *Carcinoplax*, however, is a G1 that is dorsoventrally flattened, slender, long, and with a characteristically rounded or slightly pointed apex (e.g., Castro, 2007, fig. 2A; 2009, fig. 1C) (stout, not dorsoventrally flattened in *P. coryphaea*); G2 with one to three spinules on its apex (e.g., Castro, 2007, fig. 2B; 2009, fig. 1D) (acuminate apex in the new species); and a greatly expanded vulva that lacks a sternal vulvar cover (e.g., Castro, 2007, fig. 11; 2009, fig. 1E) (smaller vulva with a sternal vulvar cover in the new species).

Pyncoplax coryphea also shares the general shape of its carapace with *Thyraplax* Castro, 2007, but the five known species of this Pacific Ocean genus have only one anterolateral tooth, although a low projection can be found in four of the species (e.g., Guinot, 1989, fig. 42, pl. 11 fig. G, as *Carcinoplax cooki*; Castro, 2007, figs. 13C, 21, 24, 26) in contrast to the two well-developed, rounded teeth in the new species. Another difference is the proportionally slender G1 of *Thyraplax*, which typically has a narrow proximal part and an acuminate apex (e.g. Guinot, 1969, fig. 77, as *C. cooki*; Castro, 2007, figs. 20C, 23D), in contrast to the proportionally short, stout, and thick G1 of *Pyncoplax*, which typically has a rounded apex (e.g., Guinot, 1969, fig. 73, as *C. surugensis*; 1989, fig. 30A, as *C. bispinosa*; Serène & Lohavanijaya, 1973, figs. 168-170, as *C. meridionalis*; Castro, 2007, fig. 16C; 2009, fig. 2C).

The new species nevertheless departs from its congeners and from the species of *Carcinoplax* and *Thyraplax* in having a complete thoracic sternal suture 7/8, and from the species of *Carcinoplax* in having complete sternal suture 6/7.

Several characters distinguish *P. coryphaea*, the only *Pycnoplax* species known from the Indian Ocean, from its six Pacific Ocean congeners. The front is concave (fig. 1A) and, in some specimens, marked by a very slight median notch, whereas the front is straight or nearly so and without a notch in all congeners (e.g., *P. latifolia*; Castro, 2007, figs. 16A, 17). The second anterolateral tooth is minute, much smaller than the first anterolateral tooth, in the new species, conspicuous and as big or even larger than the first (e.g., *P. victoriensis*; Ahyong, 2008, fig. 15B) in all congeners. Thoracic sternal suture 7/8 is complete in the new species but incomplete in its congeners. The P5 dactylus of the new species is spinous, similar to *P. aspera*, which is armed with small teeth, but smooth in the remaining five species of *Pyncoplax*.

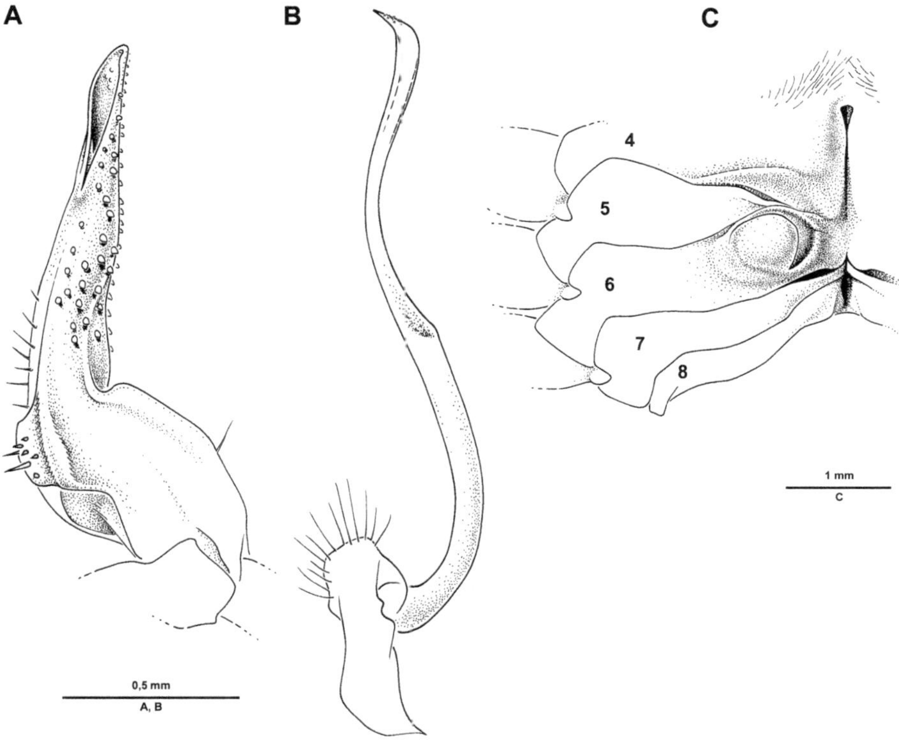

Fig. 2. *Pycnoplax coryphaea* n. sp. A, B, male holotype (cl 4.5 × cw 5.7 mm), Mozambique Channel (MNHN-IU-2010-1534). A, left G1; B, left G2; C, female paratype (cl 7.1 × cw 9.4 mm), Mozambique Channel (MNHN-IU-2010-1535), ventral view of posterior portion of thoracic sternum showing right vulva.

The G1 (fig. 2A) of the new species is unique among congeners by being straight and armed by short, thick denticles that extend along most of its length. In contrast, the G1 is curved or slightly curved, with minute denticles only on the distal part in *P. aspera* (see Castro, 2009, fig. 2C) and *P. latifolia* (see Castro, 2007, fig. 16C); conspicuously stout, curved, and with small to large denticles on the distal part as in *P. bispinosa* (see Guinot, 1989, fig. 30A, as *Carcinoplax bispinosa*) and *P. surugensis* (see Guinot, 1989, fig. 31A, as *C. surugensis*); and curved with small denticles on the distal part in *P. meridionalis* (see Serène & Lohavanijaya, 1973, figs. 168-170, as *C. meridionalis*) and *P. victoriensis*. The apex of the G2 is acuminate in the new species (fig. 2B) as in *P. surugensis* (see Guinot, 1989, fig. 31B, as *C. surugensis*) in contrast to leaf-like in *P. aspera* (see Castro, 2009, fig. 2D), armed with two denticles in *P. bispinosa* (see Guinot, 1989, fig. 30B, as *C. bispinosa*), *P. latifolia* (see Castro, 2007, fig. 16D), and *P. victoriensis*, and several minute denticles in *P. meridionalis* (see Serène & Lohavanijaya, 1973, figs. 171-173, as *C. meridionalis*).

The vulva of the new species departs from that of all congeners by being covered by a thin and flat sternal vulvar cover on the outer, posterior portion and extending over most of the opening (fig. 2C), in contrast to posterior but triangular and salient in *P. aspera* (see Castro, 2009, fig. 2E), spherical, thick and on the outer margin in *P. bispinosa* (see Castro, 2007, fig. 15), *P. meridionalis*, and *P. victoriensis* (see Castro, 2007, fig. 18), and similar as in these three species but thinner in *P. latifolia*. There is no evidence of a developed vulvar cover in *P. surugensis* (see Castro, 2007, fig. 14).

Pycnoplax coryphaea is so far known only from the southwestern Indian Ocean, whereas its congeners are exclusively from the Pacific Ocean: *P. aspera*, *P. bispinosa*, and *P. surugensis* variously distributed in the western and southwestern Pacific, *P. latifolia* in Japan, and *P. meridionalis* and *P. victoriensis* in Australia and New Zealand. The seven species have been collected at depths that typically exceed 100 m and extending to as deep as 1045 m in the case of *P. meridionalis* (see Castro, 2007, 2009; Ahyong, 2008).

Penial protection is similar among the seven species of *Pycnoplax*: an expanded, triangular or almost quadrate episternite 7 under which the penis is sheltered after it exits the coxal gonopore. In *P. coryphaea* the penis is proportionally long and slender along its entire length. In the remaining species the penis is broadly expanded along the base and ends as a short, small papilla that enters the G1 basal portion.

Pycnoplax coryphaea appears to be a small-sized species from the basis of the five known specimens. All specimens are adults, the male holotype (cl 4.5 × cw 5.7 mm) having a calcified G1 and one of the small females (cl 5.7 × cw 7.4 mm) being ovigerous. The largest specimen of *P. coryphaea* was a female, cl 7.1 × cw 9.4 mm. *Pycnoplax latifolia* also seems to be a small-sized species, with a mean size of cl 5.3 × cw 6.7 mm for the 8 specimens known and the largest specimen cl 7.0 × cw 8.9 mm (Castro, 2007: 667); no ovigerous females are known, however. *Pycnoplax aspera* is perhaps another small-size congener. The mean size of the 14 specimens known is cl 6.0 × cw 7.5 mm and the largest specimen (cl 8.2 × cw 10.4 mm) is the only ovigerous female known (Castro, 2009: 953).

ACKNOWLEDGEMENTS

The MAINBAZA expedition, a joint project of MNHN (P. Bouchet principal investigator) and Instituto Español de Oceanografia (A. Ramos principal investigator), and the MIRIKY expedition formed part of a cluster of Mozambique-Madagascar expeditions (La Planète Revisitée/Our Planet Reviewed) funded by the Total, Prince Albert II of Monaco, and Stavros Niarchos foundations, and conducted by MNHN and Pro-Natura International (PNI). My gratitude goes to H. Komatsu for his invitation to participate in the Homage to Masatsune Takeda Volume, J.-F. Dejouannet (MNHN) for the preparation of the figures, L. Corbari and P. Martin-Lefèvre (MNHN) for making available the material, and to S. Ahyong (Australian Museum) and T. Naruse (University of the Ryukyus) for their valuable comments to the manuscript.

REFERENCES

AHYONG, S. T., 2008. Deepwater crabs from seamounts and chemosynthetic habitats off eastern New Zealand (Crustacea: Decapoda: Brachyura). Zootaxa, **1708**: 1-72.

CASTRO, P., 2007. A reappraisal of the family Goneplacidae MacLeay, 1838 (Crustacea, Decapoda, Brachyura) and revision of the subfamily Goneplacinae, with the description of 10 new genera and 18 new species. Zoosystema, **29**(4): 609-774.

— —, 2009. Two new species of *Carcinoplax* H. Milne Edwards, 1852, and *Pycnoplax* Castro, 2007, from the western Pacific, and a description of the female of *Thyraplax truncata* Castro, 2007 (Crustacea, Decapoda, Brachyura, Goneplacidae). Zoosystema, **31**(4): 949-957.

GUINOT, D., 1969. Recherches préliminaires sur les groupements naturels chez les crustacés décapodes brachyoures, VII. Les Goneplacidae (suite). Bulletin du Muséum national d'Histoire naturelle, Paris, (2) **41**: 507-528, pl. 2.

— —, 1989. Le genre *Carcinoplax* H. Milne Edwards, 1852 (Crustacea, Brachyura: Goneplacidae). In: J. FOREST (ed.), Résultats des Campagnes MUSORSTOM, **5**. Mémoires du Muséum national d'Histoire naturelle, Paris, (A) **144**: 265-345.

— —, 1990. Crustacea Decapoda : le genre *Psopheticus* Wood-Mason, 1892 (Goneplacidae), In: A. CROSNIER (ed.), Résultats des campagnes MUSORSTOM, **6**. Mémoires du Muséum national d'Histoire naturelle, Paris, (A) **145**: 331-367.

GUINOT, D. & B. RICHER DE FORGES, 1981a. Crabes de profondeur, nouveaux ou rares, de l'Indo-Pacifique (Crustacea, Decapoda, Brachyura) (première partie). Bulletin du Muséum national d'Histoire naturelle, Paris, (4), (A) **2**: 1113-1153 [dated 1980, published 1981].

— — & — —, 1981b. Crabes de profondeur, nouveaux ou rares, de l'Indo-Pacifique (Crustacea, Decapoda, Brachyura) (deuxième partie). Bulletin du Muséum national d'Histoire naturelle, Paris, (4), (A) **3**: 227-260.

HAAN, W. DE, 1833. Crustacea, In: P. F. VON SIEBOLD (ed.), Fauna japonica sive descriptio animalium, quae in itinere per Japoniam, jussu et auspiciis superiorum, qui summum in India Batava Imperium tenent, suscepto, annis 1823-1830 collegit, notis, observationibus et adumbrationibus illustravit: i–xvii, i–xxxi, ix–xvi, 1-243, pls. A–J, L–Q, 1-55, circ. tab 2. (Lugduni-Batavorum [= Leiden]). [Published 1833-1850].

MANNING, R. B. & L. B. HOLTHUIS, 1988. South African species of the genus *Geryon* (Crustacea, Decapoda, Geryonidae). Annals of the South African Museum, **98**: 77-92.

MILNE EDWARDS, H., 1852. Observations sur les affinités zoologiques et la classification naturelle des Crustacés. Annales des Sciences naturelles, Zoologie, (3) **18**: 109-166, pl. 3, 4. (Second part in Annales des Sciences naturelles, Zoologie [Paris] **20**: 163-228, pls. 6-11 [1853]).

RATHBUN, M. J., 1923. Report on the Brachyrhyncha, Oxystomata and Dromiacea. Report on the crabs obtained by the F.I.S. "Endeavour" on the coasts of Queensland, New South Wales, Victoria, South Australia and Tasmania. In: Biological Results of the Fishing Experiments Carried Out by the F.I.S. "Endeavour" 1904-14, **5**(3): 95-156, pls. 16-42. (Department of Trade and Customs, Sydney).

— —, 1932. Preliminary descriptions of new species of Japanese crabs. Proceedings of the Biological Society of Washington, **45**: 29-38.

SERÈNE, R. & P. LOHAVANIJAYA, 1973. The Brachyura (Crustacea: Decapoda) collected by the Naga Expedition, including a review of the Homolidae. In: E. BRINTON, W. A. NEWMAN & W. S. WOOSTER (eds.), Scientific Results of Marine Investigations of the South China Sea and the Gulf of Thailand, 1959-1961. Naga Report, **4**(4): 1-187.

STEBBING, T. R. R., 1923. Crustacea of Natal. Fisheries and Marine Biological Survey, **3**: 1-16 [dated 1922, published 1923].

First received 11 August 2011.
Final version accepted 5 October 2011.

A NEW SPECIES OF *PEDROPLAX* (DECAPODA, BRACHYURA, GONEPLACIDAE) FROM THE CORAL SEA, AND A NEW RECORD OF *PEDROPLAX MEGALOPS* FROM NORTH-WESTERN AUSTRALIA

BY

PETER J. F. DAVIE[1])

Queensland Museum, P.O. Box 3300, South Brisbane, Queensland, 4101, Australia

ABSTRACT

A second species of *Pedroplax* Ng & Komai, 2011, *P. takedai* sp. nov., is described from the Chesterfield Reefs, Coral Sea. It is distinguished from its only other congener, *P. megalops* (Takeda, 1989), by differences in carapace morphology and shape of the male G1. *Pedroplax megalops* is recorded from north-western Australia, marking a significant southerly range extension and the first record of this species from the Indian Ocean.

RÉSUMÉ

Une seconde espèce de *Pedroplax* Ng & Komai, 2011, *P. takedai* sp. nov., est décrite des récifs de Chesterfield, dans la mer de Corail. Elle se distingue de la seule autre espèce connue *P. megalops* (Takeda, 1989) par la morphologie de la carapace et la forme du G1. *Pedroplax megalops* est signalé du nord-ouest de l'Australie, ce qui constitue une extension méridionale significative de la répartition de l'espèce et son premier signalement dans l'océan Indien.

INTRODUCTION

Ng & Komai (2011) recently erected a new genus for an unusual goneplacid crab that was first described from southern Japan by Takeda (1989) as *Psopheticus megalops*. This species had been previously known only from one small male and one female. Although Takeda (1989) placed his new distinctive species in *Psopheticus* Wood-Mason, 1892, he did comment (1989: 174) that "...inclusion to the genus...is rather arbitary [sic] and tentative". Castro

[1]) e-mail: Peter.Davie@qm.qld.gov.au

(2007: 739) excluded it from *Psopheticus* even suggesting it was unlikely to be a member of the Goneplacidae MacLeay, 1838, sensu stricto. Ng et al. (2008: 80) retained it in the Goneplacidae as an incerta sedis. Finally Ng & Komai (2011) re-examined the type material, as well as some new samples from Japan, and created the new genus *Pedroplax* to receive it, while confirming its natural monophyletic grouping with the Goneplacidae.

The abbreviations G1 and G2 refer to the male first and second gonopods, respectively, while P2–P5 refers to the first to fourth ambulatory legs. Specimens examined are deposited in the Queensland Museum (QM), and the Muséum national d'Histoire naturelle, Paris (MNHN). Measurements provided are of the carapace width and length, respectively.

TAXONOMY

Family GONEPLACIDAE MacLeay, 1838
Pedroplax Ng & Komai, 2011

Pedroplax Ng & Komai, 2011: 61-62.
Michaelia Števčić, 2011: 128, 136.

Type species. — *Psopheticus megalops* Takeda, 1989, by original designation and monotypy. Gender feminine.

Remarks. — Ng et al. (2008: 81) discussed the problematic generic and familial status of *Psopheticus megalops* Takeda, 1989, noting that it would need to be referred to its own genus (see also Castro, 2007). Ng & Komai (2011) subsequently re-examined the types as well as fresh material from Japan, and established the new genus *Pedroplax* for it, and also confirmed that it was a goneplacid. Števčić (2011) independently named a new genus, *Michaelia*, for the species (although he probably had no material), and also recognised a new subfamily in the Goneplacidae, Michaeliinae Števčić, 2011. By coincidence, both Ng & Komai's (2011) and Števčić's (2011) papers describing the new genus in question were published on 30 June 2011. Both names must thus be regarded as simultaneously published, and in such a situation, the Principle of First Reviser would normally need to be invoked to fix the valid name (ICZN Article 24.2). However in the present situation this is not necessary as *Michaelia* Števčić, 2011, is a junior homonym of *Michaelia* Berlese, 1884, *Michaelia* Haller, 1884, and *Michaelia* Trouessart, 1884, all of which are acarines (Arachnida). With regard to the use of Michaeliinae,

Števčić (2011) did not elaborate why he considered a subfamily is needed, and following the valuable discussion of Ng & Komai (2011) I can see no good reason to recognise a higher taxonomic grouping above genus.

Pedroplax takedai sp. nov.
(figs. 1, 2A–F)

Material examined. — Holotype, male (4.5 × 4.0 mm), Chesterfield Reefs, Coral Sea, 19°14′96″S, 158°30′21″E, CORAIL 2, station DW69, Warén dredge, 30-52 m depth, coll. P. Davie & B. Richer de Forges, 25 July 1988 (MNHN). Paratypes, 2 females (3.9 × 3.5, 3.0 × 2.7 mm), Chesterfield Reefs, 19°21′65″S, 158°52′69″E, CORAIL 2, station DW35, Warén dredge, 52 m depth, coll. P. Davie & B. Richer de Forges, 23 July 1988 (QM-W29096).

Description of male. — Carapace sub-hexagonal, c. 1.1 times broader than long (fig. 1A). Dorsal surface glabrous, microscopically granular; granules relatively more prominent towards frontal and lateral margins; gastro-cardiac groove shallow, obscure. Front entire, median part shallowly concave, short sulcus separating it from preorbital lobe; preorbital lobe rounded, sloping slightly posteriorly. Orbit short, subcircular; eyes large, mobile, protruding beyond orbit, ocular peduncle very short, dorsal lateral prolongation distinctively granular. Supraorbital margin entire, not separated from suborbital margin by distinct exorbital angle or tooth. Suborbital margin evenly concave, entire; inner orbital lobe broad, somewhat flattened. Anterolateral margin short, weakly concave, lateral tooth prominent, sublobiform, directed laterally, weakly raised dorsally. Gastric region between anterolateral teeth broadly swollen; similarly moderately swollen transversely at level of cardiac region. Posterolateral margins distinctly convergent posteriorly, weakly convex. Posterior carapace margin straight to weakly concave. Hepatic, pterygostomian, subhepatic regions appearing smooth but covered with low micoscopic granules. Antennule relatively large, fossa asymmetrically triangular, palp folding transversely. Basal antennal segment mobile, rectangular, placed in orbit; flagellum relatively short, as long as width of orbit. Epistome wide, broadly triangular anteromedially; posterior margin medially triangular with a low lobe, separated from bilobed lateral portion by deep concavity. Endostomial ridges well developed.

Third maxillipeds subrectangular, nearly touching basally but diverging distally, leaving moderate gape; merus about as long as wide, anterolateral angle rounded, not auriculiform; ischium subrectangular, very shallow, barely discernible sulcus. Exopod slender, reaching anterior margin of merus, with prominent tooth on inner subdistal margin; flagellum long (fig. 1E).

Chelipeds heterochelous in adult, asymmetrical; surfaces almost glabrous, smooth (fig. 1A, D). Merus short; small tooth on subproximal dorsal margin;

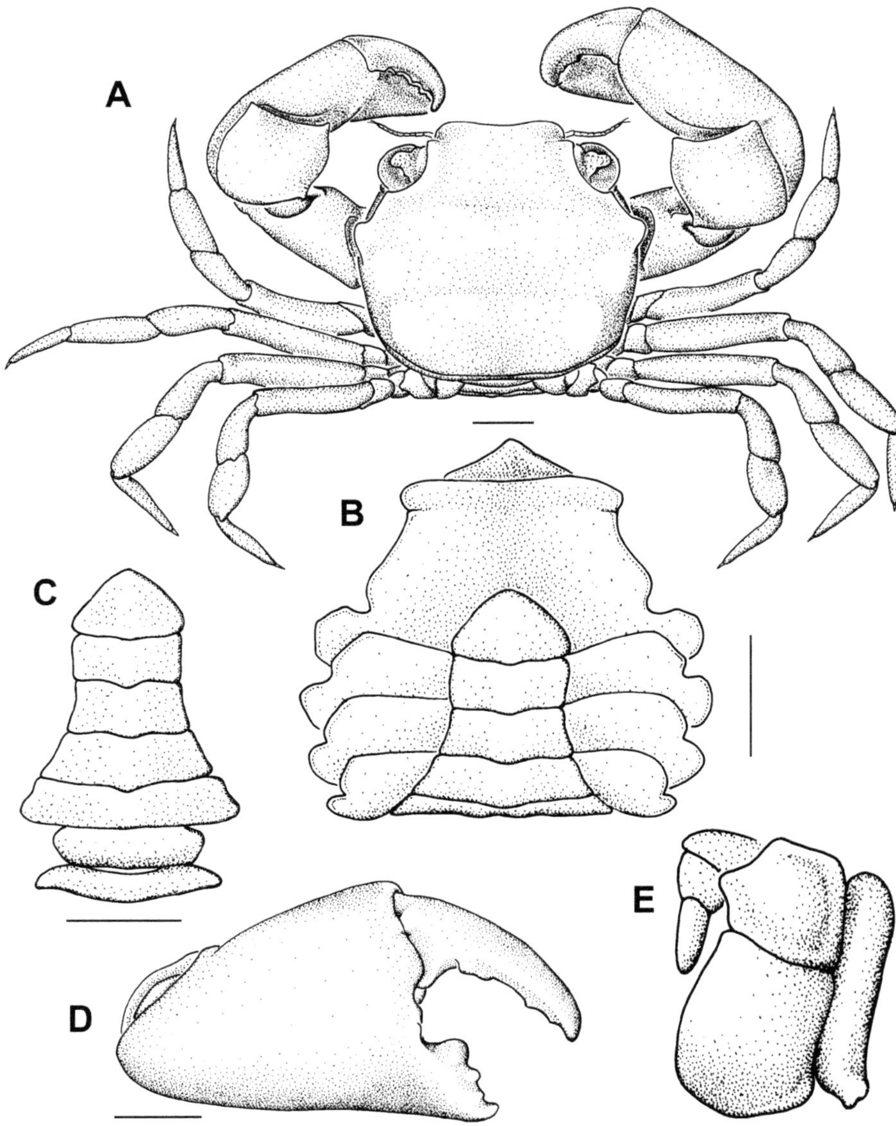

Fig. 1. *Pedroplax takedai* sp. nov., holotype male (4.5 × 4.0 mm) (MNHN), Chesterfield Reefs, Coral Sea. A, dorsal view of whole crab; B, anterior thoracic sternum, and abdominal somites 3-6 with telson; C, male abdomen; D, outer view of left chela; E, left third maxilliped. Scale line = 1 mm.

margins minutely granular. Carpus with broad, bluntly pointed lobiform tooth on inner angle. Chelae stout; fingers relatively long, with dactylus longer than upper margin of palm. Right chela largest, pollex with strongly developed basal tooth on cutting margin; dactylus relatively narrow, curved, leaving broad

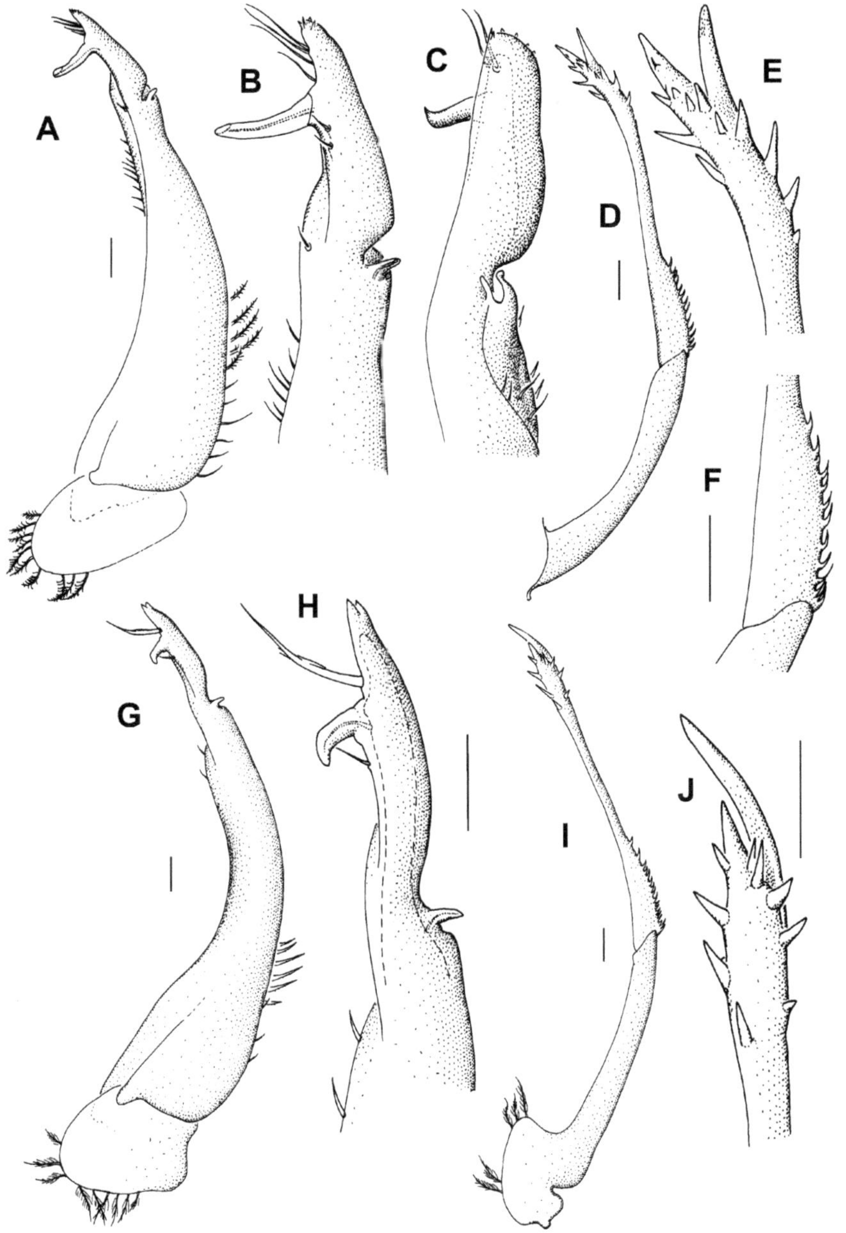

Fig. 2. A–F, *Pedroplax takedai* sp. nov., holotype male (4.5 × 4.0 mm) (MNHN). A, left G1; B, magnification of distal part of left G1; C, distal part of left G1 in sternal view; D, left G2; E, magnification of distal part of left G2; F, magnification of medial section of left G2. G–J, *Pedroplax megalops* (Takeda, 1989), male (5.6 × 4.8 mm) (QM-W29097) G, left G1; H, magnification of distal part of left G1; I, left G2; J, magnification of distal part of left G2. Scale line = 0.1 mm.

basal gape left when closed (fig. 1D). Left chela with fingers more laterally flattened; inner edges with uneven cutting teeth; no gape when closed.

P2–P5 slender, articles smooth, with scattered plumose setae; second leg longest (fig. 1). Meri slender, dorsal margins minutely serrated, without trace of distal or subdistal angle; P4 merus 3.8 times longer than wide. Propodi laterally flattened with weak longitudinal median sulcus; propodus of fourth leg ovate, subfoliaceous. Dactyli slender, gradually tapering distally; fringed with sparse short setae; dactylus of fourth leg laterally flattened, subspatuliform, broader medially.

Surface of thoracic sternites appearing almost smooth but covered with low, small granules. Sternites 1, 2 completely fused, forming prominent triangular plate, separated from sternite 3 by shallow but distinct suture (fig. 1B). Sternites 3, 4 completely fused except for lateral sutures. Press button of abdominal locking mechanism present as rounded tubercle on anterior half of sternite 5. Penis coxal, opening to base of G1 through narrow channel formed by episternite 7, sternite 8.

Sterno-abdominal cavity relatively broad; male abdomen reaching to level between posterior edges of coxae of chelipeds (fig. 1B). Abdomen relatively broad, all somites and telson free. Somite 1 longitudinally narrow, not reaching to bases of coxae of last ambulatory legs; somite 2 narrower, exposing large part of sternite 8; somite 3 transversely trapezoidal, almost reaching coxae of P5, but sternite 8 narrowly visible; somites 4, 5 trapezoidal; somite 6 rectangular, lateral margins almost parallel; telson triangular, distally rounded.

G1 (fig. 2A–C) relatively stout, recurved, distal quarter tapering to relatively sharp tip, separated from main stem by noticeable indentation bearing 2 tiny fleshy palps; single stout, narrow, subdistal lobe directly inwards at right angle to axis, slightly curved distally, slightly longer than terminal distal portion; 3 or 4 conspicuous long, simple, setae between subdistal lobe and apex. G2 slender about 1.3 times length of G1, median part dilated with numerous short spines in long row on outer margin, distal part elongated, tip recurved, bifurcated into 2 stout equal length, divergent spines; with a number of smaller but prominent subdistal spines (fig. 2D–F).

Females. — Similar to male. Chelae homochelous; less robust, relatively more slender. Abdomen rounded, covering entire thoracic sternum with all somites free. Telson broadly triangular with lateral margins weakly sinuous but not contiguous with margins of somite 6. No visible abdominal locking press-button. Vulvae prominent, occupying most of anterior half of sternite 6, without visible sternal vulvar cover.

Habitat. — Found living associated with coarse white coral sand, that contained large percentages of foraminiferans (*Marginopora*) and bleached *Halimeda* pieces (see Richer de Forges et al., 1988).

Distribution. — Only known from the lagoon of the Chesterfield Reefs system, Coral Sea.

Etymology. — It is with great pleasure that I name this new species after Dr Masatsune Takeda, in honour of his immense contribution to our understanding of the taxonomy of the Brachyura.

Remarks. — *Pedroplax takedai* sp. nov. is very similar in general appearance to *P. megalops* but can be separated by a suite of characters. 1) The carapace regions are generally not as pronounced as *P. takedai* and in particular the gastro-cardiac groove is shallow and obscure in contrast to a deeper more distinctly defined H-shaped groove in *P. megalops*. 2) The anterolateral teeth of *P. takedai* are relatively flat and not directed anteriorly to form a distinct anterior sulcus as they do in *P. megalops* (cf. present fig. 1A with Ng & Komai, 2011: fig. 3A). 3) The ambulatory legs of *P. takedai* are relatively shorter, e.g., P4 merus is 3.8 times longer than wide versus 4.9 times in *P. megalops*. 4) The larger male chela of adult *P. takedai* is relatively more robust and rounded, with a prominent enlarged tooth at the base of the fixed finger, and the dactylus is relatively narrower, and arched, such that there is a wide basal gape when the fingers are closed. In *P. megalops*, both fingers are more flattened and close against one-another (cf. Ng & Komai, 2011: fig. 3D). It is unlikely that this different chela morphology is dependent on the stage of maturation, because the present holotype male of *P. takedai* is somewhat smaller than the *P. megalops* males that were examined (4.5 mm carapace breadth versus 5.6 mm). 5) Most significantly, both species differ markedly in gonopod morphology. The G1 of *P. takedai* is slightly stouter, but in particular is broader and more evenly tapering over its apical quarter (cf. present figs 2A, B with 2G, H with Ng & Komai, 2011: fig. 3H). In addition, while both species have a remarkable large subapical palp on the G1, in *P. takedai*, this is larger and slightly curved distally, whereas in *P. megalops* it is shorter and apically hooked down. Similarly *P. takedai* has several long simple setae towards the apex, whereas *P. megalops* has a single, much longer, stouter, and more prominent bristle. The G2 also differs between species (fig. 2D, E, I, J). In *P. takedai*, the elongated distal half is stouter and the recurved tip is bifurcated into two stout, equal length, diverging spines; whereas in *P. megalops* the distal stem is narrower, with the apical portion more swollen and not recurved, and the terminal bifurcation consists of a shorter inner spine lying adjacent to a much longer and recurved outer spine.

As remarked by Ng & Komai (2011) for *P. megalops*, ovigerous females of *P. takedai* also have a small number of large eggs. The larger female had 23 eggs, while the smaller had 13 eggs. Ng & Komai (2011) suggest that this may indicate a partial or even fully abbreviated larval development. While this is possible, such a case would typically imply a relatively narrow range of endemism. Contrary to this, *P. megalops* is here reported to have a broad distribution from Japan to northern Western Australia in the Indian Ocean. It would be interesting to assess the genotype of *P. megalops* to understand the extent of genetic flow. However, both species of this genus are known from very few specimens, and thus would appear to be rare, at least using current collecting methods.

Pedroplax megalops (Takeda, 1989)
(figs. 2G–J, 3)

Psopheticus megalops Takeda, 1989: 172, figs. 16, 17; Castro, 2007: 738, 739.
"*Psopheticus*" *megalops*: Ng et al., 2008: 80.
Pedroplax megalops: Ng & Komai, 2011: 62-66, figs. 1-3.
Michaelia megalops: Števčić, 2011: 136.

Material examined. — Male (5.6 × 4.8 mm), NW Biodiversity Survey (SS2007/05), Stn 196, near Ashmore Reef, northwestern Australia, 123.5964°E 12.4324°S–123.5978°E 12.4344°S, 111 m depth, Sherman Sled tow, coll. P. Davie et al., 7 July 2007 (QM-W29097).

Diagnosis. — Carapace subhexagonal 1.1-1.2 times broader than long. Dorsal surface glabrous; H-shaped gastro-cardiac groove deep. Front entire, medially concave, slightly protruding low lateral lobe. Orbit short, sub-circular; eyes large, mobile, protruding beyond orbit, ocular peduncle very short. Supraorbital margin entire, lacking exorbital angle or tooth. Suborbital margin entire, with low inner sublobiform tooth. Anterolateral margin short, slightly concave, lateral tooth prominent, blunt, curving obliquely anteriorly and upwards. Posterolateral margins convergent posteriorly; region behind anterior part of posterolateral margin weakly swollen, with low rounded crest posteriorly. Posterior carapace margin straight to slightly concave. Hepatic, pterygostomian, subhepatic regions covered with numerous minute low smooth granules. Antennule large, fossa asymmetrical with outer margin longer than inner, distal segments folding transversely. Basal antennal segment mobile, rectangular; placed in orbit; flagellum relatively short, as long as width of orbit. Third maxillipeds subrectangular, small basal gape widening anteriorly; merus about as long as wide, anterolateral angle rounded, not auriculiform; ischium subrectangular, very shallow, barely discernible sulcus. Exopod slender, reaching anterior margin of merus, with prominent tooth on inner subdistal margin; flagellum long. Chelipeds homochelous, asymmetrical; glabrous, smooth.

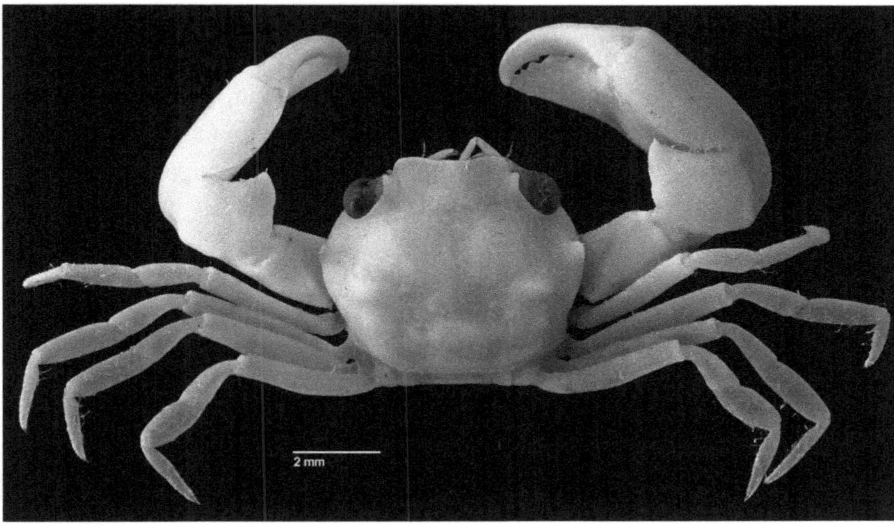

Fig. 3. *Pedroplax megalops* (Takeda, 1989), male (5.6 × 4.8 mm) (QM-W29097), from near Ashmore Reef, northwestern Australia.

Merus short, with low rounded tooth on subproximal dorsal margin. Carpus with large, obtuse lobiform tooth on inner angle. Chela stout, fingers laterally flattened, longer than palm, cutting edges with uneven cutting teeth. Ambulatory legs slender, smooth; P3 longest. Meri slender, P4 merus 4.9 times longer than wide; dorsal margins without trace of distal or subdistal angle. Propodi laterally flattened with median sulcus, that of P5 ovate, subfoliaceous. Dactyli slender, tapering distally; that of P5 laterally flattened, subspatuliform, broadest medially. G1 relatively stout, weakly curved outwards, distal quarter tapering to sharp tip, armed with large subapical palp hooked downwards apically, a single, much longer, stouter, and more prominent bristle subapically (fig. 2G, H). G2 slender about 1.3 times length of G1, median part dilated with numerous short spines on outer margin, distal part elongated, apically swollen, tip not recurved, terminally bifurcated with shorter innner spine lying adjacent to much longer recurved outer spine; 4 or 5 subdistal spines (fig. 2I, J).

Remarks. — Although this species has been previously known only from a relatively small area off southern Japan, and the present record marks a remarkably large range extension into the Indian Ocean, the male specimen examined here agrees in all major respects with the material examined and illustrated by Ng & Komai (2011). As mentioned earlier, it is interesting how a species with such few, large eggs, and potentially abbreviated development can occur across such a large range. This perhaps suggests that it will eventually prove to occur in many intervening locations. Precise habitat data has not been presented for the Japanese material, but it seems that all samples have been collected by dredging in coarse sediments associated with coral reefs.

Distribution. — Previously only from southern Japan, the present record extends its range to north-western Australia, in the Indian Ocean. It has been dredged from depths of 25 to 101 m, with the depth now increased slightly to 111 m.

ACKNOWLEDGEMENTS

Thanks are due to Bertrand Richer de Forges for inviting me on the CORAIL 2 expedition to the Chesterfield Islands, and to CSIRO and Gary Poore for the invitation to take part in the NW Biodiversity Survey to north-western Australia. The photograph of *Pedroplax megalops* was kindly taken by Geoff Thompson, and the figures of *P. takedai* were drawn by Michele Baker, and to both I am very grateful. Peter Ng kindly read the draft and offered me valuable nomenclatural advice, particularly in regard to the homonymy of *Michaelia*. Finally a big thank-you to Hironori Komatsu and Peter K. L. Ng for the invitation to contribute to the current tribute volume for Dr. Takeda.

REFERENCES

BERLESE, A., 1884. Acari, Myriapoda et Scorpiones, hucusque in Italia reperta, **16**(6): (no page number). (Padova).

CASTRO, P., 2007. A reappraisal of the family Goneplacidae MacLeay, 1838 (Crustacea, Decapoda, Brachyura) and revision of the subfamily Goneplacinae, with the description of ten new genera and eighteen new species. Zoosystema, **29**(4): 609-773.

HALLER, G., 1884. Beschreibung einiger neuen Milben. Archiv für Naturgeschicthe, **50**(1): 217-236.

MACLEAY, W. S., 1838. On the brachyurous decapod Crustacea brought from the Cape by Dr. Smith. In: A. SMITH (ed.), Illustrations of the zoology of South Africa; consisting chiefly of figures and descriptions of the objects of natural history collected during an expedition into the interior of South Africa, in the years 1834, 1835, and 1836; fitted out by 'The Cape of Good Hope Association for Exploring Central Africa': together with a summary of African zoology, and an inquiry into the geographical ranges of species in that quarter of the globe. Invertebrate, **4** [1849]: 53-71, pls. 2, 3. (Smith, Elder & Co., London).

NG, P. K. L., D. GUINOT & P. J. F. DAVIE, 2008. Systema brachyurorum: Part I. An annotated checklist of extant brachyuran crabs of the world. Raffles Bulletin of Zoology, (Supplement) **17**: 1-286.

NG, P. K. L. & T. KOMAI, 2011. A new genus for *Psopheticus megalops* Takeda, 1989 (Crustacea: Brachyura: Goneplacidae). Zootaxa, **2934**: 61-67.

RICHER DE FORGES, B., C. CHEVILLON, P. LABOUTE, G. BARGIBANT, J.-L. MENOU & P. TIRARD, 1988. La campagne CORAIL 2 sur le plateau des îles Chesterfield (N.O. "Coriolis" et N.O. "Alis", 18 juillet au 6 août 1988). Rapports scientifiques et techniques Sciences de la Mer Biologie marine, **50**: 1-68.

ŠTEVČIĆ, Z., 2011. Addition to the reclassification of brachyuran crabs (Crustacea: Decapoda: Brachyura). Part I. New taxa. Natura Croatica (Fauna Croatica), **20**(1): 125-139.

TAKEDA, M., 1989. Shallow-water crabs from the Oshima Passage between Amami-Oshima and Kakeroma-jima Islands, the northern Ryukyu Islands. Memoirs of the National Science Museum, Tokyo, **22**: 135-184, pl. 4.

TROUESSART, E.-L., 1884. Les Sarcoptides plumicoles. Journal de Micrographie, **8**: 527-532, 572-579.

WOOD-MASON, J., 1892. Illustrations of the zoology of the Royal Indian Marine Surveying Steamer *Investigator*, under the command of Commander A. Carpenter, R.N., D.S.O., and Commander R. F. Hoskyn, R.N. Part I. Crustaceans: pls. 1-5. [five unnumbered pages for Explanation of Plates]. (Office of the Superintendent of Government Printing, Calcutta).

First received 2 September 2011.
Final version accepted 24 January 2012.

FIRST ZOEA OF *PORCELLANOPAGURUS TRUNCATIFRONS* TAKEDA, 1981 (DECAPODA, ANOMURA, PAGURIDAE) DESCRIBED FROM LABORATORY-HATCHED MATERIAL

BY

YOSHIHISA FUJITA[1,2,3])

[1]) University Education Center, University of the Ryukyus, 1 Senbaru, Nishihara, Okinawa 903-0213, Japan

[2]) Marine Learning Center, 2-95-101 Miyagi, Chatan, Okinawa 904-0113, Japan

ABSTRACT

The first zoea of *Porcellanopagurus truncatifrons* Takeda, 1981, is described and illustrated from laboratory-hatched material. The zoea of *P. truncatifrons* differs from that of *P. edwardsi*, the only other species of *Porcellanopagurus* for which larval morphology is known, in the armature of the antennule, antenna, abdomen and telson. The zoea of *Porcellanopagurus* is distinguishable from those of other pagurid genera by a combination of characters of the carapace and appendages.

RÉSUMÉ

La première zoé de *Porcellanopagurus truncatifrons* Takeda, 1981, est décrite et figurée à partir d'un matériel éclos en laboratoire. La zoé de *P. truncatifrons* diffère de celle de *P. edwardsi*, seule autre espèce of *Porcellanopagurus* dont la morphologie larvaire soit connue, par l'armature de l'antennule, de l'antenne, de l'abdomen et du telson. La zoé de *Porcellanopagurus* se distingue de celle des autres genres de Pagurides par la combinaison des caractères de la carapace et des appendices.

INTRODUCTION

The pagurid hermit crab genus *Porcellanopagurus* Filhol, 1885, is known to have unique body form and behavior, with species possessing a short and symmetrical abdomen that utilizes bivalve shells as protective covering (McLaugh-

[3]) e-mail: galatheids@yahoo.co.jp

lin, 2000a). The genus presently includes 13 species from the Indo-Pacific region (McLaughlin, 2000a; McLaughlin et al., 2010), with three recorded from Japanese waters: *P. japonicus* Balss, 1913, *P. nihonkaiensis* Takeda, 1985, and *P. truncatifrons* Takeda, 1981 (Takeda, 1981, 1985; McLaughlin, 2000a).

The only known studies on the larvae of *Porcellanopagurus* species are by Roberts (1972) and McLaughlin (2000b) for *P. edwardsi* Filhol, 1885, both of which are based on wild-caught plankton material. The aim of this study is to describe and illustrate the first zoeal morphology of *P. truncatifrons* obtained from laboratory-hatched material, and to compare it with that of *P. edwardsi*.

MATERIAL AND METHODS

An ovigerous female of *Porcellanopagurus truncatifrons* Takeda, 1981, was collected at 8 m depth in Cape Maeda, Okinawa Island on 12 August 2003, transported to the laboratory of the University of the Ryukyus and maintained in 1.4-liter plastic aquaria, containing filtered seawater, until hatching occurred. After hatching, larvae were fixed and preserved in 50% ethylene glycol. Observations and illustrations of larvae were made using a NIKON OPTIPHOT-2 binocular stereomicroscope equipped with a drawing tube. Five specimens were measured and dissected for setal observations. Body parts are described progressing in an anterior to posterior direction, appendages from endopod to exopod, and segments and setae from proximal to distal. The long plumose natatory setae on the exopods of maxillipeds are drawn truncated. Terminology generally follows that of Ingle (1991) but usage of "zoeas" (as plural) follows that recommended by Clark et al. (1998). Carapace length (CL) was measured from the tip of the rostrum to the posterior midpoint of the carapace; total length (TL) from the rostral tip to the posterior median margin of the telson, excluding posterior processes.

Dissected and whole larvae as well as a parental female are deposited in the Ryukyu University Museum, Fujukan (RUMF), Okinawa, under the registration numbers RUMF-ZC-01578 (first zoeas) and RUMF-ZC-01577 (spent female), respectively.

DESCRIPTION

Porcellanopagurus truncatifrons Takeda, 1981
(figs. 1, 2)

First zoea

Size. CL 1.30-1.41 mm, TL 2.35-2.43 mm.

Carapace (fig. 1A, B). Dorsal surface with carina running from basal part of rostrum to dorsoposterior end of carapace, posteriorly produced into acute spine on each side, without setae; rostrum elongated, spine-like, extending to level of apex of antennal scaphocerite; eyes sessile.

Antennule (fig. 2A). Unsegmented, elongate rod-like; faintly bilobed, terminally with 3 aesthetascs and 2 or 3 simple setae; subterminal margin with 1 long plumose seta.

Antenna (fig. 2B). Biramous; protopod with 1 serrated spine on distoventral margin, that spine shorter than endopod; endopod fused to protopod, with 2 long plumose setae terminally; scaphocerite (exopod) ending in long, acute

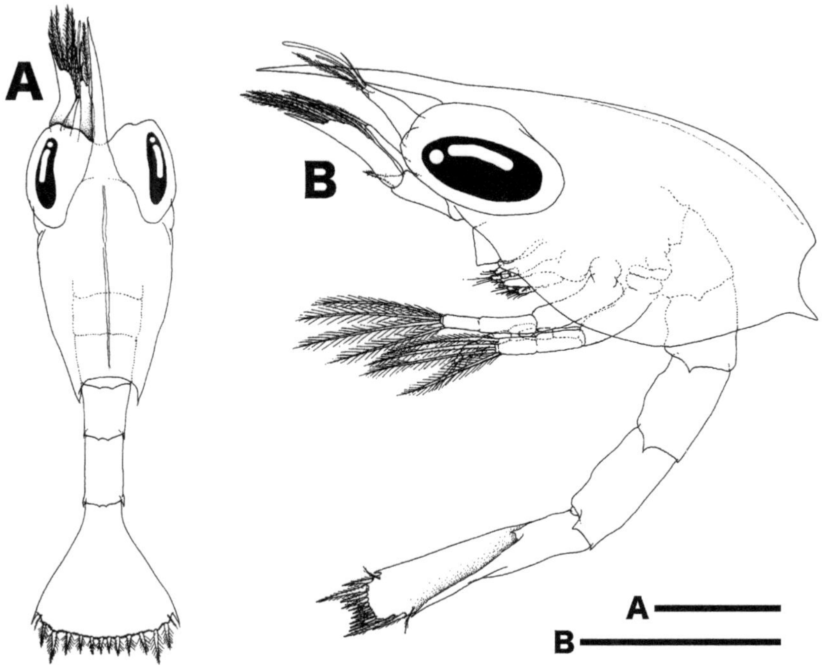

Fig. 1. First zoea of *Porcellanopagurus truncatifrons* Takeda, 1981. A, dorsal view of carapace, abdomen and telson; B, lateral view of entire animal. Scale bars = 0.5 mm.

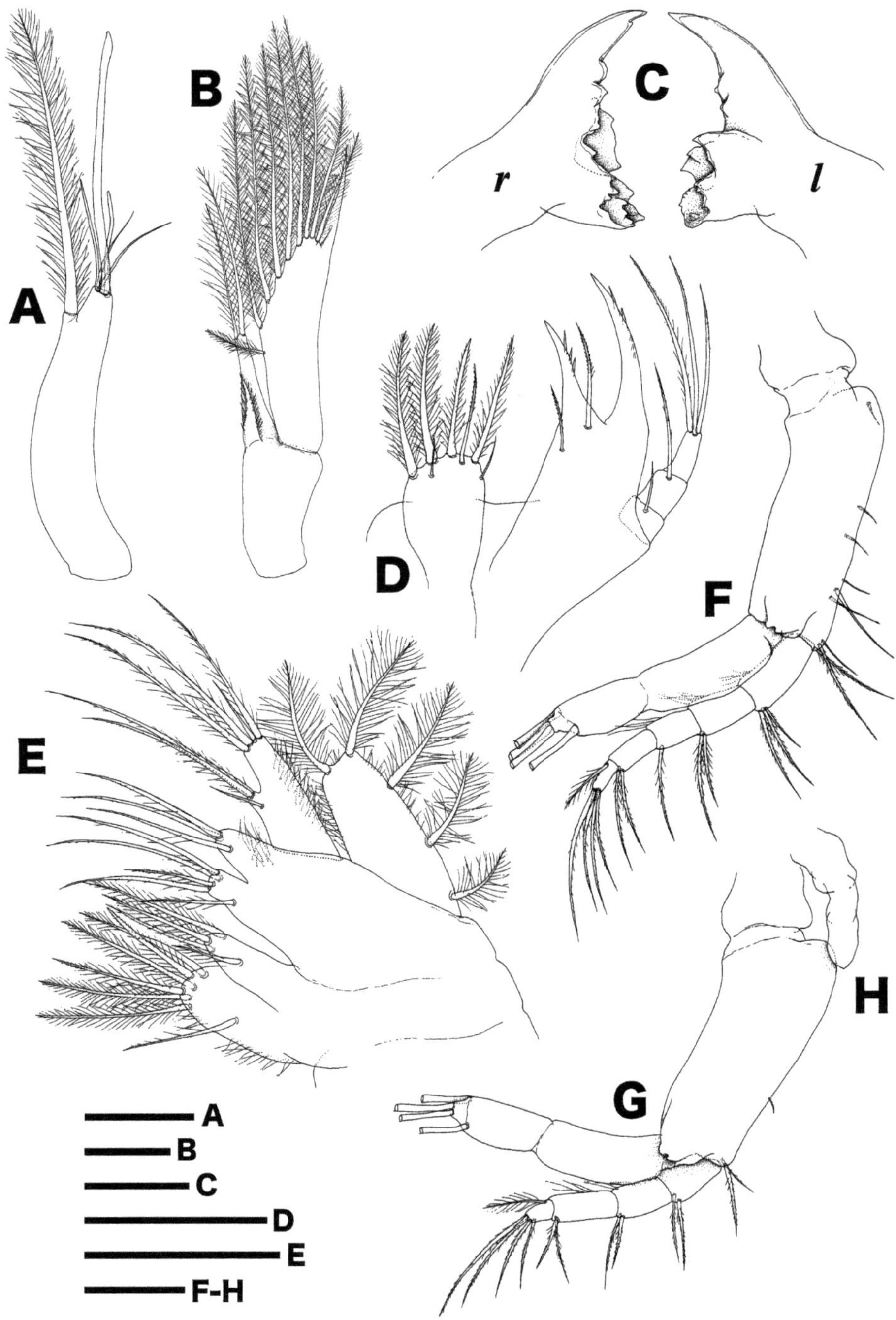

Fig. 2. First zoea of *Porcellanopagurus truncatifrons* Takeda, 1981. A, antennule; B, antenna; C, mandible (r, right side; l, left side); D, maxillule; E, maxilla; F, first maxilliped; G, second maxilliped; H, third maxilliped. Scale bars = 0.1 mm.

process on distolateral margin, mesial margin with 9 plumose setae and 1 simple seta, dorsal and ventral surfaces smooth.

Mandibles (fig. 2C). Asymmetrically dentate; incisor and molar processes distinct, incisor processes with strong teeth, molar processes serrate or spinose; palp absent.

Maxillule (fig. 2D). Coxal endite with 7 simple/plumodenticulate setae; basial endite with 2 cuspidate and 2 plumodenticulate setae; endopod 3-segmented, proximal segment with 1 short terminal simple seta, penultimate segment with 1 terminal plumose seta, ultimate segment with 3 terminal simple/plumose setae.

Maxilla (fig. 2E). Coxal and basial endites bilobed, with 7-8 + 4 and 5 + 3-4 plumose/plumodenticulate setae, respectively; endopod unsegmented but bilobed, with 3 subdistal and 4 distal plumose/plumodenticulate setae (seta formula 3 + 4); scaphognathite (exopod) with 5 marginal plumose setae. Microtrichia present on proximal lobe of coxal endite, distal lobe of basial endite and outer surface of endopod.

First maxilliped (fig. 2F). Coxa without setae; basis with 9 simple/plumose setae (seta formula, 1 + 1 + 1 + 3 + 3) on ventral margin; endopod 5-segmented, first to fourth segments with 3, 2, 1, 2 simple/sparsely plumose setae on posterior margin, respectively, first to third segments with microtrichia on dorsodistal margin, fifth segment with 4 plumodenticulate setae distally, 1 plumose seta at dorsoproximal angle; exopod indistinctly 2-segmented, distally with 4 long plumose natatory setae.

Second maxilliped (fig. 2G). Coxa without setae; basis with 1 + 2 ventrodistal setae; endopod 4-segmented, with setation of 2, 2, 2, 4 + I (I = dorsoproximal plumose seta), second and third segments each with microtrichia on dorsal margin; exopod indistinctly 2-segmented, distally with 4 long plumose natatory setae.

Third maxilliped (fig. 2H). Small unsegmented bud.

Pereiopods. Not apparent.

Abdomen (fig. 1A, B). Five somites, somite 5 longest; somites 2-5 each with pair of small median spines on posterodorsal margin and pair of posterolateral spines; pleopods absent.

Telson (fig. 1A, B). Broadened posteriorly, subtriangular in dorsal view, as long as broad, smooth on dorsal and ventral surfaces; posterior margin with 16 (8 + 8) processes, most lateral process an unarticulated spine, flanked by second short plumose seta (= anomuran hair), remaining outgrowths articulated, plumose processes; small anal spine present.

Colour in life. Not recorded.

DISCUSSION

Our knowledge on larvae of *Porcellanopagurus* is poor. Roberts (1972) described the prezoea and four zoeal stages of *P. edwardsi* from plankton samples collected in New Zealand. McLaughlin (2000b) subsequently described the megalopal and first crab stages of the same species, also collected from offshore New Zealand. As such, within the genus, the morphology of the first zoeal stage of *P. truncatifrons* can only be compared with that of *P. edwardsi*. Although the zoea of *P. truncatifrons* is generally similar to that of *P. edwardsi*, there are several morphological characters that distinguish *P. truncatifrons* from *P. edwardsi*. These are as follow (table I): (1) the antennule bears three terminal aesthetascs (two in *P. edwardsi*), (2) the antennal scaphocerite (exopod) has nine plumose setae (ten in *P. edwardsi*), (3) there are only two spines on the posterodorsal margins of the third to fifth abdominal somites (four spines in *P. edwardsi*), and (4) the posterior plumose processes of the telson are very short, the longest process being much shorter than the maximum telson width (relatively much longer in *P. edwardsi*, the longest one longer than half the maximum telson width).

In the family Paguridae Latreille, 1802, more than 70 genera have been described (McLaughlin et al., 2010). Komai & Konishi (2003) compared 10 morphological characters of the first zoeas among 13 pagurid genera with known larval morphology: *Anapagurus* Henderson, 1886, *Cestopagurus* Bouvier, 1897, *Discorsopagurus* McLaughlin, 1974, *Labidochirus* Benedict, 1892, *Lithopagurus* Provenzano, 1968, *Munidopagurus* A. Milne-Edwards & Bouvier, 1893, *Pagurus* Fabricius, 1775, *Parapagurodes* McLaughlin & Haig, 1973, *Phimochirus* McLaughlin, 1981, *Porcellanopagurus*, *Propagurus* McLaughlin & de Saint Laurent, 1998, *Pylopaguropsis* Alcock, 1905, and *Spiropagurus* Stimpson, 1858. For *P. truncatifrons*, the 10 diagnostic characters of the first zoea are as follow: (1) the antennal protopod is armed with one spine, (2) the antennal endopod bears two terminal long plumose setae, (3) the antennal scaphocerite has nine plumose setae, (4) the maxillular endopod is 3-segmented with a setation of 1, 1, 3, (5) the maxillar endopod possesses seven setae, (6) the maxillar scaphognathite has five marginal plumose setae, (7) the third maxilliped is uniramous, (8) the pereopodal buds have not appeared, (9) the pleopods are absent, and (10) the telson is narrow and triangular in shape. These features agree well with those of *P. edwardsi*, except for the number of marginal plumose setae on the antennal scaphocerite (ten instead of nine; see Roberts, 1972; table I in present study).

TABLE I

Comparison of morphological characters of first zoeas of *Porcellanopagurus truncatifrons* Takeda, 1981, and *P. edwardsi* Filhol, 1885. Abbreviations: a, aesthetascs; cs, cuspidate seta; ps, plumose seta; s, seta other than cs and ps; sp, spine; I, subterminal long plumose seta on antennule, or dorsoproximal plumose seta on endopods of maxillipeds

Species Reference	*P. truncatifrons* Present study	*P. edwardsi* Roberts (1972)
Antennule: protopod	3a + (2-3)s + I	2a + 2s + I
Antenna:		
protopod	1sp	1sp
endopod	2s	2s
scaphocerite	9ps + 1s	10ps
Maxillule:		
coxal endite	7s	7s
basial endite	2cs + 2s	2cs + 2s
endopod	1s, 1s, 3s	1s, 1s, 3s
Maxilla:		
coxal endite	(7-8 + 4)s	(8 + 4)s
basial endite	(5 + 3-4)s	(5 + 4)s
endopod	(3 + 4)s	(3 + 4)s
scaphognathite	5ps	5ps
First maxilliped:		
coxa	naked	naked
basis	(1 + 2 + 3 + 3)s	(1 + 2 + 3 + 3)s
endopod	3s, 2s, 1s, 2s, 4s + I	3s, 2s, 1s, 2s, 4s + I
Second maxilliped:		
coxa	naked	naked
basis	(1 + 2)s	(1 + 2)s
endopod	2s, 2s, 2s, 4s + I	2s, 2s, 2s, 4s + I
Third maxilliped:	uniramous bud	uniramous bud
Abdomen:		
posterodorsal spines on each third to fifith somites	2 (1 pair)	4 (2 pairs)
Telson:		
posterior processes	8 + 8	8 + 8
length of plumose processes	short	long

Overall the larvae of *Porcellanopagurus* appears closest to that of *Cestopagurus*, with the same armature on the antenna, maxillule, and maxilla, but it differs from the latter in that the telson is armed with 16 (8 + 8) instead of 14 (7 + 7) pairs of posterior marginal processes (see Dechancé, 1961, as *Catapaguroides timidus* (Roux, 1830); Komai & Konishi, 2003). In addition, de Saint Laurent (1969) provided a short description of the first zoea of *Acanthopagurus dubius* (A. Milne-Edwards & Bouvier, 1900). Morphologically, the first zoeal of *A. dubius* clearly differs from that of the two *Porcellanopagurus* species in having long posterolateral and posterodorsal spines on the carapace (see de Saint Laurent, 1969, figs. 17, 18). In contrast, the first zoeas of *P. edwardsi* and *P. truncatifrons* have distinctly shorter posterolateral carapace spines and lack posterodorsal carapace spines. Therefore, *Porcellanopagurus* can be distinguished from all other 13 pagurid genera by this combination of zoeal characters.

ACKNOWLEDGEMENTS

Many thanks to my wife Akane Fujita of the University of the Ryukyus (Faculty of Medicine) for her assistance in the fieldwork. Thanks are also due to anonymous reviewers for reviewing my manuscript: to Masayuki Osawa (Shimane University) for his valuable comments and suggestions on the early draft of the manuscript. This study is dedicated to Dr. Masatsune Takeda who has described the hermit crab species in this study and made great contributions to our knowledge of decapod crustaceans in general.

REFERENCES

CLARK, P. F., D. K. CALAZANS & G. W. POHLE, 1998. Accuracy and standardization of brachyuran larval descriptions. Invertebrate Reproduction and Development, **33**: 127-144.

DECHANCÉ, M., 1961. Sur la morphologie externe des larves de pagurides. I. *Catapaguroides timidus* (Roux). Crustaceana, **2**: 53-67.

INGLE, R., 1991. Larval stages of northeastern Atlantic crabs: an illustrated key: 1-363. (Chapman and Hall, London).

KOMAI, T. & K. KONISHI, 2003. Further note on *Propagurus miyakei* (Baba, 1986) (Decapoda: Anomura: Paguridae), with description of its first zoea. Bulletin of Marine Science, **72**: 853-869.

MCLAUGHLIN, P. A., 2000a. Crustacea: Decapoda: species of *Porcellanopagurus* Filhol and *Solitariopagurus* Türkay (Paguridae), from New Caledonia area, Vanuatu and the Marquesas: new records, new species. In: A. CROSNIER (ed.), Résultats des Campagnes MUSORSTOM, **21**. Mémoires du Muséum national d'Histoire naturelle, Paris, **184**: 389-414.

— —, 2000b. Megalopal and first crab stages of *Porcellanopagurus edwardsi* Filhol, 1885 (Crustacea: Decapoda: Anomura: Paguridea). Proceedings of the Biological Society of Washington, **113**: 721-730.

MCLAUGHLIN, P. A., T. KOMAI, R. LEMAITRE & D. L. RAHAYU, 2010. Annotated checklist of anomuran decapod crustaceans of the world (exclusive of the Kiwaoidea and families Chirostylidae and Galatheidae of the Galatheoidea) Part I — Lithodoidea, Lomisoidea and Paguroidea. In: E. Y. L. MARTYN & S. H. TAN (eds.), Checklists of anomuran decapod crustaceans of the world (exclusive of the Kiwaioidea and families Chirostylidae and Galatheidae of the Galathoidea) and marine lobsters of the world. Raffles Bulletin of Zoology, (Supplement) **23**: 5-107.

ROBERTS, P. E., 1972. Larvae of *Porcellanopagurus edwardsii* Filhol, 1885 (Crustacea: Decapoda: Paguridae) from Perseverance Harbour, Campbell Island. Journal of the Royal Society of New Zealand, **2**: 383-391.

DE SAINT LAURENT, M., 1969. Révision des genres *Catapaguroides* et *Cestopagurus* et description de quarte genres nouveaux. III. *Acanthopagurus* de Saint Laurent (Crustaces Decapodes Paguridae), Bulletin du Muséum national d'Histoire naturelle, Paris, (2) **41**: 731-742.

TAKEDA, M., 1981. A new hermit crab of the genus *Porcellanopagurus* from the Ogasawara Islands. Bulletin of the Biogeographical Society of Japan, **36**: 8-13.

— —, 1985. Occurrence of a new hermit crab of the genus *Porcellanopagurus* in the Sea of Japan. Memoirs of the National Science Museum, Tokyo, **18**: 141-144.

First received 12 October 2011.
Final version accepted 21 December 2011.

A NEW SPECIES OF *ANISOMYSIS* HANSEN, 1910 (MYSIDA, MYSIDAE) FROM INSHORE WATERS OF A SMALL ISLAND IN THE ANDAMAN SEA

BY

YUKIO HANAMURA[1,3]) and ISAO TSUTSUI[2])

[1]) National Research Institute of Fisheries Science, Fuku-ura 2-12-4, Kanazawa-ku, Yokohama, Kanagawa 236-8648, Japan

[2]) Japan International Research Center for Agricultural Sciences, Owashi 1-1, Tsukuba, Ibaraki 305-8686, Japan

ABSTRACT

A new species of *Anisomysis* Hansen, 1910 (Mysida, Mysidae) collected from inshore waters of Racha Yai Island in the Andaman Sea, eastern Indian Ocean, is reported upon. The new species, *Anisomysis* (*Paranisomysis*) *takedai*, shows a close affinity to *A.* (*P.*) *minicoyensis* Biju et al., 2006. The new species can be distinguished from *A.* (*P.*) *minicoyensis* by having a short rostrum barely reaching the base of the antennular peduncle, a comparatively broader antennal scale, a proportionately longer second article of the male fourth pleopodal exopod, and a larger number of setae on the telson.

INTRODUCTION

Species of *Anisomysis* Hansen, 1910 are small- to medium-sized animals of the Mysida Haworth, 1825 distributed in temperate to tropical waters of the Indo-West Pacific. They often form aggregations around submarine rocks, seaweed beds or sea surface and are assumed to play a significant role in the marine food web (cf., Müller, 1993; Fukuoka et al., 2002). The genus currently contains some 56 species (Murano & Fukuoka, 2003; Anderson, 2010), although the taxonomic position of *Anisomysis* (*Javanisomysis*) *gutzui* Băcescu, 1992 is controversial. Murano & Fukuoka (2003) reported that this species should be excluded from *Anisomysis* since the male fourth

[3]) Corresponding author; e-mail: hanamura@affrc.go.jp

pleopod is remarkable by having an un-segmented exopod as opposed to three articles in all other members of the genus. Consequently, Murano & Fukuoka (2003) considered *Javanisomysis* Băcescu, 1992 to be a monotypic genus and these authors subsequently recognized three subgenera including their new taxon within *Anisomysis*; i.e., *Anisomysis*; *Paranisomysis* Băcescu, 1973, and *Pseudanisomysis* Murano & Fukuoka, 2003 (see also Sawamoto & Fukuoka, 2005).

To date, approximately 10 species of *Anisomysis* have been recorded from South East Asian coastal waters (Sawamoto & Fukuoka, 2005). However, the mysid diversity in this region is still poorly understood due to limited research in terms of frequency and surveyed areas.

A notable number of specimens of *Anisomysis* were captured during seaweed and crustacean surveys on Racha Yai Island, a small island in the southern Andaman Sea (7°36′N 98°22′E), eastern Indian Ocean. This paper reports a new species belonging to the subgenus *Paranisomysis* of the genus *Anisomysis* collected from the sublittoral of this island.

Mysid specimens were fixed in approximately 70% ethanol upon sampling and preserved prior to the taxonomic analysis. Body length (BL), used as the standard, is measured from the apex of the rostrum to the posterior end of the telson without apical setae. For setal terminology, Watling (1989) is followed. Descriptions were based on adult specimens of both sexes unless reported otherwise. The specimens treated here are housed in the National Museum of Nature and Science, Tokyo (NSMT).

TAXONOMIC NOTE

Family MYSIDAE Haworth, 1825
Subfamily MYSINAE Haworth, 1825
Genus *Anisomysis* Hansen, 1910
Anisomysis (Paranisomysis) takedai sp. nov.

(figs. 1-3)

Material examined. — Holotype (NSMT-Cr 21977), male (BL 5.3 mm), sublittoral around seaweed bed in Siam Bay (7°36′30″N 98°22′26″E), Racha Yai Island, Thailand, hand net during free diving, 4 Feb. 2008, coll. K. Hamano & I. Tsutsui. Paratypes (NSMT-Cr 21978), 12 males (BL 3.2-4.9 mm), 41 females (BL 4.0-5.5 mm), 6 ovigerous females (BL 4.1-5.2 mm), 1 juvenile (BL 2.5 mm), other data same as above.

Description. — Body somewhat slender and fragile (fig. 1a).

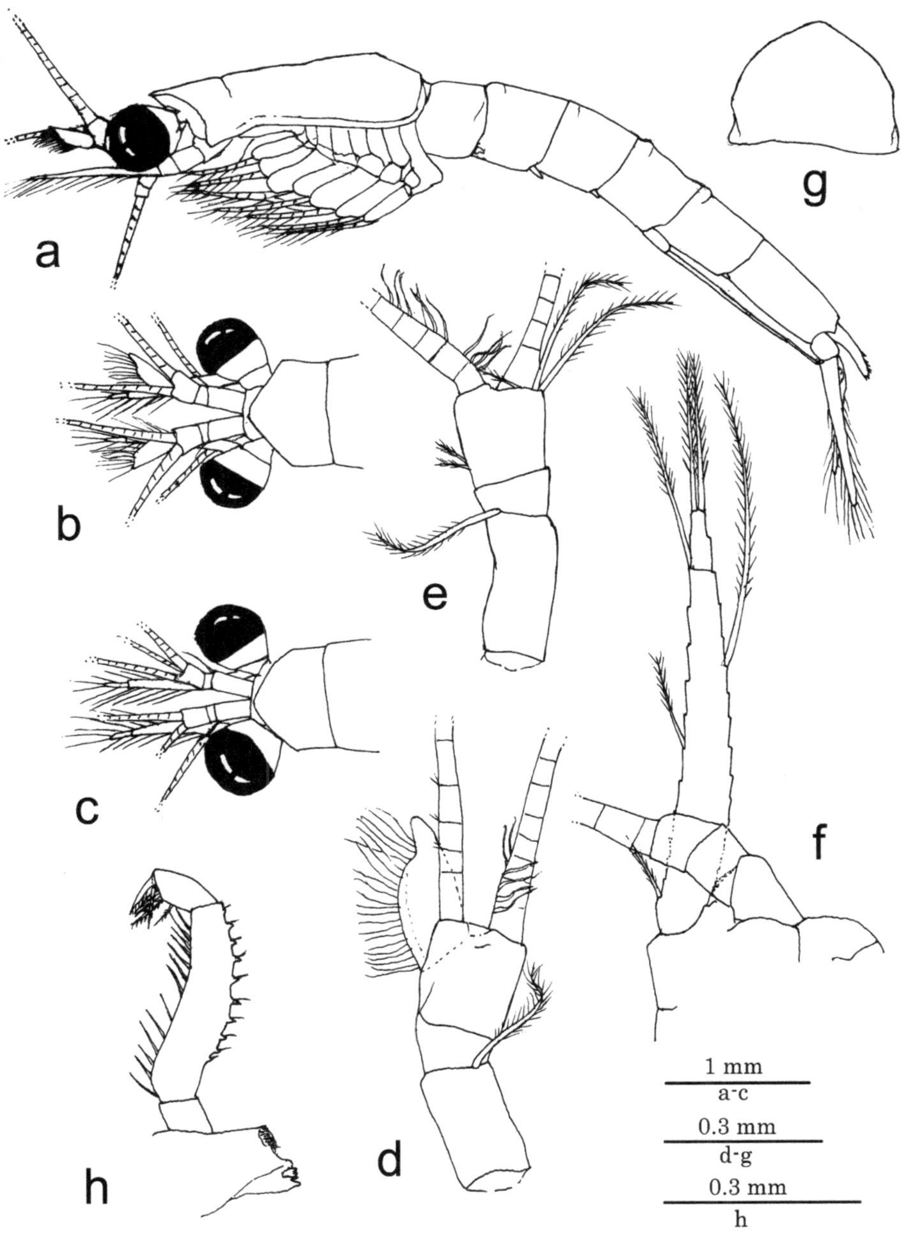

Fig. 1. *Anisomysis* (*Paranisomysis*) *takedai* sp. nov., holotype male (BL 5.3 mm) (a, b), paratype male (BL 4.9 mm) (d, f, h), paratype female (BL ca. 5.0 mm) (c, e, g): a, entire body of male, lateral view (abdomen slightly deformed due to damage); b, anterior part of carapace and cephalic appendages of male, dorsal view; c, same of female; d, right antennule of male, dorsal view; e, left antennule of female, dorsal view; f, right antenna, ventral view; g, labrum, ventral view; h, right mandible, external view.

Carapace (fig. 1a–c) with anterior dorsal margin produced into subtriangular rostrum with obtuse to moderately pointed apex reaching basal part of antennular peduncle; lateral margin of rostrum nearly straight or gently convex, basically leaving eye stalk uncovered; cervical sulcus distinct at anterior 1/3 of carapace; posterior dorsal margin excavate, leaving last 3 thoracic somites uncovered.

Eye (fig. 1a–c) globular and large relative to its body size, cornea well pigmented, distinctly larger than eye stalk and almost as large as anterior carapace depth.

Antennule sexually dimorphic (fig. 1d, e). Male peduncle more robust than female one, first article slightly shorter than combined length of second and third articles, possessing 1 or 2 long setae near distolateral end (slightly deformed in figure due to artifact); second article distinctly shorter than broad; third article broader than basal 2 articles and as long as or slightly longer than its width, sub-oval male lobe with distinct excavation near distomesial part, bearing long and dense slender setae along mesial margin. Female peduncle with first article longer than combined length of distal 2 articles, with long seta at distolateral corner (slightly deformed due to artifact); second article distinctly shorter than broad; third article as long as or slightly longer than broad, possessing a few long setae at distomesial corner.

Antennal scale of male (fig. 1f) lanceolate and curving outwards, extending beyond apex of male lobe approximately by length of its distal lobe, 7.9-8.1 times as long as broad, armed with long setae all around margins, distal lobe demarcated by suture and its length slightly more than 2 times longer than width across suture. Female antennal scale proportionately longer than that of male, 8.9-9.3 times as long as broad, extending beyond end of antennular peduncle by slightly less than half of its length.

Labrum (fig. 1g) with anteriorly margin slightly produced, subtriangular, without median spine or tooth. Mandibular palp (fig. 1h) with second article distinctly curving outwards, possessing usually 9 or 10 somewhat short, flagellate denticles along mesial margin, becoming longer and stouter towards mid-length and largest denticle 1/5-1/3 length of width across corresponding part of palp, lateral margin of palp fringed with long setae along entire margin; third article flexing outwards and about 1/3 length of second, distal margin truncated, bearing seta at distomesial part, rather stout seta/claw arising next to innermost seta, and also with several setae of subequal length. Other mouth parts as illustrated (fig. 2a–d).

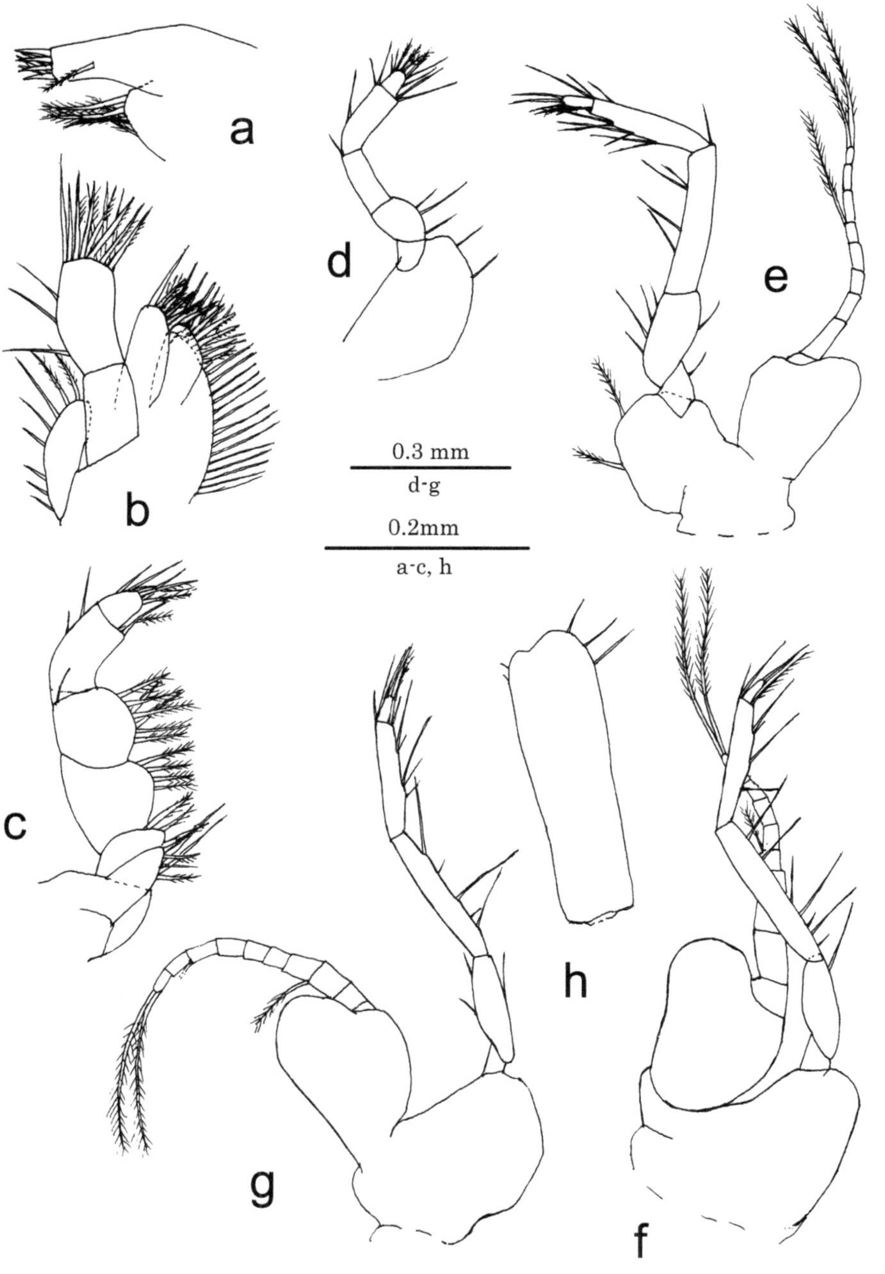

Fig. 2. *Anisomysis* (*Paranisomysis*) *takedai* sp. nov., paratype male (BL 4.9 mm) (a–d, f–h), paratype female (BL ca. 5.0 mm) (e): a, right maxillule, external view; b, right maxilla, external view; c, d, right endopod of first and second thoracopods, posterior view; e, right fourth thoracopod, anterior view; f, g, right sixth and seventh thoracopods, posterior view; h, right penis.

Third to eighth thoracopods (fig. 2e–g) similar in shape; exopods with basal plate rounded at distolateral corner, following flagellum composed of 8 or 9 articles; endopods proportionately stout, merus consistently longer than carpopropodus. Penis (fig. 2h) cylindrical, approximately 3 times as long as broad, bearing a few setae distally.

Abdomen (fig. 1a) with first 5 somites subequal in length, sixth somite 1.19-1.31 times as long as fifth (slightly deformed in holotype specimen due to damage).

Male first to third, and fifth pleopods (fig. 3a–c, e) unsegmented and rudimentary, first pleopod with well developed side lobe, slightly shorter than second one; second pleopod with marked side lobe; third and fifth pleopods subequal in length and appreciably shorter than second, side lobe indistinct. Male fourth pleopod (fig. 3d) bi-ramous: endopod unsegmented and rudimentary, bearing a few short setae along margins: exopod slender and elongated, reaching basal 1/3-1/2 of telson; first article 2.33-2.80 times as long as second and again 1.23-1.40 times longer than combined length of distal 2 articles; second article 1.00-1.20 times as long as third, latter with slightly longer attenuated seta and also basally stout incised seta with setulles. Female pleopods (fig. 3f–h) unsegmented and rudimentary, length decreasing in order of second, first, third, and fourth ~ fifth.

Uropod (fig. 3i) with endopod approximately 0.8 times as long as exopod, bearing about 10 short smooth setae along lateral margin; exopod entire and slender, slightly curving outwards, about 10 or more times longer than broad.

Telson (fig. 3j) 0.59-0.60 times as long as sixth abdominal somite, 1.42-1.57 times longer than basal width, distinctly divided into 2 parts by deep constriction around distal 2/7; proximal part 1.13-1.16 times as long as basal width, bearing commonly 6, occasionally 5, spinous setae anterior to narrowest part, median seta(e) usually longer than those placed on anterior and posterior edges; posterior part of telson distinctly shorter than its width, constriction 68-75% width of posterior lobe and again 33-37% of maximal width at base, apical margin truncated, possessing 9 or 10 somewhat wide, lanceolate setae of subequal length, and lateral margin with 4 or 5 somewhat narrow setae, increasing its length posteriorly.

Colour. — Semi-transparent when alive, although dark-brownish or blackish chromatophores present around mouthparts, basal part of some thoracopods and along basal part of female marsupium in preserved specimens.

Etymology. — This species is named in honour of Dr. Masatune Takeda, who contributed substantially to the taxonomy of the Indo-West Pacific Crustacea.

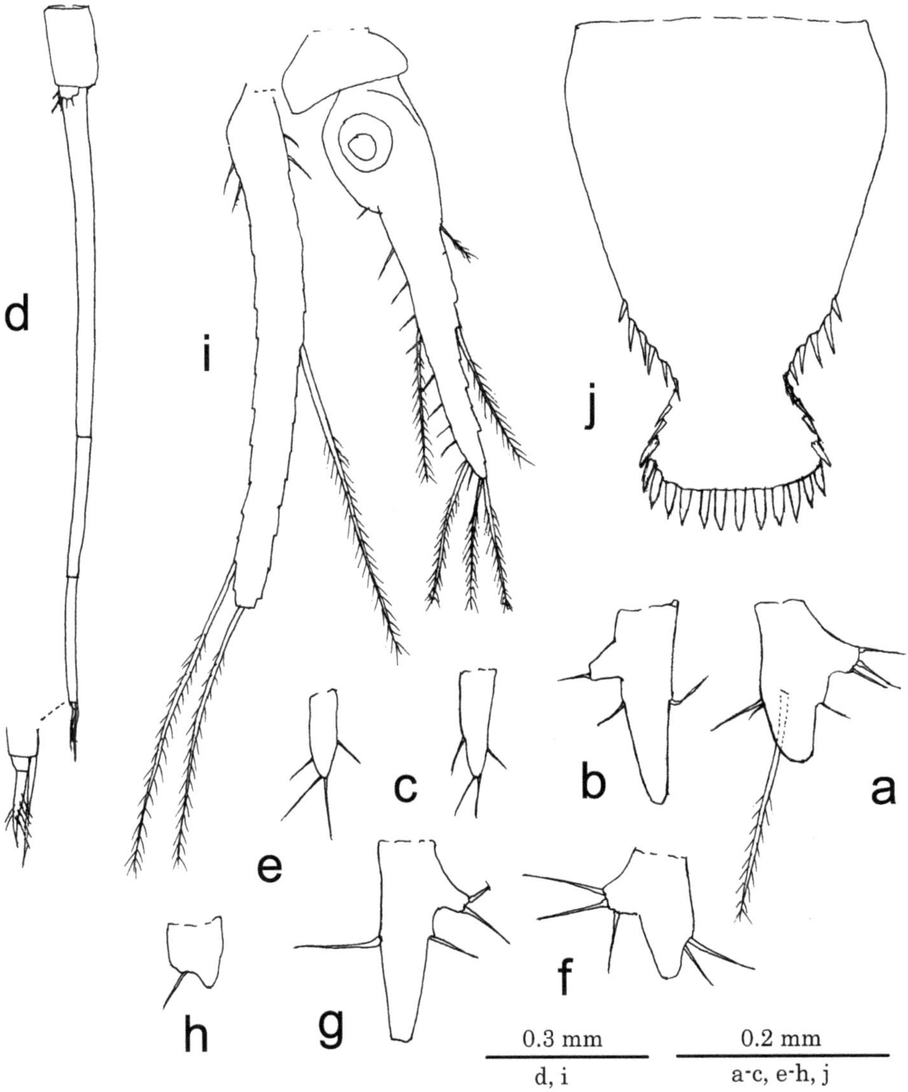

Fig. 3. *Anisomysis* (*Paranisomysis*) *takedai* sp. nov., paratype male (BL 4.9 mm) (a–e, i), paratype female (BL ca. 5.0 mm) (f–h, j): a–e, right first to fifth pleopods of male; f–h, right first, second and fourth pleopods of female; i, right uropod, ventral view; j, telson, dorsal view.

Remarks. — Our specimens are characterised by a telson composed of two parts by a deep subapical constriction as well as a truncated posterior margin and the second article of the mandibular palp possessing flagellate denticles. The following five species are known to share these characters; *A.* (*P.*) *lamellicauda* Hansen, 1912; *A.* (*P.*) *marisrubri* Băcescu, 1973; *A.* (*P.*)

TABLE I

Morphological features of *Anisomysis* (*Paranisomysis*) *minicoyensis* Biju et al., 2006 and *A.* (*P.*) *takedai* sp. nov.

Characters	*Anisomysis* (*P.*) *minicoyensis*	*Anisomysis* (*P.*) *takedai* sp. nov.
Body length	(male) 5.3-6.2 mm; (female) 5.6-6.3 mm	(male) 4.0-4.5 mm; (female) 4.0-5.5 mm
Rostrum	(male) triangular with obtuse to moderately pointed apex, extending mid-length of 1st article of antennular peduncle in male; (female) overreaching base of antennular peduncle; lateral margin nearly straight	(male & female) triangular with obtuse to moderately pointed apex barely reaching base of antennular peduncle; lateral margin straight or gently rounded
Antennule	(male) 1st article of peduncle subequal to combined length of anterior 2 articles; (female) 1st article of peduncle longer than rest of articles	(male) 1st article of peduncle slightly shorter than combined length of anterior 2 articles; (female) 1st article of peduncle longer than rest of articles
Antennal scale	(male) 9.7× longer than broad; (female) 10.7× longer than broad	(male) 7.9-8.1× broad; (female) 8.9-9.3× broad
Mandibular palp	(male & female) 2nd article with 10 or 11 flagellate denticles, less developed; 3rd article 1/3 length of 2nd one	(male & female) 2nd article with 9 or 10 flagellate denticles, less developed; 3rd article about 1/3 length of 2nd one
Male fourth pleopodal exopod	reaching mid-length of telson; 1st article ca. 3.2× longer than 2nd and again 1.5× longer than 2nd + 3rd; 2nd article 0.9× longer than 3rd	reaching basal 1/3-1/2 length of telson; 1st article 2.33-2.80× than 2nd and again 1.23-1.40× longer than 2nd + 3rd; 2nd 1.0-1.2× longer than 3rd
Telson	(male & female) 1.4× longer than basal width; proximal part ca. 1.0× longer than broad; width across constriction ca. 70% width of distal lobe; 26 or 27 marginal setae	(male & female) 1.42-1.57× longer than basal width; proximal part ca. 1.13-1.16× longer than broad; width across constriction ca. 68-75% width of distal lobe and 34-37% width of broadest part at base; 28-31 marginal setae

ohtsukai Murano, 1994; *A.* (*P.*) *arabica* Wooldridge & Victor, 2004; and *A.* (*P.*) *minicoyensis* Biju et al., 2006.

Among these species, *A.* (*P.*) *takedai* from the Andaman Sea shows a closer resemblance to *A.* (*P.*) *minicoyensis* recorded from off south-western India. *Anisomysis* (*P.*) *takedai*, however, has: 1) a relatively short rostrum barely reaching to the base of the antennular peduncle in both sexes (vs. overreaching base (females) and/or reaching mid-length (males) of the first article of the antennular peduncle in *A.* (*P.*) *minicoyensis*); 2) a comparatively shorter antennal scale less than 9.5 times as long as broad (vs. more than 9.5 times as long as broad); 3) a proportionately shorter first article of the male fourth pleopodal exopod and its second article longer than third article (vs. first article comparatively long and second article shorter than third); 4) a comparatively longer first article of the male antennular peduncle and 5) a telson with 28-31 marginal setae (vs. 26 or 27 setae). These morphological differences (table I) warrant the separation of the two populations at the species rank.

Distribution. — Known only from the type locality, Siam Bay in Racha Yai Island. The types were collected on an open fine-sand beach with a gentle slope, and small-scale aggregations of *A.* (*P.*) *takedai* were often observed around seaweed beds of brown algae, *Turbinaria* spp., growing onto corals, *Acropora* spp., which are predominant in the sublittoral zone of the bay.

ACKNOWLEDGEMENTS

We express our appreciation to Dr. Kaoru Hamano, National Research Institute of Fisheries and Coastal Environments of Inland Sea, Onomichi, for her help in the field. We also thank Dr. Koichi Konishi, National Research Institute of Fisheries Science, Yokohama, for technical support in the preparation of figures. This study was supported in part by research grants from JSPS for the senior author (Nos. 20570101 and 22405030) and from the Toyota Foundation for the second author (D09-R-0648).

REFERENCES

ANDERSON, G., 2010. Mysida Classification, January 20, 2010. http://peracarida.usm.edu/MysidaTaxa.pdf. (accessed on 11 March 2011)

BĂCESCU, M., 1973a. Contribution à la connaissance des Mysidés benthiques de la mer Rouge. Rapports et Procès-verbaux des Réunions Commission Internationale pour l'Exploration Scientifique de la Mer Méditerranée, **21**(9): 643-646.

— —, 1973b. *Anisomysis levi* n. sp. from the Red Sea and the dichotomic key of the species belonging to the genus, with description of a new taxon, *Paranisomysis* n. sg. Revue Roumaine de Biologie, Série de Zoologie, **18**(3): 173-180.

— —, 1992. *Javanisomysis gutzui* n. sg., n. sp., mysidacé grégaire des eaux Indonésiennes. Revue Roumaine de Biologie, Série de Biologie Animale, **37**(2): 79-86.

BIJU, A., S. U. PANAMPUNNAYIL & M. P. PRABHAKARAN, 2006. Mysidacea (Crustacea) from the Minicoy lagoon (Lakshadweep, India) with description of a new species of *Anisomysis*. Marine Biology Research, **2**(4): 291-295.

FUKUOKA, K., Y. HANAMURA & M. MURANO, 2002. Supplementary information on the taxonomy of six species of *Anisomysis* (Crustacea: Mysidacea: Mysidae). Plankton Biology and Ecology, **49**(1): 38-43.

HANSEN, H. J., 1910. The Schizopoda of the Siboga Expedition. Siboga Expeditie, **37**: 1-123, pls. 1-16.

— —, 1912. The Schizopoda. Memoirs of the Museum of Comparative Zoology at Harvard College, **35**(4): 173-296.

HAWORTH, A. H., 1825. A new binary arrangement of the macrurous Crustacea. Philosophical Magazine and Journal of Natural Philosophy, Chemistry and Arts, **65**(323): 183-184.

MÜLLER, H.-G., 1993. World Catalogue and Bibliography of the Recent Mysidacea: 1-491. (Laboratory for Tropical Ecosystems Research and Information Service, Vetzler).

MURANO, M., 1994. *Anisomysis* (*Paranisomysis*) *ohtsukai* (Crustacea, Mysidacea), a new mysid from Japan. Bulletin of the National Science Museum, Tokyo, (A) **20**(4): 155-159.

MURANO, M. & K. FUKUOKA, 2003. A systematic study of the genus *Anisomysis* (Crustacea: Mysida: Mysidae), with descriptions of six new species. Bulletin of the National Science Museum, Tokyo, (A) **29**(2): 65-102.

SAWAMOTO, S. & K. FUKUOKA, 2005. Lists of mysid species and references for their identification in Southeast Asian waters. Bulletin of Institute of Oceanic Research and Development, Tokai University, **26**: 79-93. [In Japanese.]

WATLING, L., 1989. A classification system for crustacean setae based on the homology concept. In: B. E. FELGENHAUER, L. WATLING & A. B. THISTLE (eds.), Functional morphology of feeding and grooming in Crustacea. Crustacean Issues, **6**: 15-26. (A. A. Balkema, Rotterdam).

WOOLDRIDGE, T. H. & R. VICTOR, 2004. Additions to the mysid fauna (Crustacea: Mysidacea) from coastal waters of Oman, including descriptions of two new species. Hydrobiologia, **511**: 247-258.

First received 16 August 2011.
Final version accepted 28 November 2011.

A NEW GENUS AND NEW SPECIES OF CHASMOCARCINID CRAB (DECAPODA, BRACHYURA, GONEPLACOIDEA) FROM SHALLOW WATERS IN JAPAN

BY

TOMOYUKI KOMAI[1,4]), PETER K. L. NG[2,5]) and YUSUKE YAMADA[3,6])

[1]) Natural History Museum and Institute, Chiba, 955-2, Aoba-cho, Chuo-ku, Chiba 260-8682, Japan

[2]) Tropical Marine Science Institute, Kent Ridge Campus, National University of Singapore, Upper Kent Ridge Road, Singapore 119260, Republic of Singapore

[3]) Ishikawa, Uruma, Okinawa 904-1115, Japan

ABSTRACT

A new genus and new species of chasmocarcinid crab, *Microtopsis takedai*, is described on the basis of four specimens from Okinawa, Ryukyu Islands, Japan. The new genus is characterized by the subtriangular merus of the third maxilliped, the presence of a row of closely set stiff setae on the cutting margin of the pollex of the minor male chela (for both chelae in the case of females), the relatively short male abdomen, and the stout male first gonopod with a distal part encircled by short to long spines.

INTRODUCTION

The systematics of the traditional Goneplacidae MacLeay, 1838, has been changed substantially over the last 30 years. Now regarded as a superfamily, Goneplacoidea, it now contains 12 families (see Karasawa & Kato, 2003a, b; Števčić, 2005, 2011; Ng & Manuel-Santos, 2007; Ng et al., 2008; Castro et al., 2010; Castro & Ng, 2010), although the status of the recently established Neommatocarcinidae Števčić, 2011, will need to be re-examined. Of these, the Chasmocarcinidae Serène, 1964, is relatively less well studied. It currently consists of three subfamilies (Chasmocarcininae Serène, 1964; Trogloplacinae

[4]) Corresponding author; e-mail: komai@chiba-muse.or.jp

[5]) e-mail: dbsngkl@nus.edu.sg

[6]) e-mail: y-yusuke@okiso.co.jp

Guinot, 1986, and Megaesthesiinae Števčić, 2005) and contains 10 genera and 37 species (updated from Ng et al., 2008). The status of a number of families (Raouliidae Števčić, 2005; Typhlocarcinodidae Števčić, 2005, and Caecopilumnidae Števčić, 2011) and their relationship to Chasmocarcinidae is still uncertain. The family is now being revised by P. Castro and the second author.

During recent SCUBA diving samplings in Okinawa Islands, Ryukyu Islands, by the third author, four specimens of an unusual chasmocarcinid were collected from shallow water, which superficially resembled species of *Camatopsis* Alcock & Anderson, 1899, and *Hephthopelta* Alcock, 1899. The structures of its third maxilliped, male abdomen and male first gonopod, however, were so distinct that it needs to be referred to a new genus, which is described here.

Specimens examined are deposited in the Natural History Museum and Institute, Chiba (CBM); and Zoological Reference Collection of the Raffles Museum of Biodiversity Museum, National University of Singapore (ZRC). Measurements provided are of the carapace width and length, respectively. The abbreviations G1 and G2 refer to the male first and second gonopods, respectively.

TAXONOMIC ACCOUNT

Family CHASMOCARCINIDAE Serène, 1964, sensu Ng et al., 2008

Subfamily CHASMOCARCININAE Serène, 1964

Microtopsis gen. nov.

[New Japanese name: Higebuto-kamoto-gani-zoku]

Type species. — *Microtopsis takedai* sp. nov., by monotypy.

Diagnosis. — Carapace (figs. 1, 2A) subtrapezoidal; front almost straight; anterolateral margin arcuate, with row of minute granules, without distinct lobes or teeth. Epistome (fig. 2C) fused with inter-antennular septum, somewhat depressed; posterior margin delimited, with shallow median cleft. Eyestalk (fig. 2A–C) filling orbit, slightly movable; cornea greatly reduced, with dark pigmentation. Antennule (figs. 1, 2B, C) with basal segment tightly fitting in fossa, peduncle not particularly enlarged but unable to fold into fossa; dorsal flagellum bearing elongated aesthetascs. Third maxilliped (fig. 3A) completely filling buccal cavern; merus subtriangular, lateral margin convex, anterolateral angle not produced. Chelipeds (figs. 1, 3B–E) subequal in length, dissimilar; cutting margin of pollex of minor cheliped in male (fig. 3D) or of

Fig. 1. *Microtopsis takedai* gen. et sp. nov., holotype, male (3.1 × 2.4 mm), CBM-ZC 10491, entire animal in dorsal view. Scale bar: 1 mm.

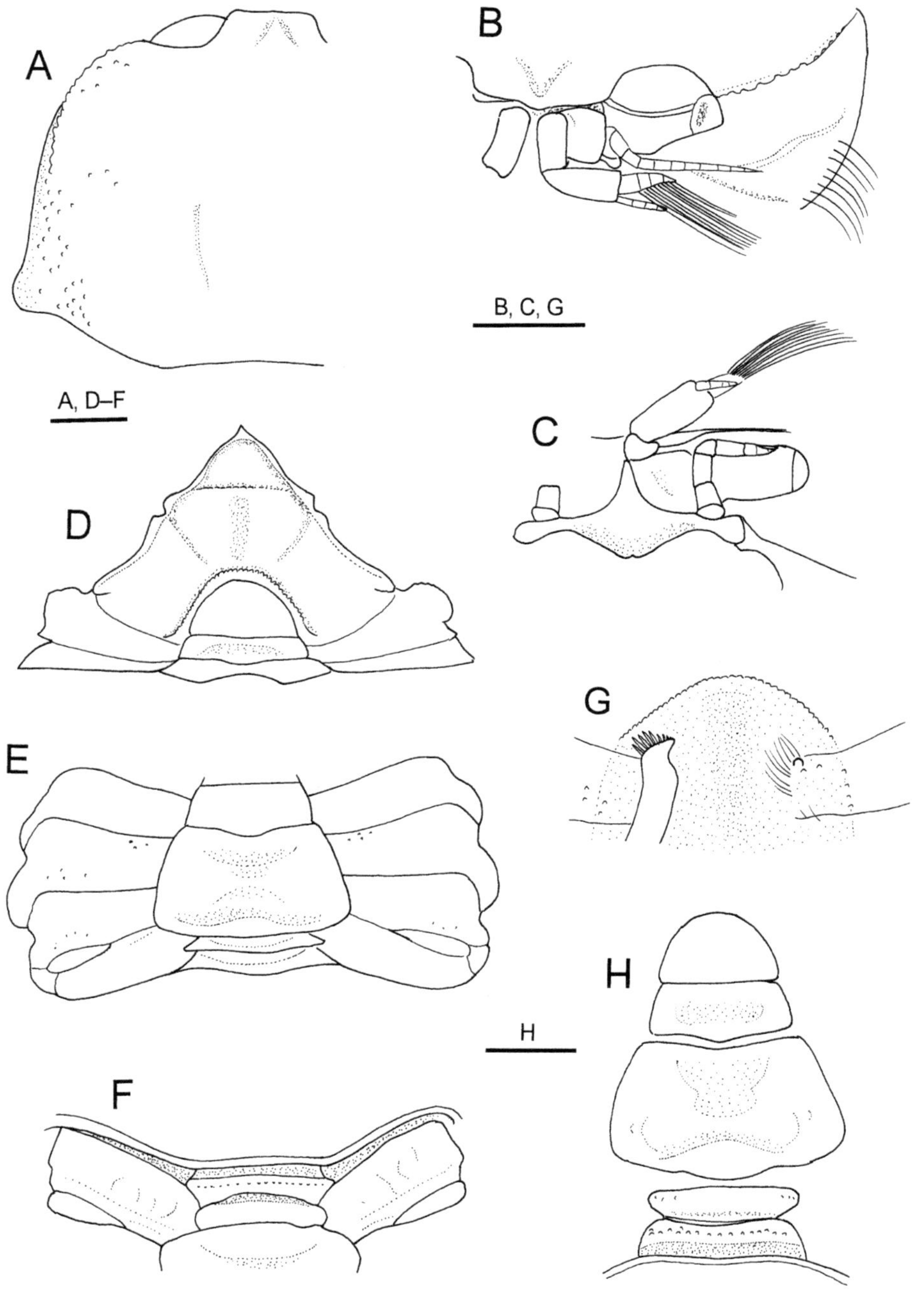

Fig. 2. *Microtopsis takedai* gen. et sp. nov., holotype, male (3.1 × 2.4 mm), CBM-ZC 10491. A, carapace and eye, left side, dorsal view (setae omitted); B, carapace and cephalic appendages, left side, anterior view; C, inter-antennular septum, epistome and left cephalic appendages, anteroventral view (antennal flagellum damaged during observation); D, anterior

both chelipeds in female (fig. 5A, B) each with row of dense stiff setae. Meri of ambulatory legs (fig. 1) smooth, unarmed. Thoracic sternites 6-8 visible from posterior view (fig. 2E, F). Male abdomen (fig. 2H) with somites 3-5 completely fused; lateral margins of fused somites 3-5 slightly convex or nearly straight. Male gonopore coxal, penis coxo-sternal, lying in a channel between sternites 7 and 8, completely covered by supplementary sternal plate between sternites 7 and 8, plate reaching halfway across exposed part of thoracic sternum (fig. 2F). G1 (fig. 4A–E) stout, distal segment slightly twisted, tip curved inwards, encircled by short to long spinules. G2 (fig. 4F) subequal in length to G1, very slender, slightly sinuous.

Etymology. — The name is derived from an arbitrary combination of *micro*, for "small", and the suffix of *Camatopsis*, which the new genus resembles. Gender feminine.

Remarks. — *Microtopsis* gen. nov. agrees well with Chasmocarcinidae in every diagnostic aspect, particularly in the presence of a male supplementary sternal plate between thoracic sternites 7 and 8, a coxo-sternal penis arrangement, and an abdomen with fused somites 3-5 (Ng et al., 2008). The so-called male supplementary sternal plate, also known as the coxo-sternal double plate, penial channel or invaginated channel, is believed to correspond to an invagination of thoracic sternite 8 and ostensibly protects the penis (see Guinot, 1986: pl. 2 figs. C, D; Davie, 1988: fig. 1F; Davie & Guinot, 1996: figs. 3G, 7). In all members of the Chasmocarcinidae as defined by Davie & Guinot (1996), the male supplementary sternal plate is distinct. The Chasmocarcinidae of Ng et al. (2008), however, may be a more heterogeneous grouping as the male thoracic sternal condition in some genera is not well understood and it is not sure if all members have the supplementary sternal plate. The supplementary sternal plate of chasmocarcinids does vary in structure. For example, in *Raoulia* Ng, 1987, the plate is actually slightly mobile (unpublished data) while in some species of *Hephthopelta* Alcock, 1899, the strongly calcified penis almost completely fills the groove between thoracic sternites 7 and 8, giving the appearance of a plate being present (fig. 7D). This condition is similar to that in another genus, *Scalopidia* Stimpson, 1858, except it is much more pronounced

---

thoracic sternum, ventral view; E, thoracic sternites 5-8 and anterior part of abdomen, posteroventral view; F, thoracic sternite 8 and abdominal somites 1 and 2, posterior view; G, anterior part of sterno-abdominal cavity, showing position of abdominal locking press button (left side), distal part of right G1 is also shown, ventral view; H, abdomen, ventral view. Scale bars: 0.5 mm.

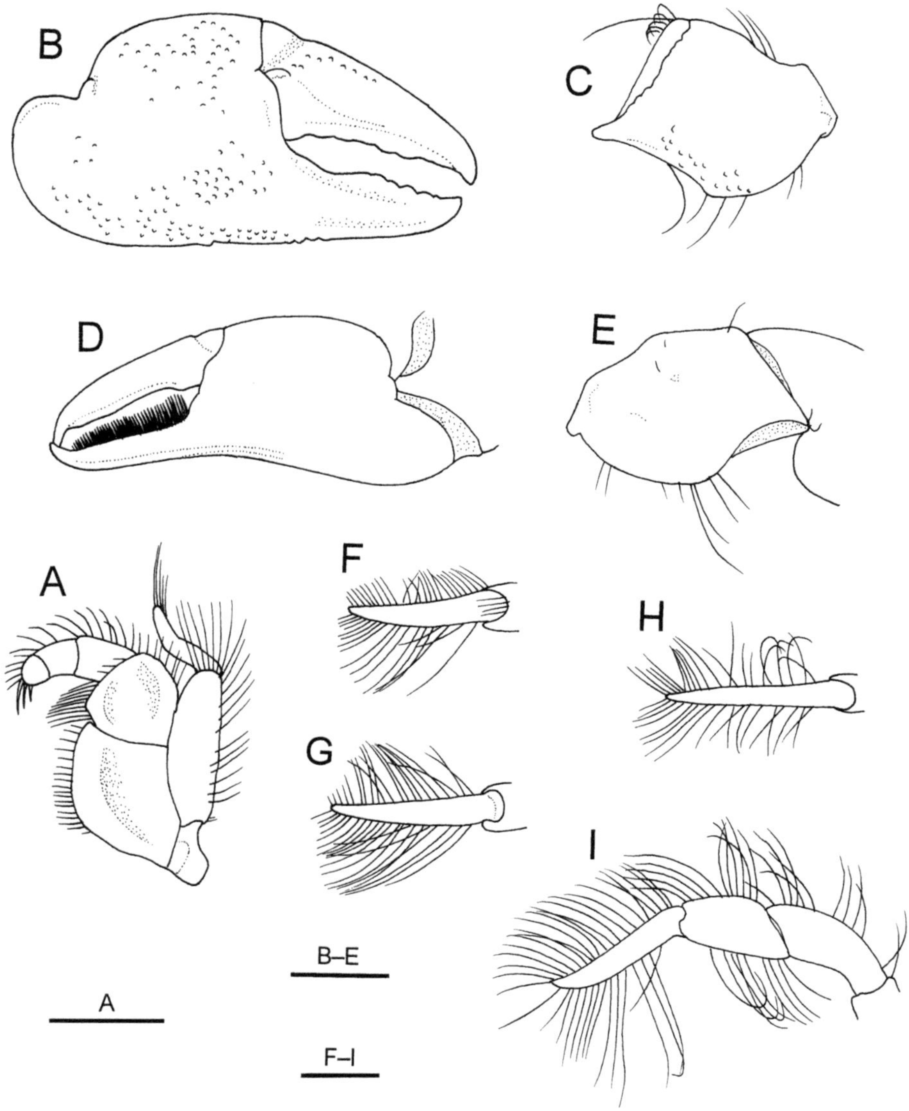

Fig. 3. *Microtopsis takedai* gen. et sp. nov., holotype, male (3.1 × 2.4 mm), CBM-ZC 10491. A, left third maxilliped, outer view; B, chela of major right cheliped, anterior (extensor) view; C, carpus of right major cheliped, dorsal view; D, chela of minor left cheliped, anterior (extensor) view; E, carpus of left minor cheliped, dorsal view; F–H, dactyli of left second to fourth pereopods (first to third ambulatory legs), extensor view; I, distal 3 segments of left fifth pereopod, dorsal view. Scale bars: 0.5 mm.

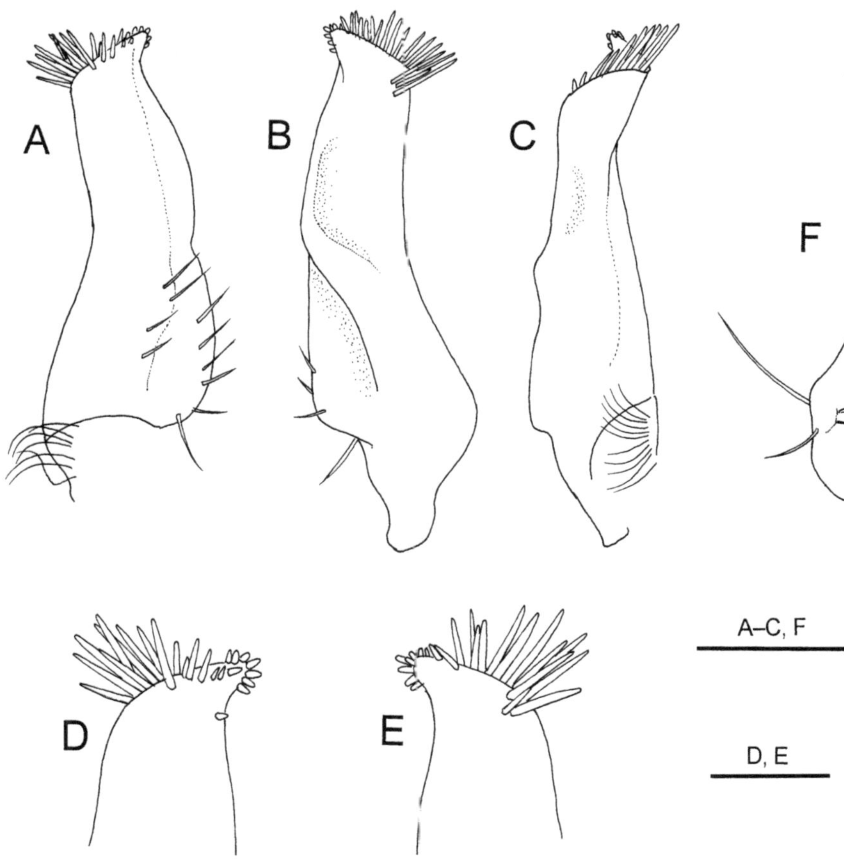

Fig. 4. *Microtopsis takedai* gen. et sp. nov., holotype, male (3.1 × 2.4 mm), CBM-ZC 10491. A, left G1, dorsal view; B, same, ventral view; C, same, lateral view; D, close up of tip of left G1, dorsal view; E, G1, ventral view; F, left G2. Scale bars: 0.25 mm for A–C, F; 0.1 mm for D, E.

in *Hephthopelta*. Ng et al. (2008: 85) nevertheless excluded *Scalopidia* from the Chasmocarcinidae and left it in its own family, Scalopididae Števčić, 2005, as although the penis was calcified, it was not covered by a sternal plate.

Microtopsis is allied to *Camatopsis* Alcock & Anderson, 1899 (type species *Camatopsis rubida* Alcock & Anderson, 1899), *Hephthopelta* (type species *Hephthopelta lugubris* Alcock, 1899), and *Chasmocarcinops* Alcock, 1900 (type species *Chasmocarcinops gelasimoides* Alcock, 1900), in the basal article of the antennular peduncle completely filling the antennular fossa that prevents the distal two articles of the peduncle and flagella from folding inside. *Camatopsis* and *Hephthopelta* have long been placed in the Goneplacidae sensu lato, while *Chasmocarcinops* was originally classified in the Pinnotheri-

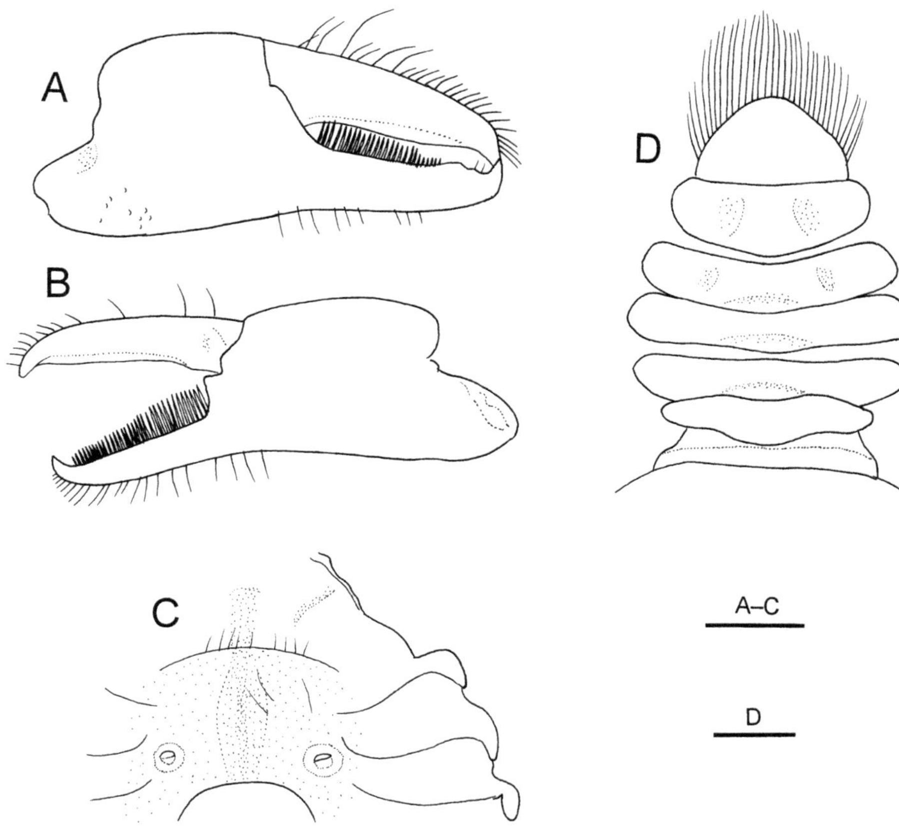

Fig. 5. *Microtopsis takedai* gen. et sp. nov., paratype, female (3.4 × 2.8 mm), CBM-ZC 10492. A, major right chela, anterior (extensor) view; B, minor left cheliped, anterior (extensor) view; C, thoracic sternum, showing position of valvae, ventral view; D, abdomen, ventral view. Scale bars: 0.5 mm.

dae (Alcock, 1900: 334; Tesch, 1918: 280) and which Serène (1964a: 266) correctly referred to Goneplacidae, and later to Chasmocarcininae (Serène, 1964b: 186; Serène & Soh, 1976: 22, 23).

Among the three allied genera, *Camatopsis* appears closest to *Microtopsis* gen. nov., shared characters being: the eyestalk is only slightly movable with cornea substantially reduced; at least minor chela (both chelae in the case of females) with a comb-like row of stiff setae on the cutting margin of the pollex (fig. 6D). In *Hephthopelta*, the eye is better developed than in *Microtopsis*

Fig. 6. *Camatopsis rubida* Alcock & Anderson, 1899, male (10.6 × 9.0 mm), ZRC 2009.347, Taiwan. A, entire animal in dorsal view; B, anterior thoracic sternum, posterior abdominal somites and telson, ventral view; C, posterior abdominal somites and groove between thoracic sternites 7 and 8, posterior view; D, chelae, outer view.

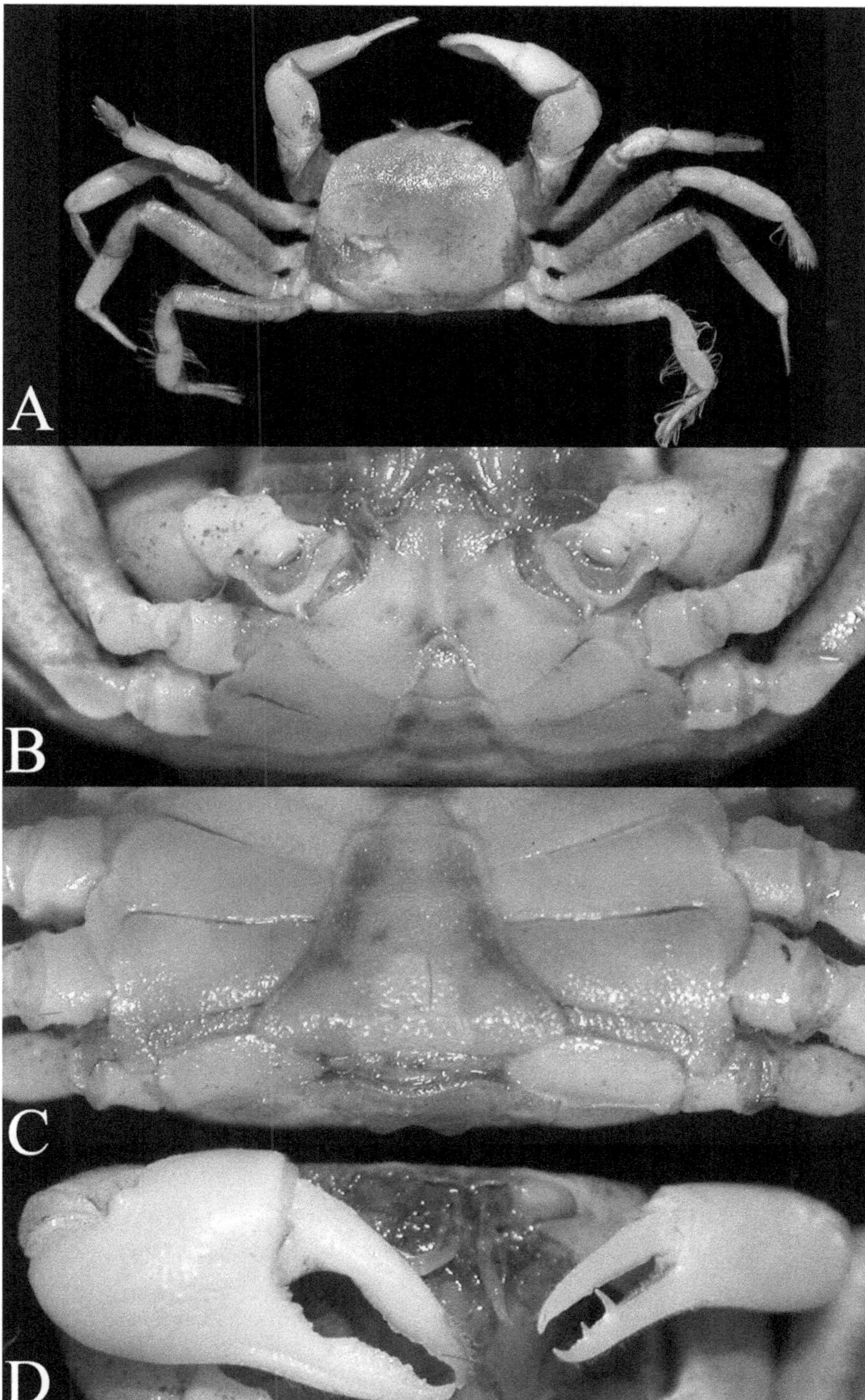
A
B
C
D

and *Camatopsis*, being more freely movable with darkly pigmented cornea, whereas in *Chasmocarcinops* the eye is completely fused to the orbit and the cornea is hardly pigmented. No comb-like setal row is present on the cutting margins of the fixed fingers of chelae in *Hephthopelta* and *Chasmocarcinops*. The chelipeds of *Camatopsis* and *Microtopsis*, however, are easy to separate as the pollex of the minor chela of *Camatopsis* has two prominent vertical spine-like teeth on the distal half of the cutting margin (fig. 6D), a feature absent in *Microtopsis* (figs. 3D, 5B).

The third maxilliped of *Microtopsis* is diagnostic in being relatively short, with the merus broadly subtriangular (fig. 3A), not distinctly quadrate or subovate like in *Camatopsis* and *Hephthopelta* (see Alcock, 1899: pl. 4 figs. 2a, b, 3a; Alcock & MacGilchrist, 1905: pl. 74 fig. 3; Serène, 1964a: figs. 16A, 17A, 18A). In fact, the merus is similar to that in *Chasmocarcinops* but in *Microtopsis*, it is proportionately broader (see Alcock & McArdle, 1903: pl. 62 fig. 2a; Serène, 1964a: fig. 20A).

The relatively short male abdomen of *Microtopsis* is also diagnostic, particularly in the shape of the fused somites 3-5, which is subtrapezoidal with lateral margins convex to nearly straight, and the telson is broad and sub-semicircular (fig. 2H). In *Camatopsis* (as well as *Hephthopelta* and *Chasmocarcinops*), the male abdomen is distinctly more slender and elongated; the fused somites 3-5 are trapezoidal with concave or sinuous lateral margins (see figs. 6C, 7D); and the telson is more elongated and triangular (figs. 6B, 7C; Serène, 1964a: figs. 16B, 18B, 20B). The broad male telson in *Microtopsis* also results in the distal part of the sterno-abdominal cavity being proportionately much broader (fig. 2G) than in *Camatopsis* (fig. 6B), *Hephthopelta* (fig. 7C) and *Chasmocarcinops*.

Compared to *Microtopsis*, *Camatopsis* and *Chasmocarcinops*, the carapace of *Hephthopelta* is the highest and most vaulted (fig. 7A, C). The carpus of the cheliped of most species of *Hephthopelta* also has a strong inner tooth and the ambulatory legs are very slender (fig. 7A; Serène, 1964b: pl. 22), while the carpus of the chelipeds in the other three genera has no tooth and the legs are proportionately stouter (cf. fig. 1 for *Microtopsis*; fig. 6A for *Camatopsis*; Alcock & MacGilchrist, 1905: pl. 74 fig. 3 for *Chasmocarcinops*). The form of the supplementary sternal plate in *Microtopsis* is noteworthy

Fig. 7. *Hephthopelta* cf. *lugubris* Alcock, 1899, male (11.9 × 10.3 mm), ZRC, Philippines. A, entire animal in dorsal view; B, anterior part of carapace, frontal view; C, anterior thoracic sternum, posterior abdominal somites and telson, ventral view; D, posterior abdominal somites and groove between thoracic sternites 7 and 8.

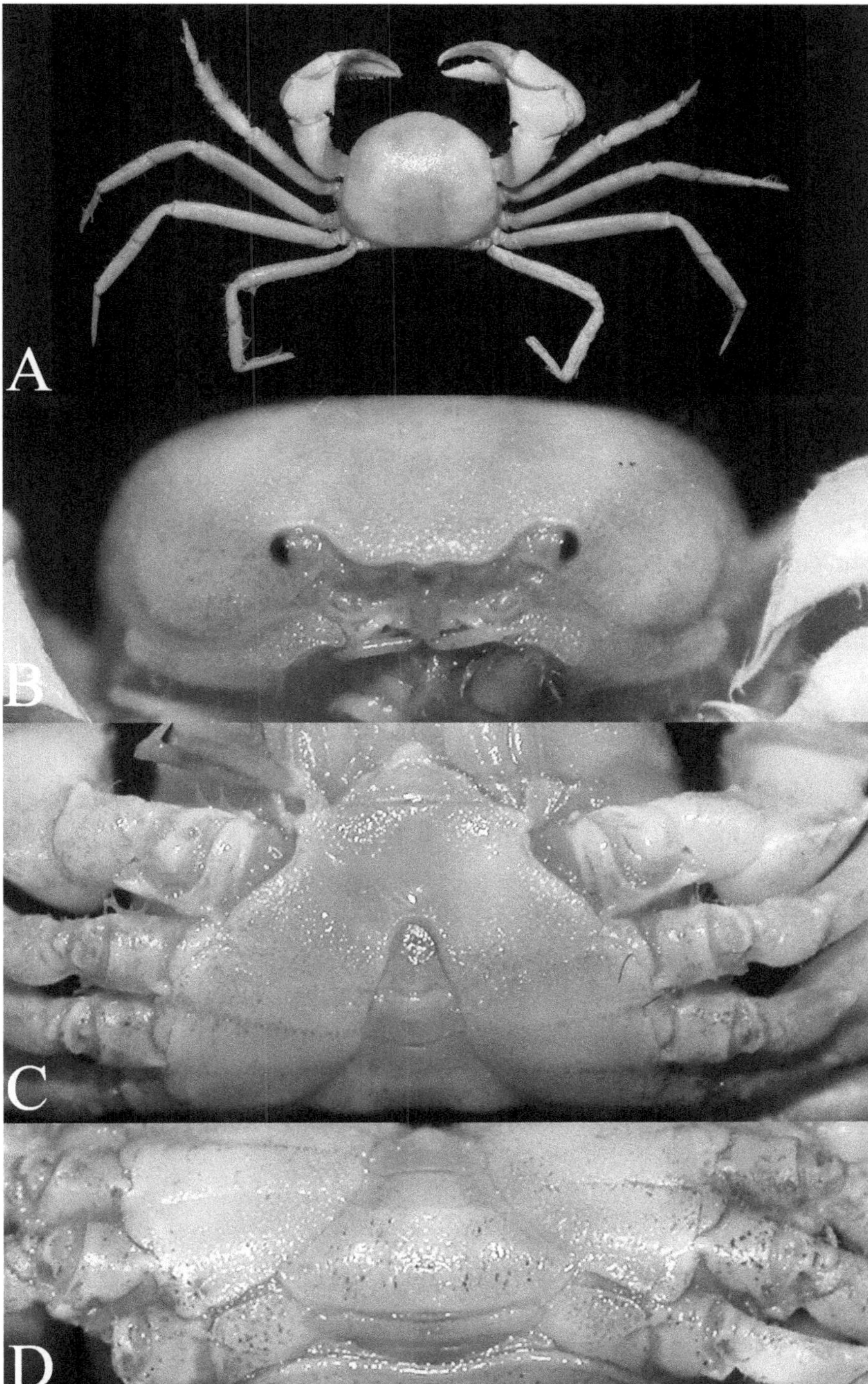
A
B
C
D

as it reaches only about halfway across the exposed part of the thoracic sternum, well before the sterno-abdominal cavity (fig. 2F). In *Camatopsis* and *Chasmocarcinops*, the supplementary sternal plate reaches all the way across the thoracic sternum (fig. 6B). As discussed earlier, in *Hephthopelta*, the supplementary sternal plate is actually not a well-defined plate and is partially mobile (fig. 7D).

The G1 of *Microtopsis* is very unusual, being short and having the distal part partially twisted, with apical margin encircled by short to long spines (fig. 4A-E). In *Camatopsis*, *Hephthopelta* and *Chasmocarcinops*, the G1 is elongated, gradually tapering to a sharp point or flared tip, and with the distal parts lined only with distinct short spines (cf. Serène, 1964a: figs. 16C, 18C, 20C; Serène & Soh, 1976: fig. 21A).

A note on *Camatopsis rubida* is perhaps necessary as the present concept of the species is unclear. Tesch (1918: 235, 236) first recognized two different forms of *Camaotpsis rubida*, referred to as "forma a" (pl. 16 fig. 3, 3a–d) and "forma b" (pl. 16 fig. 3e–i). He commented that it was possible that these two forms represented variants of the same species and refrained from naming a second species. His figures and description provide evidence that the material actually consisted of two species. Tesch (1918: 237) realized that perhaps the most inexplicable character was the form of the merus of the third maxilliped; quadrate and broad in "forma a" (Tesch, 1918: pl. 16 fig. 3b) and narrow in "forma b" (Tesch, 1918: pl. 16 fig. 3g). Tesch's "forma a" agrees very well with the type description and figures of *Camatopsis rubida* (see Alcock, 1900: 329; Alcock & MacGilchrist, 1905: pl. 74 fig. 3). It is likely that his "forma b" specimens are either small individuals of *Chasmocarcinops gelasimoides*, or a species allied to it, having a similar third maxilliped structure. The meri of the ambulatory legs of *Chasmocarcinops gelasimoides*, however, is gently serrated (Alcock, 1900: 335; Alcock & McArdle, 1903: pl. 62 figs. 2, 3), but completely smooth in *Camatopsis rubida* (see Alcock & MacGilchrist, 1905: pl. 74 fig. 3; Tesch, 1918: pl. 16 fig. 3, 3e). Tesch (1918: 236, pl. 16 fig. 3c) described the margin before the two spines on the pollex of the minor chela of his "forma a" as armed with "very minute, closely-crowded serrulations" but this is not accurate as it is in fact lined with short stiff setae, giving the illusion of dentition. The margin is actually smooth. He described the same condition for his "forma b" (Tesch, 1918: 236, pl. 16 fig. 3h) but in a different arrangement.

Microtopsis takedai sp. nov.
[New Japanese name: Higebuto-kamato-gani]
(figs. 1-5)

Material examined. — Holotype: male (3.1 × 2.4 mm) (CBM-ZC 10491), Awa, Nago, Okinawa Island, Ryukyu Islands, 12 m, coll. Y. Yamada, SCUBA diving, 17 April 2010. Paratypes: 1 female (3.4 × 2.8 mm) (CBM-ZC 10492), Kyoda, Nago, Okinawa Island, 30 m, coll. Y. Yamada, SCUBA diving, 15 April 2011; 1 male (damaged), 1 female (3.6 × 2.7 mm) (ZRC), Awa, Nago, Okinawa Island, 13 m, coll. Y. Yamada, SCUBA diving, 2 January 2011.

Comparative material. — *Camatopsis rubida* Alcock & Anderson, 1899: 1 male (10.6 × 9.0 mm) (ZRC 2009.347), Tahsi, northeastern Taiwan, 28 August 2000; 1 female (8.2 × 7.3 mm) (ZRC 2009.348), Tahsi, northern Taiwan, 1 December 1999, coll. J.-W. Lin.

Hephthopelta cf. *lugubris* Alcock, 1899: 1 male (11.9 × 10.3 mm) (ZRC), Maribohoc Bay, Panglao, Bohol, Philippines, 100-200 m, coll. J. Arbasto, November 2003–April 2004; 1 female (6.2 × 5.4 mm) (CBM-ZC 8770), RV *Hakuho-maru*, KH02-4 cruise, stn SA-2, Sulu Sea, Philippines, 10°36.48′N 121°31.56′E, 362-372 m, coll. S. Ohtsuka, beam trawl, 8 December 2002.

Chasmocarcinops gelasimoides Alcock, 1900: 4 males (largest 10.2 × 9.1 mm) (ZRC 2000.1341), Johore Shoals, 01°20.081′N 104°23.11′E, Singapore, dredged by D. Wee et al., 14 July 1993.

Description. — Males. Carapace (figs. 1, 2A, B) slightly vaulted, subtrapezoidal in general outline in dorsal view, 1.3 times wider than long, greatest width at posterolateral angles. Dorsal surface strongly arched anteriorly and posteriorly along longitudinal axis, transversely flat, almost glabrous, microscopically granular, with coarse granules laterally; regions not well defined, except for indistinct shallow longitudinal cardiac grooves. Front narrow (0.17 times as wide as carapace), anterior margin slightly deflexed, nearly straight, rounded laterally; dorsal surface with posteriorly bifurcate shallow groove medially. Supraorbtal margin smooth, continuous with infraorbital margin; infraorbital margin smooth, slightly sinuous. Anterolateral margin arched, lined by coarse granules. Lateral carapace margins slightly diverging posteriorly, posterolateral angles slightly produced. Posterior carapace margin nearly straight. Pterygostomial region with distinct groove extending posteriorly from suborbital region.

Inter-antennular septum fused with epistome (fig. 2C), narrow, not reaching frontal margin of carapace. Epistome slightly depressed; posterior margin distinct, slightly produced medially, with minute median cleft. Endostome shallow, smooth.

Eyestalks (fig. 2A–C) filling fully orbit, but slightly movable, with sharp carina on anterior surface dorsally; cornea small, located terminally, dark pigment visible inner side. Antennular peduncle (figs. 1, 2B, C) with first article moderately large, subquadrate, completely filling antennular fossa such that distal 2 articles cannot be folded; second article articulated at dorsomedial angle of basal article; third article longer than penultimate article; dorsal flagellum with elongated aesthetascs, distinctly longer than flagellum. Antennal peduncle (fig. 2B, C) with first article fused with carapace; flagellum slightly overreaching lateral end of eye when extended laterally.

Third maxilliped (fig. 3A) filling buccal cavern when closed. Ischium with lateral and mesial margins subparallel, lateral margin concave, proximal margin slightly oblique; outer surface with shallow obliquely longitudinal groove medially. Merus subtriangular, distinctly shorter than ischium; lateral margin gently convex, mesial margin obtusely angular; outer face with shallow grooves adjacent to lateral and mesial margins. Distal 3 segments moderately robust; carpus articulated at distomesial angle of merus; dactylus short, with some short stiff setae distally. Exopod relatively broad, not reaching distal margin of merus, unarmed on inner margin; flagellum with 2 articles.

Chelipeds (figs. 1, 2B–E) subequal in length, heteromorphic. Major (right) cheliped large. Ischium short, unarmed, with sparse setae. Merus with outer surface bearing granules ventrally; dorsal surface rounded, unarmed; ventral surface smooth, nearly flat; inner face with row of plumose setae along ventral margin. Carpus short, rhomboidal in dorsal view; dorsal surface generally smooth, but with some granules adjacent to inner margin; inner margin obtusely angular, without conspicuous spine or tooth. Chela subtriangular in general outline in outer view with rounded margins, about twice as long as high. Palm 1.2 times higher than long (measured along dorsal margin); dorsal surface rounded; outer face glabrous, sparsely granular; inner face weakly convex, glabrous, with patch of granules at base of pollex; ventral margin (including fixed finger) nearly straight, non-carinate. Tips of fingers crossing when closed. Pollex nearly straight with slightly curving tip, cutting margin faintly dentate. Dactylus 1.3 times longer than palm, almost straight, terminating in blunt tip; dorsal surface with sparse granules, non-carinate; outer and inner faces nearly smooth; cutting margin faintly dentate. Minor (left) cheliped with ischium and merus similar to those of major cheliped. Carpus slightly narrower than on major cheliped. Chela about half of that of major chela in height, ventral margin notably sinuous, bluntly carinate. Palm slightly narrowing distally; dorsal surface rounded; outer and inner faces smooth, with sparse short setae. Fingers strongly compressed, crossing at tips. Pollex with tip upturned, cutting margin smooth, with prominent row of dense stiff setae filling hiatus between dactylus. Dactylus slender, about 1.2 times longer than palm; dorsal margin non-carinate; tip curving downwards; cutting margin with thin edge, nearly smooth.

Ambulatory legs (fig. 1) moderately slender, fourth pereopod (third leg) longest, about twice length of carapace width; fifth pereopod (fourth leg) reaching to midlength of dactylus of fourth pereopod when fully extended laterally. Meri smooth, unarmed. Carpi slightly shorter than propodi. Dactyli (fig. 3F–I) longer than propodi, slightly recurved (second and third pereopods), nearly straight (fourth pereopod) or sinuously curved with tip directed outward (fifth pereopod). Brush-like setal row present on dactylus and/or propodus.

Thoracic sternum (fig. 2D–F) noticeably broadened posteriorly; surface microscopically granular, with scattered minute setae. Sternites 1 and 2 completely fused, forming triangular plate with margins bordered by carina. Sternite 3 hexagonal, with shallow median groove, separated from sternite 4 by strongly oblique, shallow sulcus. Sternite 5 broad. Sternites 5-8 visible from posterior view, all nearly perpendicular to horizontal plane of carapace, each part on either side of sterno-abdominal cavity subrectangular in general outlines. Sutures separating sternites distinct, interrupted medially. Sternite 8 with narrow supplementary sternal plate, completely covering penis; not reaching to sterno-abdominal cavity. Press-button abdominal locking mechanism on sternite 5 papilla-like, located adjacent to suture between sternite 4, partially obscured by tuft of short setae (fig. 2G). Episternites on sternites 4-6 developed as subsemicircular plates; that on sternite 7 small, rounded. Sterno-abdominal cavity deep, extending to anterior margin of sternite 4, anterior part corresponding to sternite 4 bordered by distinct carina (fig. 2G).

Abdomen (fig. 2H) consisting of 4 freely articulated somites with somites 3-5 completely fused. Somites 1 and 2 very short longitudinally, transversely carinate, distinctly narrower than fused somites 3-5. Fused somites 3-5 widest, occupying about one-third of width of thoracic sternite 7, trapezoidal in shape, wider than long; outer face shallowly depressed medially. Somite 6 trapezoidal, much wider than long. Telson subsemicircular, slightly wider than long.

Male gonopore coxal. Penis long, concealed in channel between thoracic sternites 7 and 8, completely covered with supplementary sternal plate which reaches about halfway across thoracic sternum (fig. 2F); tip of penis slightly protruding from gonopore opening on sternum (coxo-sternal arrangement). G1 (fig. 4A–E) stout, reaching beyond suture between thoracic sternites 4 and 5 (fig. 2G), slightly twisted, weakly curved outward, outer face with obliquely longitudinal carina; proximomesial part bearing several minute spiniform setae; short distal part curving inwards, terminating in blunt apex, with several terminally blunt, very short spines at apex and numerous subterminal long spines surrounding distal part; basal segment large, squarish. G2 (fig. 4F) very slender, subequal in length to G1, gently curved inward, terminating into acute apex, distal half inserted in G1 in situ.

Females. Generally similar to males in non-sexual characters. Chelipeds slightly dissimilar (fig. 5A, B). Major chela (fig. 5A) with slightly sinuous ventral margin; palm with scattered granules ventrally, otherwise nearly smooth; pollex with row of stiff setae on proximal three-quarters of smooth cutting margin; dactylus about 1.5 times longer than palm measured along dorsal margin; cutting margin with thin edge. Minor cheliped (fig. 5B) similar to that of male. Thoracic sternite 4 with low transverse carina bearing sparse setae (fig. 5C); vulva moderately large, flat, located mesial to end of sutures between sternites 5/6 and 7/8 (fig. 5C). Abdomen (fig. 5D) moderately broad, somites 3 and 4 widest; somite 1 transversely carinate; lateral margins of somites 3-6 strongly convex. Telson roundly triangular, 1.7 times wider than long, margins fringed with long setae.

Color in life. — Body and appendages generally pale gray, except for brownish pterygostomial regions of carapace and inner faces of carpi of chelipeds.

Distribution. — Known so far only from Nago, Okinawa Island, Ryukyu Islands, 12-30 m.

Ecology. — Although chasmocarcinids generally inhabit soft bottoms at sublittoral to bathyal depths, the present new species occurs in shallow water at depths of 12-30 m. Individuals were found burying in mud. They were collected using a small sieve. No association with other invertebrates was observed.

Etymology. — It is our pleasure to dedicate this new species to Dr. Masatsune Takeda in recognition of his immense contributions to the taxonomy of decapod crustaceans, particularly Brachyura.

ACKNOWLEDGEMENTS

The second author's visit to Japan was funded by a 2011 NUS-JSPS (National University of Singapore–Japan Society for the Promotion of Science) exchange program.

REFERENCES

ALCOCK, A., 1899. An account of the deep-sea Brachyura collected by the Indian Marine Survey Ship Investigator: i–iv, 1-85, pls. 1-4. (Trustees of the Indian Museum, Calcutta).

— —, 1900. Materials for a carcinological fauna of India. No. 6. The Brachyura Catometopa or Grapsoidea. Journal of the Asiatic Society of Bengal, Calcutta, **69**(3): 279-486.

ALCOCK, A. & A. R. ANDERSON, 1899. Natural history notes from H.M. Royal Indian marine survey ship 'Investigator', Commander T.H. Heming R.N., commanding. Series 3, number 2. An account of the deep sea Crustacea dredged during the surveying season of 1897-98. Annals and Magazine of Natural History, (7) **3**: 1-27, 278-292.

ALCOCK, A. & A. C. MACGILCHRIST, 1905. Illustrations of the Zoology of the Royal Indian Marine Survey Ship Investigator, under the command of Captain T.H. Heming, R.N. (Retired). Crustacea — Part XI: pls. 68-76. (Office of the Superintendent of Government Printing, Calcutta).

ALCOCK, A. & A. F. MCARDLE, 1903. Illustrations of the Zoology of the Royal Indian Marine Survey Ship Investigator, under the command of Commander T.H. Heming, R.N. Crustacea – Part X: pls. 61-67. (Office of the Superintendent of Government Printing, Calcutta).

CASTRO, P., D. GUINOT & P. K. L. NG, 2010. A new family for *Sotoplax robertsi* Guinot, 1984, with a diagnosis and key to the Goneplacoidea MacLeay, 1838 (Crustacea: Decapoda: Brachyura). Zootaxa, **2356**: 36-56.

CASTRO, P. & P. K. L. NG, 2010. Revision of the family Euryplacidae Stimpson, 1871 (Crustacea: Decapoda: Brachyura: Goneplacoidea). Zootaxa, **2375**: 1-130.

DAVIE, P. J. F., 1988. A new genus and species of goneplacid (Crustacea: Brachyura) from Queensland, Australia. Memoirs of the Queensland Museum, **25**(2): 259-264.

DAVIE, P. J. F. & D. GUINOT, 1996. Two new freshwater species in *Australocarcinus* Davie, with remarks on Trogloplacinae Guinot and Goneplacidae MacLeay (Crustacea: Decapoda: Brachyura). Memoirs of the Queensland Museum, **39**(2): 277-287.

GUINOT, D., 1986. Description d'un Crabe cavernicole aveugle de Nouvelle-Bretagne (Papouasie Nouvelle-Guinée), *Trogloplax joliveti* gen. nov., sp. nov., et établissement d'une sous-famille nouvelle, Trogloplacinae subfam. nov. Comptes Rendus de l'Académie des Sciences (Paris), (3) **303**(8): 307-312.

KARASAWA, H. & H. KATO, 2003a. The family Goneplacidae MacLeay, 1838 (Crustacea: Decapoda: Brachyura): systematics, phylogeny and fossil records. Paleontological Research, **7**(2): 129-151.

— — & — —, 2003b. The phylogeny, systematics and fossil record of the Goneplacidae MacLeay (Crustacea, Decapoda, Brachyura) revisited. Contributions in Zoology to the Mesozoic and Cenozoic Decapod Crustaceans, **72**(2-3): 147-152.

MACLEAY, W. S., 1838. On the Brachyurous Decapod Crustacea. Brought from the Cape by Dr. Smith. In: A. SMITH, Illustrations of the Zoology of South Africa; consisting chiefly of figures and descriptions of the objects of natural history collected during an expedition into the interior of South Africa, in the years 1834, 1835, and 1836; fitted out by 'The Cape of Good Hope Association for Exploring Central Africa': together with a summary of African Zoology, and an inquiry into the geographical ranges of species in that quarter of the globe, published under the Authority of the Lords Commissioners of Her Majesty's Treasury, Invertebratae. IV [1849]: 53-71, pls. 2, 3. (Smith, Elder & Co., London).

NG, P. K. L., 1987. The Indo-Pacific Pilumnidae II. A revision of the genus *Rhizopa* Stimpson, 1858 and the status of the Rhizopinae Stimpson, 1858 (Crustacea: Decapoda: Brachyura). Indo-Malayan Zoology, **4**(1): 69-111, pl. 1.

NG, P. K. L., D. GUINOT & P. J. F. DAVIE, 2008. Systema Brachyurorum: Part I. An annotated checklist of extant brachyuran crabs of the world. Raffles Bulletin of Zoology, (Supplement) **17**: 1-286.

NG, P. K. L. & M. R. MANUEL-SANTOS, 2007. Establishment of the Vultocinidae, a new family for an unusual new genus and new species of Indo-West Pacific crab (Crustacea: Decapoda: Brachyura: Goneplacoidea), with comments on the taxonomy of the Goneplacidae. Zootaxa, **1558**: 39-68.

SERÈNE, R., 1964a. Goneplacidae et Pinnotheridae récoltés par le Dr. Mortensen. Papers from Dr. Th. Mortensen's Pacific Expedition 1914-1916, part 80. Videnskabelige Meddelelser fra Dansk Naturhistorisk Forening I København, **126**: 181-282, pls. 16-24.

— —, 1964b. Redescription du genre *Megaesthesius* Rathbun et définition des Chasmocarcininae, nouvelle sous-famille des Goneplacidae (Decapoda Brachyura). Crustaceana, **7**(3): 175-187, figs. 1-6.

SERÈNE, R. & C. L. SOH, 1976. Brachyura collected during the Thai-Danish expedition (1966). Phuket Marine Biological Center Research Bulletin, **12**: 1-37.

ŠTEVČIĆ, Z., 2005. The reclassification of Brachyuran Crabs (Crustacea: Decapoda: Brachyura). Natura Croatica (Fauna Croatica), **14**(1): 1-159.

— —, 2011. Addition to the reclassification of brachyuran crabs (Crustacea: Decapoda: Brachyura). Part I. New taxa. Natura Croatica (Fauna Croatica), **20**(1): 125-139.

STIMPSON, W., 1858. Prodromus descriptionis animalium evertebratorum in expeditione ad Oceanum Pacificum Septentrionalem missa, C. Ringgold et Johanne Rodgers ducibus, observatorum et descriptorum. Pars V. Crustacea Ocypodoidea. Proceedings of the Academy of Natural Sciences of Philadelphia, **10**: 93-110 [39-56].

TESCH, J. J., 1918. Decapoda F. (Decapoda Brachyura continued) Goneplacidae and Pinnotheridae. Siboga Expéditie, **39**c[1]: 149-296.

First received 25 August 2011.
Final version accepted 28 November 2011.

EBALIA TAKEDAI, A NEW SPECIES OF LEUCOSIID CRAB (DECAPODA, BRACHYURA) FROM PANGLAO, PHILIPPINES

BY

HIRONORI KOMATSU[1,4]), MARIVENE R. MANUEL-SANTOS[2,5]) and PETER K. L. NG[3,6])

[1]) Department of Zoology, National Museum of Nature and Science, 4-1-1 Amakubo, Tsukuba, Ibaraki 305-0005, Japan

[2]) Zoology Division, National Museum, P. Burgos St., P.O. Box 2659, Manila 1000, Philippines

[3]) Tropical Marine Science Institute and Department of Biological Sciences, National University of Singapore, Lower Kent Ridge Road, Singapore 119260, Republic of Singapore

ABSTRACT

A new species of the family Leucosiidae, *Ebalia takedai*, is described and illustrated on the basis of four ovigerous females from Panglao, Philippines. *Ebalia takedai* n. sp. resembles *E. longispinosa* Ihle, 1918, *E. sakaii* Takeda & Miyake, 1972 and *E. weberi* Komatsu & Takeda, 2007, in the shape and armature of the carapace, mouth parts and cheliped, but can be distinguished from them by having two spines on the anterolateral margin of carapace.

INTRODUCTION

The genus *Ebalia* Leach, 1817 (type species *E. bryerii* Leach, 1817, a subjective synonym of *Cancer tumefactus* Montagu, 1808) currently contains 70 species (Ng et al., 2008). In addition to their list, *E. weberi* Komatsu & Takeda, 2007, and *E. koyo* Komatsu, 2011, have been added to *Ebalia*, whilst *E. yokoyai* Sakai, 1965, was transferred to *Tanaoa* Galil, 2003, by Komatsu et al. (2005). The taxonomy of *Ebalia* is unstable and needs a major revision (see Ihle, 1918; Serène & Soh, 1976; Naruse & Ng, 2006). Among the collection of leucosiid crabs collected by the PANGLAO BIODIVERSITY

[4]) Corresponding author; e-mail: h-komatu@kahaku.go.jp

[5]) e-mail: ladomedaeus@hotmail.com

[6]) e-mail: dbsngkl@nus.edu.sg

PROJECT from the central Philippines, were four ovigerous females referred to an undescribed species of *Ebalia*. In this paper, we place them to a new species. The description and illustration of this new species forms the basis of the present paper.

The specimens examined are preserved in the crustacean collection of the National Museum, Manila, Philippines (NMCR); the Zoological Reference Collection of the Raffles Museum of Biodiversity Research, National University of Singapore (ZRC); Muséum national d'Histoire naturelle, Paris, France (MNHN); National Museum of Nature and Science, Tokyo, Japan (NSMT); and the Museum of New Zealand Te Papa Tongarewa (NMNZ). The size of the specimen is indicated by the carapace width (including spines) and length (including spines) in millimeters. Abbreviations used in text are: G1 = male first pleopod; G2 = male second pleopod; ovig. = ovigerous; stn = station.

TAXONOMY

Family LEUCOSIIDAE Samouelle, 1819

Ebalia takedai n. sp.

(figs. 1, 2)

Material examined. — Holotype, ovig. female (4.5 × 3.3 mm), NMCR-15019, stn T39, Cervera shoal, West Pamilican Island, Pangalao, 09°30.1′N 123°50.4′E, muddy sand, 100-138 m, trawl, coll. by PANGLAO BIODIVERSITY PROJECT, 6 July 2004.

Paratypes: 1 ovig. female (4.4 × 3.2 mm), MNHN-IU-2009-825; 1 ovig. female (4.4 × 3.4 mm), ZRC 2011.1022; 1 ovig. female (4.7 × 3.4 mm), NSMT-Cr 22010. Same data as holotype.

Comparative material examined. — *Ebalia longispinosa* Ihle, 1918: 1 female (6.1 × 5.1 mm), NSMT-Cr 5294, northeast of Tanega-shima Island, Osumi Islands, Japan, 30°46.1′N 131°08.1′E, 45 m, dredge, coll. M. Imajima & M. Takeda, 17 June 1975; 1 ovig. female (6.8 × 5.5 mm), NSMT-Cr 5294, northeast of Tanega-shima Island, Osumi Islands, Japan, 30°45.9′N 131°11.2′E, 70 m, dredge, coll. M. Imajima & M. Takeda, 17 June 1975.

Ebalia sakaii Takeda & Miyake, 1972: 1 male (4.0 × 3.2 mm), 2 ovig. females (4.3 × 3.5 mm, 4.4 × 3.4 mm), NSMT-Cr 11062, Hyotanse Bank, Izu Islands, Japan, 34°22.35′N 139°05.26′E, 109 m, dredge, coll. T. Okutani, 11 September 1989.

Ebalia weberi Komatsu & Takeda, 2007: holotype male (4.1 × 3.7 mm), NMNZ CR. 011251, between Meteorological Station and Hutchinson Bluff (ca. 29°10′S 177°50′W), Raoul I., Kermadec Is., New Zealand, beam trawl, 146-110 m, coll. RV Acheron, 4 April 1973.

Description of female. — Carapace (fig. 1) pentagonal in general outline, 1.3-1.4 times broader than long, covered with closely set, flattened, round granules, granules on spines modified into spinules. Front strongly produced,

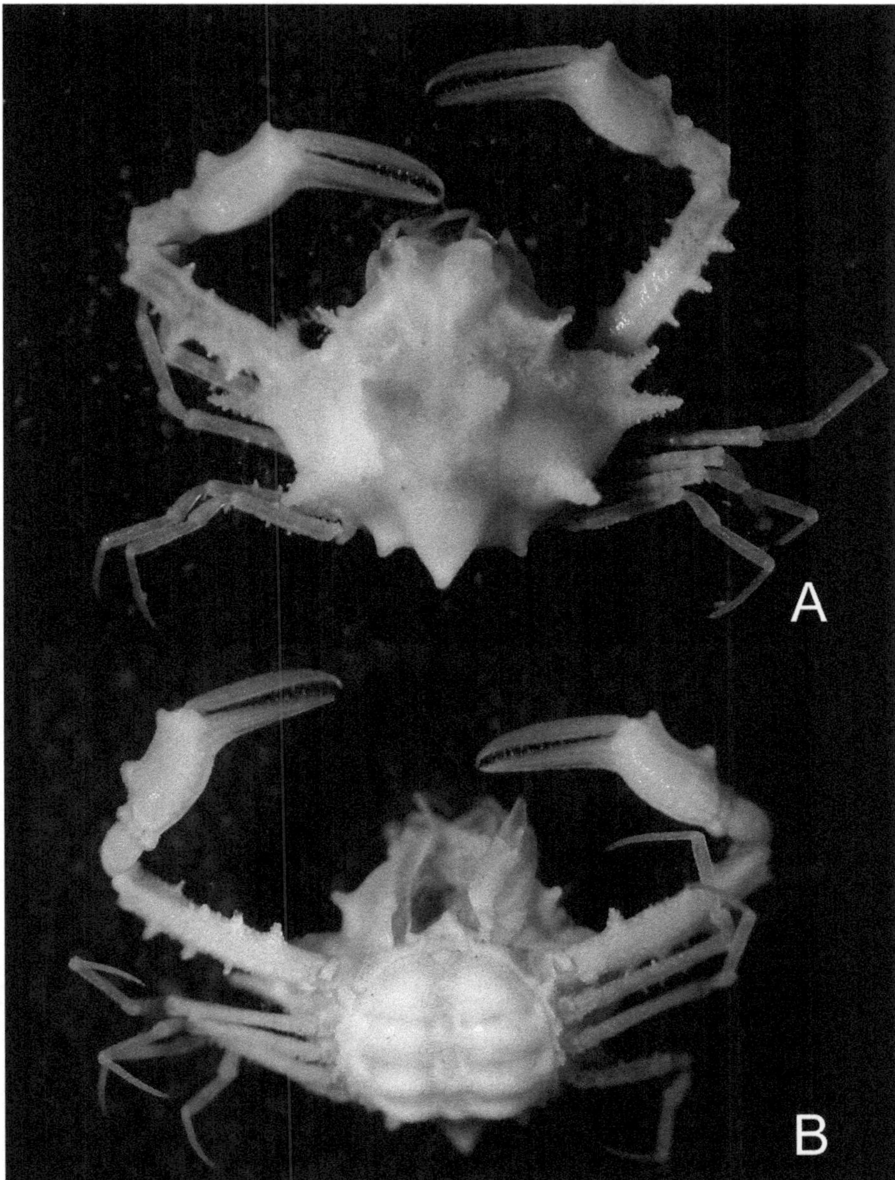

Fig. 1. *Ebalia takedai* n. sp., holotype ovig. female (4.5 × 3.3 mm), NMCR-15019. A, dorsal view; B, ventral view.

deeply concave medially, with pair of low, subdistal tubercles on both corners; margin broadly V-shaped. Mesogastric region gently raised longitudinally; gastro-cardiac region raised, not separated from each other, with pair of gastric spines and 1 shorter cardiac spine, spines erect. Intestinal region subconical, posteriorly projecting beyond posterior margin. Posterior margin transversely straight, with small triangular tubercle on each side. Hepatic region weakly demarcated, margin with short spine. Pterygostomian margin with short spine, shorter than hepatic spine. Branchial region convex, expanded, sloping anterolaterally, with 2 long spines on anterolateral margin and 1 long, submarginal spine on metabranchial region, second anterolateral spine longest.

Ocular peduncle very short. Orbit with 2 straight fissures on dorsal roof, with V-shaped notch on infraorbital margin; orbital hiatus filled with basal article of antenna. Antennule folded into oblique fossa; basal article occupying ventral half of fossa. Basal article of antenna subrectangular, filled in orbital hiatus. Afferent channel with V-shaped notch on anterior margin.

Mandible (fig. 2A, B) well calcified; cutting edge triangular in outline, pointed medially; endopod palp 3-segmented, terminal article fringed with short setae. Maxillule (fig. 2C): coxal endite subcylindrical, with short setae on tip; basial endite triangular, with stout, thin setae on mesial margin; endopod absent. Maxilla missing. First maxilliped (fig. 2D): coxal endite semiglobular, with dense setae; basial endite lobular, triangular, fringed with long, plumose setae; endopod lobular, longitudinally expanded, fitting in efferent channel, fringed with short setae around apex; exopod longitudinally filiform, bearing flagellum with some long, terminal setae. Second maxilliped (fig. 2E): endopod with long, simple setae along inner margins of ischium and merus, densely setose around outer margin of propodus, dactylus fringed with stout setae around tip; exopod filiform, with long setae on distal portion of mesial margin, bearing flagellum with some terminal setae.

Third maxilliped (fig. 2F, G) covered with flattened, round granules; basis fused with ischium, but with suture on internal surface; ischium longitudinally convex in lateral 0.7, with low subdistal tubercle on outer surface; merus slightly bent dorsally in situ, as long as ischium along mesial margin, concave in distal 0.7, with subproximal tubercle on outer surface; exopod fringed with short setae along lateral margin; internal exopodal ridge vestigial.

Cheliped (fig. 1) slender, 1.4-1.6 times as long as carapace, measured along outer margin from merus to tip of movable finger, entirely covered with flattened, round granules except on fingers; coxal condyle very small, rounded in female; merus flattened on dorsal surface, triangular in cross section, with 4 or 5 spines on outer margin, with 2 or 3 spines plus some additional spinules on inner margin; carpus short, with small triangular process on outer surface; palm convex dorsally, with small median tubercle, outer margin with triangular

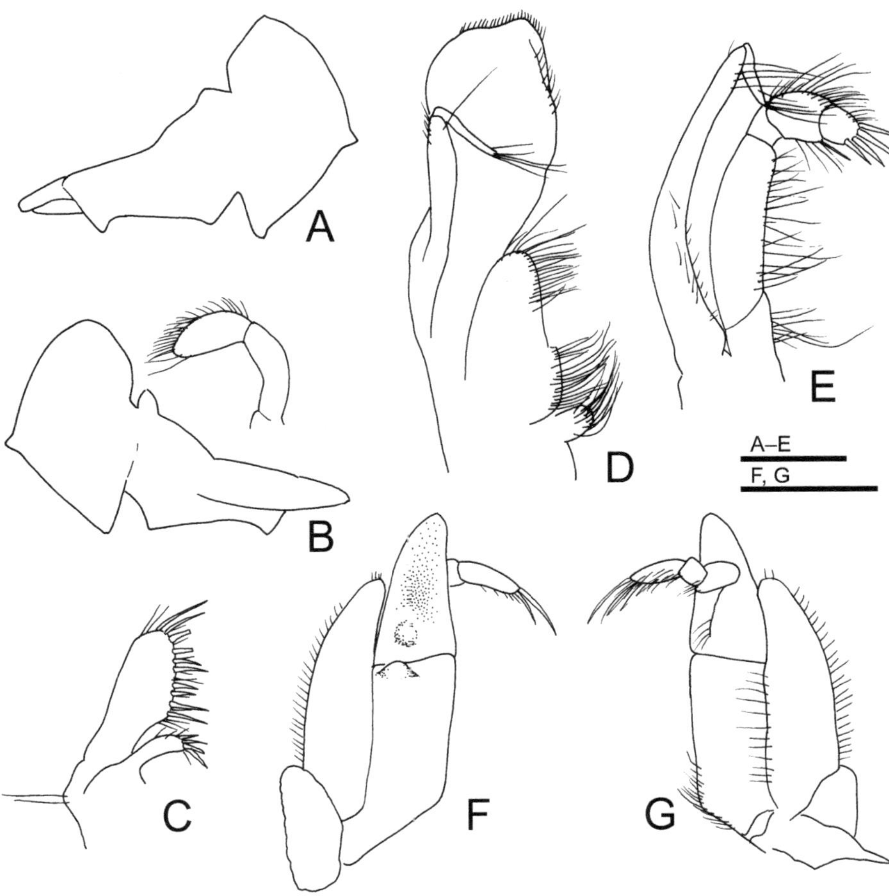

Fig. 2. *Ebalia takedai* n. sp., holotype ovig. female (4.5 × 3.3 mm), NMCR-15019. A, mandible, external view; B, same (palp detached), internal view; C, maxillule, external view; D, first maxilliped, external view; E, second maxilliped, external view; F, third maxilliped, external view; G, same, internal view. Scales: A–E = 0.25 mm; F, G = 0.5 mm.

tooth on proximal 0.3; fingers slender, straight, 1.1-1.4 times as long as palm along outer margin; both cutting edges of fingers finely dentate, crossing at tip.

Ambulatory legs (fig. 1) similar in shape, gradually decreasing in length from first to fourth, covered with flattened, round granules except dactyli; coxal condyles reduced in female; meri subcylindrical, with some spinules on dorsal and ventral surfaces; carpi and propodi subcylindrical; dactyli subconical, slightly incurved, with small dactylo-propodal locks on proximal borders of dorsal surfaces, inner and outer margins rimmed with microscopic granules.

Thoracic sternites (fig. 1B) covered with flattened, round granules of various sizes; first to third sternites fused together; abdominal cavity reaching to buccal cavern; sutures between thoracic sternites 4/5, 5/6, 6/7 and 7/8 interrupted medially; median longitudinal suture absent.

Abdomen (fig. 1B) covered with flattened, round granules; first somite completely concealed beneath carapace; second somite short, transversely subrectangular; third to sixth somites completely fused together, ovate, moderately convex ventrally, concave and glabrous between somites, with median sulcus; telson tongue-shaped, fringed with short setae around tip, directed anterodorsally in situ.

Colouration in life. — Whole body white or light brown, irregularly speckled with orange or brown spots on behind frontal region to cardiac region, metabranchial region and pereiopods.

Remarks. — *Ebalia takedai* n. sp. belongs to a group of species that includes *E. longispinosa* Ihle, 1918, *E. sakaii* Takeda & Miyake, 1972, and *E. weberi* Komatsu & Takeda, 2007. All possess a pentagonal carapace; a pair of gastric and a cardiac tubercles or spines; posteriorly projecting, subconical intestinal region; reduced endopod of the maxillule; reduced internal exopodal ridge of the third maxilliped; tuberculate or spinate outer margin of the chelipedal merus; presence of a triangular lobe on the outer margin of chelipedal palm; a median tubercle on the dorsal surface of chelipedal palm; relatively long and slender chelipedal fingers and a female abdomen composed of a separate second but fused third to sixth somites. However, *Ebalia takedai* can easily be distinguished from these species by having two spines on each anterolateral margin of carapace (versus four in *E. longispinosa* and two or three small triangular lobes in *E. sakaii* and *E. weberi*).

Ebalia takedai n. sp. most closely resembles *E. longispinosa* Ihle, 1918, in the spiny carapace and pereiopods, but in addition to the total number of anterolateral spines of the carapace (four versus eight), it can be readily distinguished by the frontal margin having a low subdistal tubercle on each corner (versus short distal spine in *E. longispinosa*).

Etymology. — This new species is dedicated to Prof. Masatsune Takeda, an honorary curator of the National Museum of Nature and Science, Tokyo, for his enormous contribution to carcinology.

Distribution. — Known only from the type locality, Panglao, Philippines, occurring at the depths of 100-138 m.

ACKNOWLEDGEMENTS

The authors sincerely thank to Prof. Masatsune Takeda for his guidance and encouragement over the years, and dedicate this new species for his 70th birthday. The Philippine work was funded by the French Government, Muséum national d'Histoire naturelle (Paris), Total Foundation, ASEAN Regional Centre for Biodiversity Conservation (Manila), National University of Singapore and the National Taiwan Ocean University. The first author acknowledges a Raffles Museum research fellowship for his stay in Singapore. We thank Tohru Naruse, Sammy de Grave and an anonymous reviewer for their critical reading and valuable comments on an earlier version.

REFERENCES

GALIL, B., 2003. Four new genera of leucosiid crabs (Crustacea: Brachyura: Leucosiidae) for three new species and nine species previously in the genus *Randallia* Stimpson, 1857, with a description of the type species, *R. ornata* (Randall, 1939). Proceedings of the Biological Society of Washington, **116**(2): 395-422.

IHLE, J. E. W., 1918. Die Decapoda Brachyura der Siboga-Expedition. III. Oxystomata: Calappidae, Leucosiidae, Raninidae. Siboga Expéditie, **39**b2: 159-322.

KOMATSU, H., 2011. Crabs dredged off the Ogasawara Islands (Crustacea, Decapoda, Brachyura). Memoirs of the National Museum of Nature and Science, Tokyo, **47**: 219-277.

KOMATSU, H., M. R. MANUEL & M. TAKEDA, 2005. A small collection of leucosiid crabs (Crsutacea: Decapoda: Brachyura) from Balicasag Island, Bohol, Philippines. Species Diversity, **10**(2): 105-123.

KOMATSU, H. & M. TAKEDA, 2007. First record of leucosiid crabs (Crustacea: Decapoda: Brachyura) from the Kermadec Islands, with description of a new species. Bulletin of the National Museum of Nature and Science, Tokyo, (A) **33**(2): 61-66.

LEACH, W. E., 1817. The zoological miscellany; being descriptions of new, or interesting animals: 1-151. (E. Nodder and Son, Covent Garden and London).

MONTAGU, G., 1808. Description of several marine animals found on the south coast of Devonshire. Transactions of the Linnean Society of London, **9**: 81-114.

NARUSE, T. & P. K. L. NG, 2006. Two new species of leucosiid crabs (Decapoda: Brachyura) from the Ryukyu Islands, Japan. Crustacean Research, **35**: 108-116.

NG, P. K. L., D. GUINOT & P. J. F. DAVIE, 2008. Systema Brachyurorum: Part 1. An annotated checklist of extant brachyuran crabs of the world. Raffles Bulletin of Zoology, (Supplement) **17**: 1-286.

SAKAI, T., 1965. The Crabs of Sagami Bay Collected by His Majesty, the Emperor of Japan: i–xvi , 1-206, 1-92, 1-3, pls. 1-100, 1 map. (Maruzen, Tokyo).

SAMOUELLE, G., 1819. The entomologist's useful compendium; or an introduction to the knowledge of British insects, comprising the best means of obtaining and preserving them, and a description of the apparatus generally used; together with the genera of Linné, and the modern method of arranging the classes Crustacea, Myriapoda, Spiders, Mites and

Insects, from their affinities and structure, according to the views of Dr. Leach. Also an explanation of the terms used in entomology; a calendar of the times of appearance and usual situations of near 3,000 species of British insects; with instructions for collecting and fitting up objects for the microscope: 1-496. (London).
SERÈNE, R. & C. L. SOH, 1976. Brachyura collected during the Thai-Danish expedition (1966). Phuket Marine Biological Center Research Bulletin, **12**: 1-37.
TAKEDA, M. & S. MIYAKE, 1972. New crabs from the sea around the Tsushima Islands. Bulletin of the National Science Museum, Tokyo, **15**(2): 253-265.

First received 30 August 2011.
Final version accepted 22 January 2012.

REMARKABLE ZOEAS OF TWO SPECIES OF DEEP-SEA SPIDER CRABS (BRACHYURA, MAJOIDEA, EPIALTIDAE, PISINAE)

BY

KOOICHI KONISHI[1,3]) and TOMOMI SAITO[2])

[1]) National Research Institute of Fisheries Science, 2-12-4 Fukuura, Yokohama, Kanagawa 236-8648, Japan

[2]) Port of Nagoya Public Aquarium, Minato-ku, Nagoya, Aichi 455-0033, Japan; present address: Usa Marine Institute, Kochi University,194 Inoshiri, Kochi 781-1164, Japan

ABSTRACT

The first zoeas of two pisinine species, *Laubierinia nodosa* (Rathbun, 1916) and *Oxypleurodon stimpsoni* Miers, 1886, are described and illustrated from laboratory-hatched materials. Among majoid larvae, the present zoeas share a salient distinguishing character in having a pair of forehead cornua on the carapace. As sometimes found in deep-sea decapod larvae, the present zoeas have some lecithotrophic features, i.e., large yolk mass within cephalothorax and reduced mouthparts.

INTRODUCTION

The classification for majoid crabs has been undergone drastic modifications in a recent decade. Martin & Davis (2001) elevated several majid subfamilies to family level, including the Pisinae. Recently, Ng et al. (2008) treated the Pisidae as a subfamily of the Epialtidae. In recent years, the epialtid genus *Sphenocarcinus* A. Milne-Edwards, 1875, has been substantially revised and redefined and most species have been transferred out to *Oxypleurodon* Miers, 1886, *Rhinocarcinus* Richer de Forges & Ng, 2009, and *Laubierinia* Richer de Forges & Ng, 2009 (Richer de Forges & Ng, 2009a, b).

The two species discussed here, have been classified in *Sphenocarcinus* at one time or another, but are now known as *Laubierinia nodosa* (Rathbun, 1916) and *Oxypleurodon stimpsoni* Miers, 1886, respectively. Ng et al. (2008)

[3]) Corresponding author; e-mail: kzoea@fra.affrc.go.jp

have commented that the present suprageneric classification based on adult morphology is still unsettled and substantial changes can be expected once more work is done. In parallel with morphological revisions of adult crabs, molecular analysis studies have been done in several majoid families, and initial indications are that many of the higher-level characters used are subject to convergence (e.g., see Hultgren & Stachowicz, 2008; Hultgren et al., 2009; Windsor & Felder, 2011).

Larval knowledge on the deep-sea spider crabs, except for snow crabs of the genus *Chionoecetes* Krøyer, 1838, important fishery species in Japan and adjacent waters, has been relatively poorly documented. This paper describes zoeal morphology of the pisinid genera, *Oxypleurodon* and *Laubierinia*, and also gives morphological notes on these deep-sea pisinine zoeas.

MATERIAL AND METHODS

Ovigerous females were collected from the waters off Makurazaki (130° 19′E 31°08′N–130°21′E 31°08′N), southern Japan, at depths of 263-285 m, on 4 June 1999 and 4 December 1999. Mean egg diameters of *Laubierinia nodosa* and *Oxypleurodon stimpsoni* were 0.95 ± 0.04 mm and 0.90 ± 0.03 mm, respectively. First zoeas of *L. nodosa* were hatched on 18 January 1999 while those of *O. stimpsoni* were obtained on 19 July 1999. Numerous yolk granules were found under the carapaces of both zoeas.

Drawings and measurements were made with a drawing tube attached to an Olympus BX51 microscope. All illustrations were made with Illustrator® 10 (Adobe Systems Inc.). Larval descriptions mostly followed the method proposed by Clark et al. (1998) and setal terminology is according to Ingle (1992). All setal arrangements are listed from proximal to distal. Roman numerals for setation denote the dorso-lateral seta. Carapace length (CL) was measured from the anterior border of the eye to the posterior border of the cephalothorax. The distance between the tip of rostral spine and the tip of dorsal spine (RDD) was also measured in zoeas. These values are shown as mean ± SD and the range is given in parenthesis. The specimens used in this study are deposited in National Museum of Nature and Science, Tsukuba.

DESCRIPTION OF FIRST ZOEAS

Laubierinia nodosa (Rathbun, 1916)
(fig. 1)

Dimensions: CL = 1.34 ± 0.05 mm (range 1.22-1.39 mm, n = 13), RDD = 2.30 ± 0.13 mm (range 2.15-2.46 mm, n = 13).

Carapace (fig. 1A): Globose, dorsal and rostral spines well developed, slightly curved, no lateral spines. Anterodorsal and posterodorsal regions of carapace each with a pair of minute simple setae. A pair of prominent anterodorsal cornua and a median tubercle. Anterior ventral margin just adjacent to maxilla indented. A large anteriorly-directed plumose seta (so-called "majid seta"), plus 3 posterior setae on ventral margin (fig. 1A′). Eyes sessile.

Antennule (fig. 1C): Uniramous, unsegmented, conical with 3 terminal aesthetascs and 2 short simple setae.

Antenna (fig. 1D): Biramous, protopod process long, pointed, with 2 rows of marginal spinules. Endopod rudiment about 1/3 length of protopod. Exopod elongated, approximately 2.5 times longer than endopod, with 1 serrate and 1 simple setae at distal one-third.

Mandible (fig. 1E): Incisor and molar processes, with rudimentary palp.

Maxillule (fig. 1F): Coxal endite with 5 simple setae. Basial endite with a small spine and 3 simple setae plus 2 plumose setae, respectively. Endopod unsegmented, with a terminal and 2 subterminal simple setae, rarely with 3 small spines plus a terminal seta (fig. 1F′).

Maxilla (fig. 1G): Coxal and basial endites slightly bilobed, with 3-4 + 2 and 3 + 2-4 setae, respectively. Endopod unsegmented, with a simple subterminal seta, rarely with a simple terminal seta and a minute subterminal spine (Fig. 1G′). Scaphognathite with 16-19 marginal plumose setae.

Maxilliped 1 (fig. 1H): Coxa without setae. Basis with 2 + 2 + 2 + 2, rarely 2 + 1 + 2 + 2, inner setae. Endopod 5-segmented with 3, 2, 1, 2, 4 + I, rarely 2, 2, 1, 2, 4 + I setae. Exopod with 4 long natatory setae.

Maxilliped 2 (fig. 1I): Coxa without setae. Basis with 1 + 1 + 1 setae. Endopod 2-segented with 0,5 setae, but distal segment indistinctly segmented. Exopod with 4 long natatory setae.

Maxilliped 3 (fig. 1J): Rudimentary biramous bud, but rarely exopodal ramus with a simple terminal seta (fig. 1J′).

Pereiopods (fig. 1K): Rudimentary buds, tip of pereiopod 1 bifurcate.

Abdomen (fig. 1A, B): 5 somites and a forked telson. A pair of short mid-dorsal setae on each somite. Somite 2 with dorsolateral spine. Somite 3 with 2

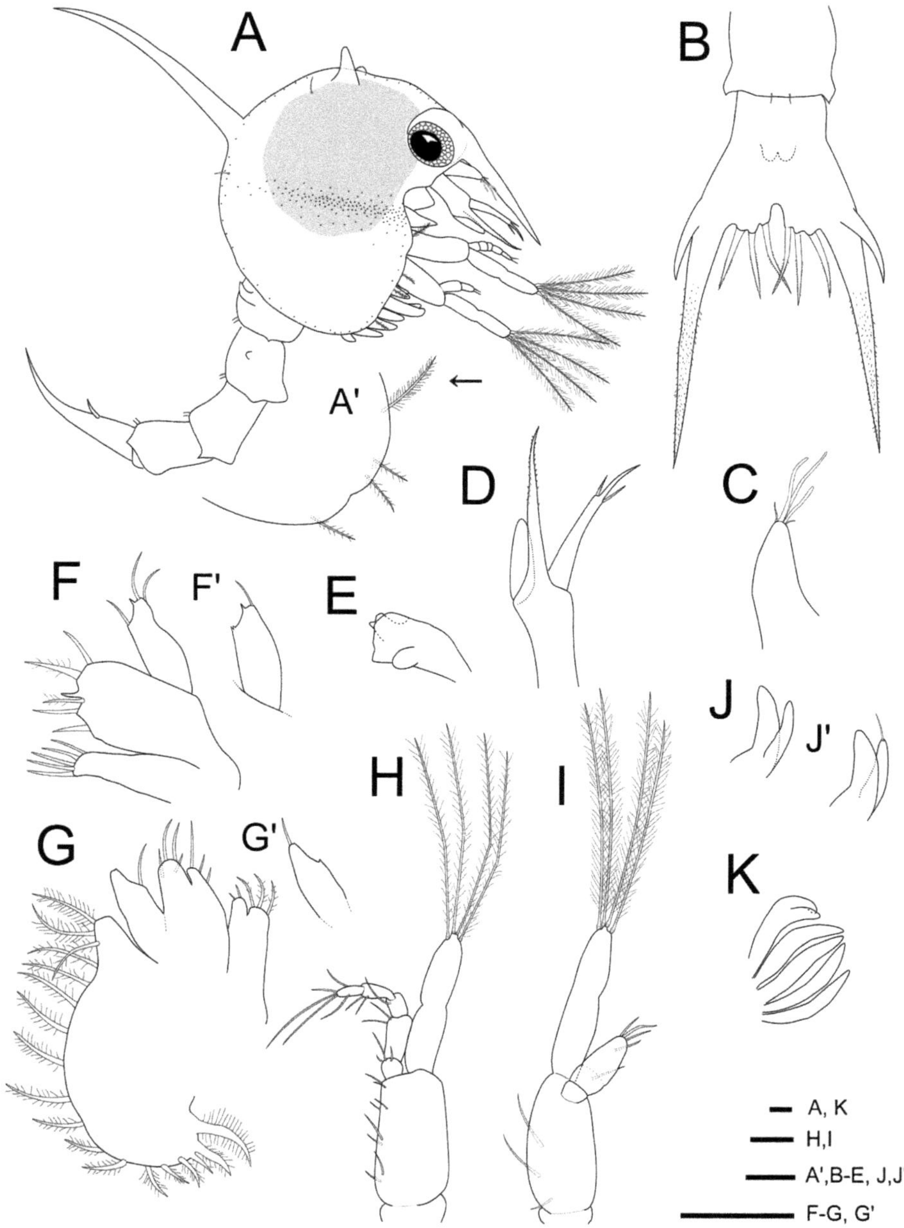

Fig. 1. *Laubierinia nodosa* (Rathbun, 1916), zoea 1. A, whole animal in lateral view; A′, antero-ventral margin of carapace showing "majid seta" (arrow); B, abdominal somite 6 and telson, dotted line showing a ventral anal opening; C, antennule; D, antenna; E, mandible; F, maxillule; F′, another type of maxillular endopod; G, maxilla; G′, another type of maxillar endopod; H, maxilliped 1; I, maxilliped 2; J, maxilliped 3; J′, another type of maxilliped 3; K, pereiopods. Dotted area shows large yolk mass within cephalothorax. Scale bars = 0.1 mm.

short lateral knobs, which in some specimens may be different in size. Telson bifurcated, with a shallow median notch and 3 pairs of stout, denticulate setae on inner margin. Both furcae with a large dorso-lateral spine proximately, covered with numerous setules.

Oxypleurodon stimpsoni Miers, 1886
(fig. 2)

Dimensions: CL = 1.30 ± 0.03 mm (range 1.25-1.34 mm), RDD = 2.04 ± 0.13 mm (range 1.85-2.24 mm, n = 12).

Carapace (fig. 2A): Globular, long dorsal and rostral spines, no lateral spines. Anterodorsal and posterodorsal regions of carapace each with a pair of minute simple setae. A pair of short frontal cornua and a small median frontal tubercle. A large anteriorly-directed plumose seta (so-called "majid seta"), with 2 or 3 posterior setae on ventral margin (fig. 2A′). Eyes sessile.

Antennule (fig. 2C): Uniramous and unsegmented, conical, with 3 aesthetascs and a short simple seta distally.

Antenna (fig. 2D): Biramous, protopod spinulate distally. Endopod about 1/3 length of protopod. Exopod with 2 subterminal spines.

Mandible (fig. 2E): Incisor and molar processes, with rudimentary palp.

Maxillule (fig. 2F): Coxal endite with 6 or 7 simple/plumose setae. Basial endite with small pointed protuberance and 4 simple setae. Endopod unsegmented, with 3 terminal setae and 2 subterminal spinules.

Maxilla (fig. 2G): Coxal and basial endite slightly bilobed, with 3 + 2-3 and 2-3 + 3 setae, respectively. Endopod unsegmented, with 2 terminal setae. Scaphognathite with 16-22 marginal plumose setae.

Maxilliped 1 (fig. 2H): Coxa without setae. Basis with 2 + 2 + 2 + 2 inner setae. Endopod 5-segmented with 3, 2, 1, 2, 4 + I setae, respectively. Exopod with 4 long natatory setae. Exopod with 4 long natatory setae.

Maxilliped 2 (fig. 2I): Coxa without setae. Basis with 1 + 1 + 1 or 1 + 1 setae. Endopod 2-segmented with 0, 5 setae, respectively, but distal segment indistinctly segmented. Exopod with 4 long natatory setae.

Maxilliped 3 (fig. 2J): Biramous rudimentary bud.

Pereiopods (fig. 2K): Rudimentary buds, tip of pereiopod 1 bifurcate.

Abdomen (fig. 2A, B): 5 somites and a forked telson. A pair of short mid-dorsal setae on each somite. Somite 2 with dorsolateral spine. Somite 3 with 2 short lateral knobs, rarely absent on one side. Telson bifurcated, with shallow median notch and 3 pairs of stout, denticulate setae on inner margin. Both furcae with a large dorso-lateral spine proximately, covered with numerous setules.

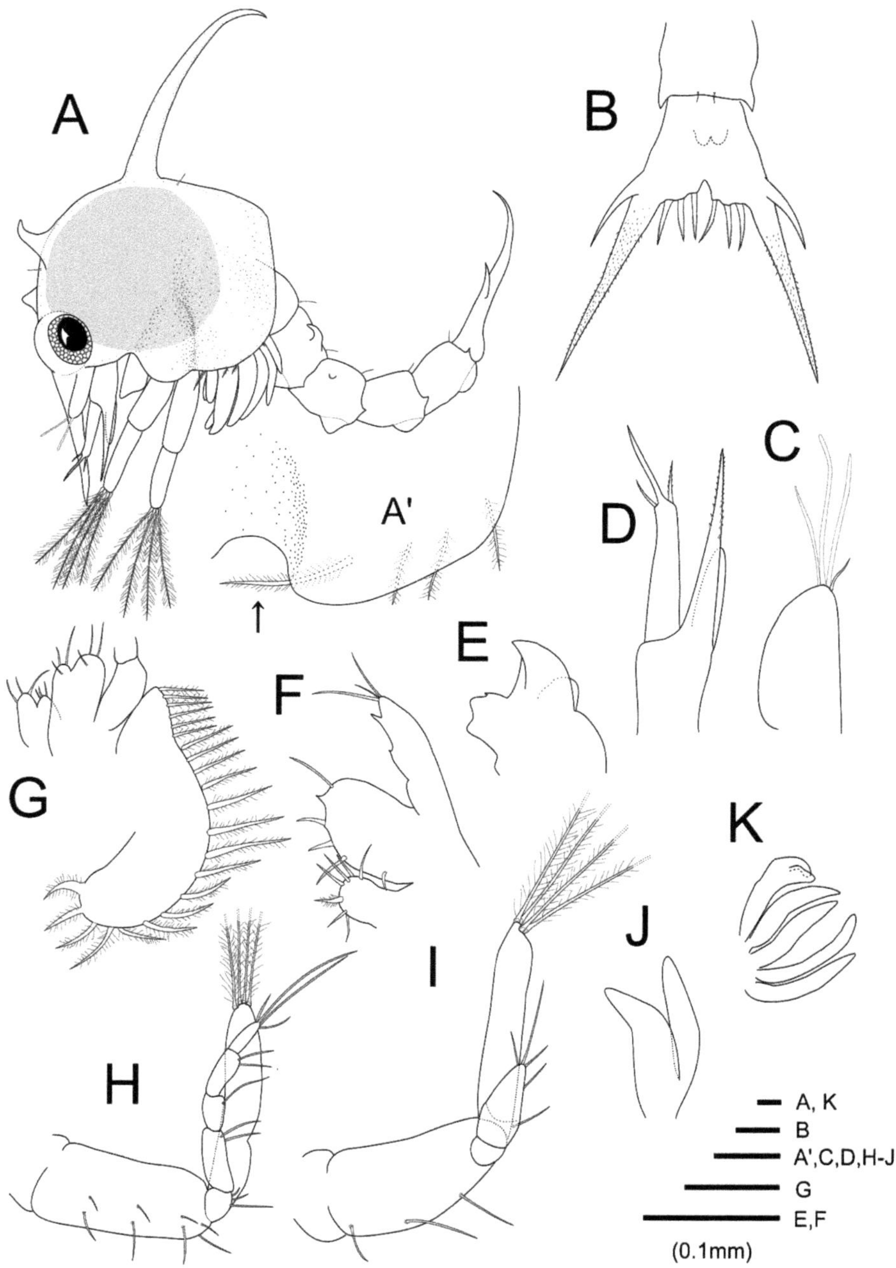

Fig. 2. *Oxypleurodon stimpsoni* Miers 1886, zoea 1. A, whole animal in lateral view; A′, antero-ventral margin of carapace showing "majid seta" (arrow); B, abdominal somite 6 and telson, dotted line showing a vental anal opening; C, antennule; D, antenna; E, mandible; F, maxillule; G, maxilla; H, maxilliped 1; I, maxilliped 2; J, maxilliped 3; K, pereiopods. Dotted area shows large yolk mass within cephalothorax. Scale bars = 0.1 mm.

DISCUSSION

The subfamily Pisinae currently contains of 46 genera (Ng et al., 2008), and larval stages have been described in 12 genera, as summarized in table I. Ingle (1979) discussed character sets for subfamilies within the Majidae sensu lato and postulated some larval features in the Pisinae as follow: 1) absence of carapace lateral spines, 2) rostral and dorsal spines of moderate length, 3) one spine on the telson fork, 4) dorsolateral processes on the 2nd abdominal somite and rarely on 3rd somite, 5) posterolateral processes on the 3rd-5th abdominal somites often short, 6) basis of the maxilliped 2 with not more than 3 setae, and 7) antennal exopod with subterminal setae. However, as workers accumulated larval descriptions over the years, these characters no longer work, and recent studies argued that there are no larval characters that can define this subfamily (e.g., Marques & Pohle, 2003). Recently Santana et al. (2006) described larvae of the pisinine *Notolopas brasiliensis* Miers, 1886, and they suggested that pisinine zoeal characters could be observed in the setation of coxal and basial endites of the maxillule (with the exception of *Pisoides* H. Milne Edwards & Lucas, 1853), and the endopod of the first maxilliped (with the exception of *Doclea* Leach, 1815) (see also Santana et al., 2004).

Table I compares the main morphological features in previously described zoeas of various pisinine genera. The first zoeas of the present two species have large yolk mass in the cephalothorax indicating they undergo lecithotrophic development. In such zoeas, lecithotrophy is invariably associated with reduced cephalothracic appendages (i.e., in the mandibles, endopods of maxillule, maxilla and maxilliped 2). Furthermore, in connection with lecithotrophy, abbreviated development is also evident in the mandibular palp at the first zoeal stage. This reduced condition has also been observed in other deep-sea pisinine species like *Goniopugettia sagamiensis* (Gordon, 1930) (see Taishaku & Konishi, 2001) and *Rochinia debilis* Rathbun, 1932 (see Komatsu & Takeda, 2003). As such, comparisons of these structures those of with normal planktotrophic species would not be phylogenetically informative. As shown in table I, the present two zoeas show a useful shared character in having a pair of forehead cornua on the carapace. At present, of course, it is unclear whether this unique and diagnostic zoeal feature is significant for future phylogenetic analyses. It is certainly a useful identification feature that separates them from other members of the subfamily. A similar character, for example, has also been found in the zoeas of the inachid *Anasimus latus* Rathbun, 1894 (Sandifer & van Engel, 1972). The zoea of *G. sagamiensis* can be also easily distinguished from the

TABLE I

Comparison of major zoeal characters in genera of the subfamily Pisinae

Genera	Carapace						Maxillule	Maxilla		Maxilliped 1	Maxilliped 2		Telson	References
	RS	DS	LS	PE	FC	ventro-lateral margin	endopod	endopod	scaphog-nathite	endopod	basis	endopod	DLS	
Anamathia	+	+	–	–	–	ms + 4	1, 6	3 + 3	14	3, 2, 1, 2, 4 + I	1, 1, 1	0, 1, 4	long	1), 2)
Doclea	–	+	–	–	–	ms + 7	1, 6	2 + 3	13	3, 2, 1, 2, 4 + I	1, 1, 1	0, 1, 3	–	3), 4), 5)
Goniopugettia	+	+	–	+	–	ms + 6	r	r	14-15	3, 2, 1, 2, 4 + I	1, 1, 1	0, 1, 3	long	6)
Herbstia	+	+	–	–	–	ms + 6	?	?	?	?	?	?	long	7)
Hyastenus	+	+	–	–	–	ms + 5	1, 6	5	11	3, 2, 1, 2, 4 + I	1, 1, 1	0, 1, 5	r/–	5), 8)
Libinia	+	+	–	–	–	ms + 7	1, 5	5	12	3, 2, 1, 2, 4 + I	1, 1, 1	0, 1, 4	short	9), 10)
Libidoclea	+	+	–	–	–	ms + 5	1, 5	6	14	3, 2, 1, 2, 4 + I	1, 1, 1	0, 1, 5	long	11)
Notolopas	+	+	–	–	–	ms + 5	1, 5	5	11	3, 2, 1, 2, 4 + I	1, 1, 1	0, 1, 4	short	12)
Phalangipus	+	+	–	–	–	ms + 4	1, 6	4-5	11-13	3, 2, 1, 2, 4 + I	1, 1, 1	0, 1, 4	short	5)
Pisa	r	+	–	–	–	ms + 6	1, 6	5	12-14	3, 2, 1, 2, 4 + I	1, 1, 1	0, 1, 5	long	13)
Pisoides	+	+	–	–	–	ms + 4	1, 4	4	9-10	3, 2, 1, 2, 4 + I	1, 1, 1	0, 1, 4	long	5), 8), 14)
Rochinia	+	+	+/–	–	–	ms + 4-5	0, 2/1, 6	3-6	11/17-18	3, 2, 1, 2, 4 + I	1/1, 1, 1	0, 1, 4-5	long	15), 16)
Scyra	+	+	–	–	–	ms + 5	1, 4	4	10-11	3, 2, 1, 2, 4 + I	1, 1, 1	0, 1, 5	long	17)
Laubierinia	+	+	–	–	+	ms + 3	r	r	16-19	2-3, 2, 1, 2, 4 + I	1, 1, 1	0, 5	long	This study
Oxypleurodon	+	+	–	–	+	ms + 3	r	r	16-22	3, 2, 1, 2, 4 + I	0/1, 1, 1	0, 5	long	This study

+: present, –: absent.

DS: dorsal spine, DLS: dorsolateral spine, FC: forehead cornua, FS: frontal spines, LS: lateral spines, ms: majid seta, r: reduced form, PE: protruding edge, RS: rostral spine.

1) Sandifer & van Engel (1972), 2) Guerao & Abelló (1996), 3) Krishnan & Kannupandi (1987), 4) Krishnan & Kannupandi (1988), 5) Terada (1983), 6) Taishaku & Konishi (2001), 7) Bourdillon-Casanova (1960), 8) Kurata (1969), 9) Boschi & Scelzo (1968), 10) Clark et al. (1998), 11) Fagetti (1969a), 12) Santana et al. (2006), 13) Rodriguez (1997), 14) Fagetti (1969b), 15) Ingle & Clark (1980), 16) Komatsu & Takeda (2003), 17) Kim & Hong (1999).

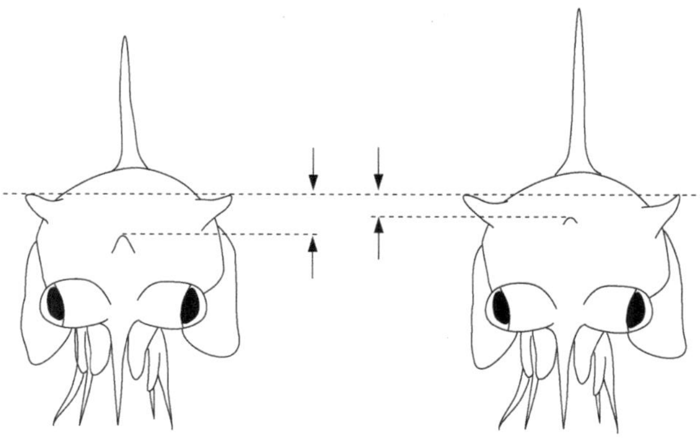

Fig. 3. Semidiagrammatic frontal view, comparing the position of forehead ornamentation between the first zoeas of *Laubierinia nodosa* and *Oxypleurodon stimpsoni*.

other pisinine members by having a protruding edge on the anterior region of the carapace (Taishaku & Konishi, 2001).

While the general morphology of the zoeas of the present two species are very similar, they can be distinguished by the relative position of anterior carapace ornamentation: in lateral view, the median tubercle and forehead cornua line up approximately on the same line in *Laubierinia nodosa*, while thet are further from each other in *Oxypleurodon stimpsoni* (fig. 3).

At the moment, the authors do not discern any consistent pattern in the morphological diversity of pisinine zoeas, which may suggest that the subfamily may be polyphyletic. The direction forward therefore is clearly to combine adult morphological, larval and molecular studies as far as possible.

ACKNOWLEDGEMENTS

We are grateful to K. Ikeue (Makurazaki City, Kagoshima Prefecture) and the crew of Nishiki-Maru for collecting the ovigerous female crabs. We dedicate this paper to Dr. Masatsune Takeda, who has been contributed so much to Japanese carcinology over the years, and who also kindly helped the authors identify the present adult material.

REFERENCES

BOSCHI, E. E. & M. A. SCELZO, 1968. Larval development of the spider crab *Libinia spinosa* H. Milne Edwards, reared in the laboratory (Brachyura, Majidae). Crustaceana, (Supplement) **2**: 170-180.

BOURDILLON-CASANOVA, L., 1960. Le méroplancton du Golfe de Marseille: les larves de Crustacé Décapodes. Recuéil des Travaux de la Station Marine d'Endoume (Marreille), **30**: 1-286.

CLARK, P. F., D. K. CALAZANS & S. S. RODORIGUES, 1998. *Libinia spinosa* H. Milne Edwards, 1834 (Crustacea: Majidae: Pisinae): a reappraisal of larval characters from laboratory reared material. Invertebrate Reproduction and Development, **33**: 145-157.

FAGETTI, E., 1969a. The larval development of the spider crab *Libidoclaea granaria* H. Milne Edwards & Lucas under laboratory conditions (Decapoda Brachyura; Majidae, Pisinae). Crustaceana, **12**: 131-140.

— —, 1969b. Larval development of the spider crab *Pisoides edwardsi* (Decapoda Brachyura) under laboratory conditions. Marine Biology, **4**: 160-165.

GUERAO, G. & P. ABELLÓ, 1996. Morphology of the prezoea and first zoea of the deep-sea spider crab *Anamathia rissoana* (Brachyura, Majidae, Pisinae). Scientia Marina, **60**: 245-251.

HULTGREN, K. M., G. GUERAO, F. P. L. MARQUES & F. P. PALERO, 2009. Assessing the contribution of molecular and larval morphological characters in a combined phylogenetic analysis of the superfamily Majoidea. In: J. W. MARTIN, K. A. CRANDALL & D. L. FELDER (eds.), Crustacean Issues **18**: Decapod Crustacean Phylogenetics: 437-455. (CRC Press, Boca Raton).

HULTGREN, K. M. & J. J. STACHOWICZ, 2008. Molecular phylogeny of the brachyuran crab superfamily Majoidea indicates close congruence with trees based on larval morphology. Molecular Phylogenetics and Evolution, **48**: 986-996.

INGLE, R. W., 1979. The larval development of the spider crab *Rochinia carpenteri* (Thomson) [Oxyrhyncha: Majidae] with a review of majid subfamilial larval features. Bulletin of the British Museum Natural History, (Zoology) **37**: 47-66.

— —, 1992. Larval stages of Northeastern Atlantic crabs. An illustrated key: i–xii, 1-363. (Natural History Museum Publications and Chapman and Hall, London).

INGLE, R. W. & P. F. CLARK, 1980. The larval and post-larval development of Gibb's spider crab, *Pisa armata* (Latreille) [family Majidae: subfamily Pisinae] reared in the laboratory. Journal of Natural History, **14**: 723-735.

KIM, D. N. & S. Y. HONG, 1999. Larval development of *Scyra compressipes* (Decapoda: Brachyura: Majidae: Pisinae) reared in the laboratory. Journal of Crustacean Biology, **19**: 782-791.

KOMATSU, H. & M. TAKEDA, 2003. First zoea of the deep-water spider crab, *Rochinia debilis* Rathbun, 1932 (Crustacea, Decapoda, Majidae), from Sagami Bay, central Japan. Bulletin of the National Science Museum, Tokyo, (A) **29**: 197-203.

KRISHNAN, T. & T. KANNUPANDI, 1987. Laboratory larval culture of a spider crab, *Doclea muricata* (Fabricius, 1787) (Decapoda, Majidae). Crustaceana, **53**: 292-303.

— — & — —, 1988. Larval development of the spider crab *Doclea gracilipes* Stimpson, 1857 (Decapoda: Brachyura, Majidae) reared in the laboratory. Journal of Crustacean Biology, **8**: 420-429.

KURATA, H., 1969. Larvae of Decapoda Brachyura of Arasaki, Sagami Bay — IV. Majidae. Bulletin of Tokai Regional Fisheries Research Laboratory, **57**: 81-127. [In Japanese with English summary.]

MARQUES, F. & G. POHLE, 2003. Searching for larval support for majoid families (Crustacea: Brachyura) with particular reference to Inachoididae Dana, 1851. Invertebrate Reproduction and Development, **43**: 71-82.

MARTIN, J. W. & G. E. DAVIS, 2001. An updated classification of the Recent Crustacea. Science Series, Natural History Museum of Los Angeles County, **39**: i–vii, 1-124.

MIERS, E. J., 1886. Part II. Report on the Brachyura collected by H.M.S. Challenger during the years 1873-76. In: Report on the Scientific Results of the Voyage of H.M.S. Challenger during the years 1873-1876 under the command of Captain George S. Nares, N.R., F.R.S. and the late Captain Frank Tourle Thomson, R.N. prepared under the Superintendence of the late Sir C. Wyville Thomson, Knt., F.R.S. & Regis Professor of Natural History in the University of Edinburgh of the civilian scientific staff on board and now of John Murray one of the naturalists of the expedition, (Zoology) **17**: i–l, 1-362, pls. 1-29. (Her Majesty's Stationary Office, London, Edinburgh and Dublin).

NG, P. K. L., D. GUINOT & P. J. F. DAVIE, 2008. Systema Brachyurorum: Part I. An annotated checklist of extant brachyuran crabs of the world. Raffles Bulletin of Zoology, (Supplement) **17**: 1-286.

RATHBUN, M. J., 1916. New species of crabs of the family Inachidae and Parthenopidae. In: Scientific results of the Philippine cruise of the fisheries Steamer "Albatross", 1907-1910. N°34. Proceedings of the United States National Museum, **50**(2135): 527-559.

RICHER DE FORGES, B. & P. K. L. NG, 2009a. New genera, new species and new records of Indo-West Pacific spider crabs (Crustacea: Brachyura: Epialtidae: Majoidea). Zootaxa, **2025**: 1-20.

— — & — —, 2009b. On the majoid genera *Oxypleurodon* Miers, 1886, and *Sphenocarcinus* A. Milne-Edwards, 1875 (Crustacea: Brachyura: Epialtidae), with descriptions of two new genera and five new species. Raffles Bulletin of Zoology, (Supplement) **20**: 247-266.

RODRIGUEZ, A., 1997. Larval and postlarval development of *Pisa tetraodon* (Pennant, 1777) (Decapoda: Majidae) reared in the laboratory. Journal of Plankton Research, **19**: 29-41.

SANDIFER, P. A. & W. A. VAN ENGEL, 1972. Larval stages of the spider crab, *Anasimus latus* Rathbun, 1894 (Brachyura, Majidae, Inachinae) obtained in the laboratory. Crustaceana, **23**: 141-151.

SANTANA, W., F. P. L. MARQUES, A. FRANSOZO & G. BERTINI, 2006. Larval development of *Notolopas brasiliensis* Miers, 1886 (Brachyura: Majoidea: Pisidae) described from laboratory reared material and a reappraisal of the characters of Pisidae. Papéis Avulsos de Zoologia, **46**: 219-232.

SANTANA, W., G. POHLE & F. MARQUES, 2004. Larval development of *Apiomithrax violaceus* (A. Milne Edwards, 1868) (Decapoda: Brachyura: Majoidea: Pisidae) reared in laboratory conditions, and a review of larval characters of Pisidae. Journal of Natural History, **38**: 1773-1797.

TAISHAKU, H. & K. KONISHI, 2001. Lecithotrophic larval development of the spider crab *Goniopugettia sagamiensis* (Gordon, 1931) (Decapoda, Brachyura, Majidae) collected from the continental shelf break. Journal of Crustacean Biology, **21**: 748-759.

TERADA, M., 1983. Larval development of four Japanese spider crabs (Brachyura, Majidae, Pisinae). Proceedings of the Japanese Society of Systematic Zoology, **25**: 17-28. [In Japanese with English summary.]

WINDSOR, A. M. & D. L. FELDER, 2011. A new species of *Hemus* (Majoidea: Majidae: Mithracinae) from the Pacific coast of Panamá, with a key to the genus. Zootaxa, **2799**: 63-68.

First received 3 August 2011.
Final version accepted 16 November 2011.

MICHELEA TAKEDA SP. NOV. (CRUSTACEA, DECAPODA, AXIIDEA, MICHELEIDAE) FROM THE SOUTH CHINA SEA

BY

WENLIANG LIU[1,2] and RUIYU LIU (J. Y. LIU)[1,3]

[1]) Institute of Oceanology, Chinese Academy of Sciences, Qingdao 266071, China

[2]) Chongxi Wetland Research Center, The State Key Laboratory of Estuarine and Coastal Research, East China Normal University, Shanghai 200062, China

ABSTRACT

A new species of the genus *Michelea* Kensley & Heard, 1991, *M. takeda*, from the Beibu Gulf (Tonkin Gulf), northern South China Sea, is described and illustrated. The new species is closely allied to *M. leura* (Poore & Griffin, 1979) but differs markedly in its wider rostrum, the lower margin of the propodus of pereopod 1 being distinctly convex and the shorter uropodal endopod. This genus is recorded for the first time from Chinese waters.

INTRODUCTION

The genus *Michelea* Kensley & Heard, 1991 is characterized by the rostrum being absent or obsolete, the carapace without a dorsolateral carina running from each side of the rostrum, the pleopods bearing marginal lamellae, at least on the lateral edge of the endopod but usually on the margins of both rami, and the uropodal rami being ovate. Eleven described species are known in the genus, from the tropical and subtropical waters of the western Pacific (Japan, Australia and New Caledonia) and the western Atlantic (Caribbean Sea) (table I).

While working on a systematic study of the axiidean fauna of the China Sea, a single specimen of an undescribed species of *Michelea* was found in collections from the Beibu Gulf, northern South China Sea. In this paper this is described as a new species, *M. takeda*, representing the first record of this genus from Chinese waters.

[3]) Corresponding author; e-mail: jyliu@qdio.ac.cn

TABLE I

Species of *Michelea* with relevant literature

Eastern Atlantic species
M. abranchiata Poore, 1997 — Caribbean Sea (Poore, 1997)
M. lamellosa Kensley & Heard, 1991 — Jamaica (Kensley & Heard, 1991)
M. vandoverae (Gore, 1987) — Eastern Florida Coast (Gore, 1987); Gulf of Mexico (Kensley & Heard, 1991); Caribbean Sea (Poore, 1997)
Western Pacific species
M. dampieri Poore, 2008 — Western Australia (Poore, 2008)
M. devaneyi Poore, 1997 — Marshall Islands (Poore, 1997)
M. hortus Poore, 1997 — Western Australia (Poore, 1997)
M. lepta (Sakai, 1987) — Okinawa, Japan (Sakai, 1987)
M. leura (Poore & Griffin, 1979) — Australia (Poore & Griffin, 1979)
M. microphylla Poore, 1997 — Australia (Poore, 1997)
M. novaecaledoniae Poore, 1997 — New Caledonia (Poore, 1997)
M. paraleura Poore, 1997 — Australia (Poore, 1997)
M. takeda sp. nov. — South China Sea (present study)

Material examined in this study was collected in the Beibu Gulf (Gulf of Tonkin) (1959-1960, 1962) and deposited in the Institute of Oceanology, Chinese Academy of Sciences, Qingdao, China (IOCAS). The drawings were made with the aid of a drawing tube mounted on a Zeiss Stemi Sv11 compound microscope. The following abbreviation is used throughout the text: cl, length of carapace.

SYSTEMATICS

Family MICHELEIDAE Sakai, 1992

Michelea Kensley & Heard, 1991

Michelea takeda sp. nov.

(figs. 1-3)

Material examined. — Holotype: Y62A-6/MBM210244, ♂ (cl, 4.0 mm), Beibu Gulf, 19.30°N 108°E, depth 107 m, bottom muddy sand, coll. Zhengang Fan, 9 January 1960.

Description of male holotype. — Adult male. Rostrum (figs. 1, 2A) small, flat, not reaching midlength of eyestalks, about 0.4 times as long as broad at base, subacute tip. Carapace without dorsolateral carina running from each side of rostrum to posterior third of gastric region, bearing 2 pairs of anterolateral transverse setal rows, cervical groove distinct, reaching 0.5 length of cephalothorax, posterodorsal margin of carapace is acutely produced.

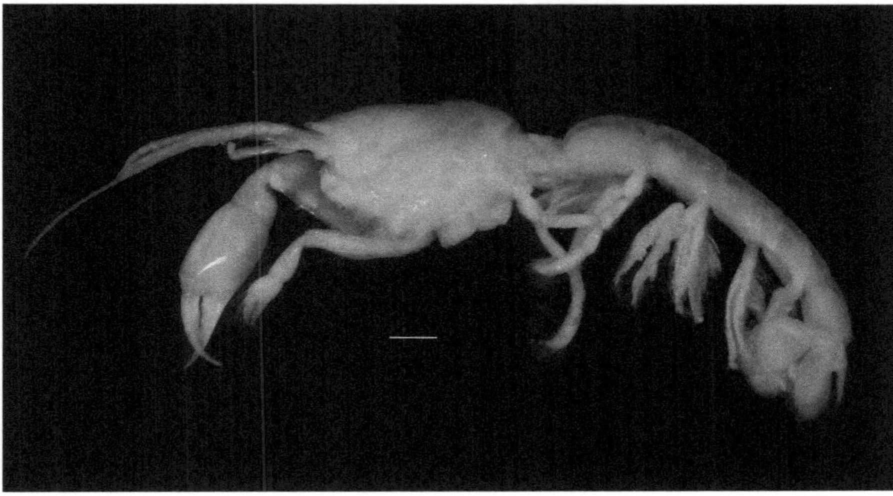

Fig. 1. *Michelea takeda* sp. nov., holotype male, Y62A-6/MBM210244, entire animal, lateral view. Scale = 1 mm.

Eyestalks (fig. 2A) slightly flattened, unarmed, and cornea is reduced into a small pigmented spot. Antennular peduncle not reaching distal margin of fourth segment of antennal peduncle. Antennal peduncle wide, unarmed; antennal scale well developed and dagger-like, located on upper surface of article 3, reaching level of proximal third of article 4.

Maxilliped 3 (fig. 2B) pediform, with sparse setae on lower margin; ischium oblong, 2.1 times as long as wide, with crista dentata on internal surface; merus 1.4 length of ischium, ventral margin with 1 small acute tooth arising distal third; carpus broadened distally, 2.8 times as long as wide, slightly longer than merus (upper margin); propodus oblong, 2.1 times as long as wide, 0.8 length of carpus; dactylus finger-like, 1.8 times as long as wide, 0.6 length of propodus. Exopod missing because of damage.

Pereopod 1 (fig. 3A) equal (punctuation should be changed). Ischium about as long as high; upper margin slightly concave and unarmed; lower margin straight, armed with a subterminally tooth (broken). Merus 2.8 length of ischium, about 1.8 times as long as high; upper margin strongly convex, unarmed; lower margin almost straight, with a sharp tooth located at midlength. Carpus triangular, about as long as high, 0.5 times as long as merus; upper margin almost straight; proximoventral margin broadly rounded, smooth. Chela heavy, palm about 1.5 times as long as high, subequal to (or slightly shorter than) merus; upper margin slightly convex, unarmed, lower margin convex, unarmed; fixed finger about 1.1 times as long as upper margin of palm, bear-

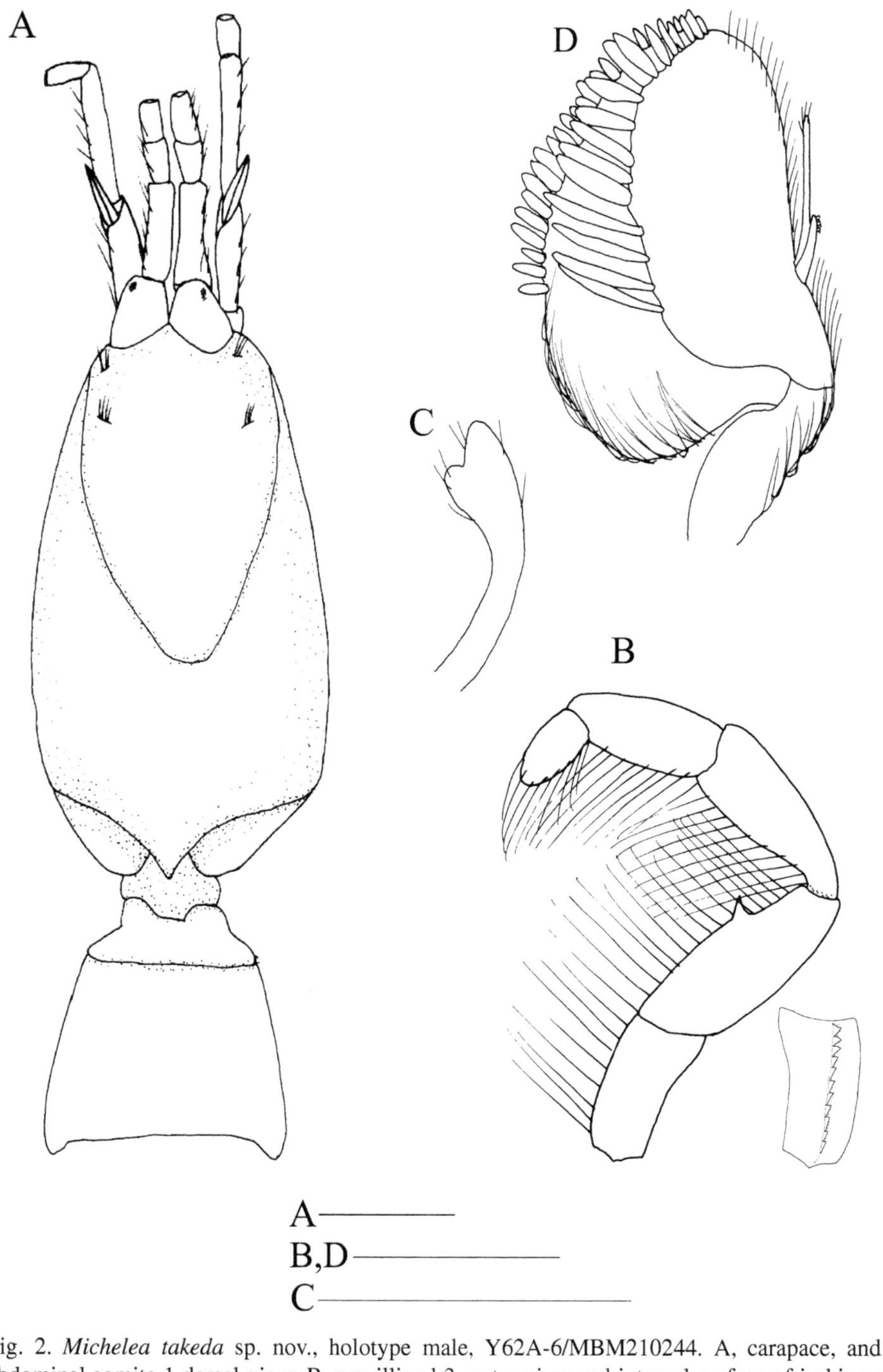

Fig. 2. *Michelea takeda* sp. nov., holotype male, Y62A-6/MBM210244. A, carapace, and abdominal somite 1 dorsal view; B, maxilliped 3, outer view and internal surface of ischium; C, male pleopod 1, posterior view; D, male pleopod 2, posterior view. Scales = 1 mm.

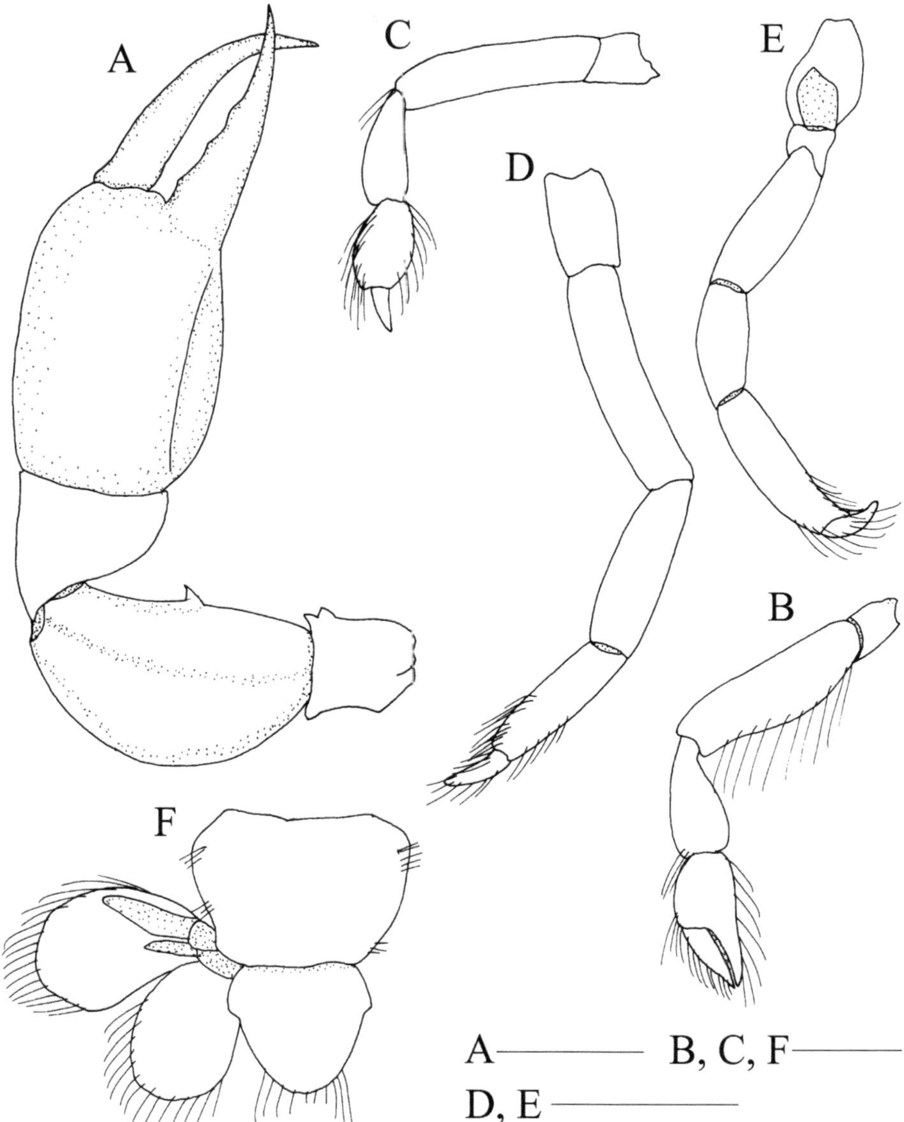

Fig. 3. *Michelea takeda* sp. nov., holotype male, Y62A-6/MBM210244. A, larger cheliped, lateral view; B–E, pereopod 2-5, lateral view; F, abdominal somite 6, telson and uropods, dorsal view. Scales = 1 mm.

ing three obtuse teeth on cutting edge; dactylus incurved distally, about as long as fixed finger, cutting edge unarmed.

Pereopod 2 (fig. 3B) chelate. Ischium unarmed. Merus 3.9 length of ischium, distoventral margin angular. Carpus broadened distally. Palm about

0.85 length of dactylus; cutting edge with fine and sharp chitinous spinules.

Pereopod 3 (fig. 3C) simple. Merus about 5.3 length of ischium. Carpus about 0.5 length of merus, broadened distally. Propodus ovate, nearly 1.4 times as long as wide, sparsely setose on both dorsal and ventral surface. Dactylus finger-like, 2.5 times as long as high, terminally acute.

Pereopod 4 (fig. 3D) longer than the third pereopod, merus about 2.3 length of ischium. Carpus 0.8 length of merus, lower margin slightly convex. Propodus shorter than carpus, dorsal and ventral surface distally with sparse long setae. Dactylus elongate, 2.8 times as long as high.

Pereopod 5 (fig. 3E) semichelate. Carpus 0.9 length of merus, slightly convex on upper margin. Propodus longer than carpus, about 1.3 length of carpus; fixed finger short, straight. Dactylus simply curving.

Abdominal somite 1 (fig. 2A) triangular posteroventrally, without setal rows; abdominal somite 2 longer than somite 3; abdominal somite 6 weakly rounded on ventral margin.

Male pleopod 1 (fig. 2C) uniramous, non-segmented, with a triangular distal blade, distal margin concave. Pleopod 2 (fig. 2D) biramous, foliaceous, endopod with 18 simple, non-articulated lamellae along distal margin to distal two thirds of outer margin, mesial magin with appendices interna and masculina, appendix masculina 2.5 length of appendix interna; exopod considerably wider than endopod, with 20 simple lamellae along distal margin to distal two thirds of outer margin, lamellae smaller than those on endopod. Pleopods 3-5 essentially similar to pleopod 2.

Telson (fig. 3F) 0.95 length of abdominal somite 6, maximum width about 1.2 times as wide as long, trapezoidal; lateral margin with distinct, angular lobe proximally; dorsal surface unarmed; posterior margin rounded, no lateral angle defined. Uropodal endopod ovate, 1.4 times as long as wide, bearing marginal seta; exopod ovate and longer than endopod, bearing marginal seta.

Gill formula is identical to *M. leura* (see Poore & Griffin, 1979).

Etymology. — The species is named in honour of Dr. Masatsune Takeda (National Museum of Nature and Science, Tokyo) for his great contribution to the world carcinology. Used as a noun in apposition.

Remarks. — *Michelea takeda* sp. nov. bears some resemblance to *M. leura* (Poore & Griffin, 1979) from Australia in the gill formula and lamellae in pleopods 2-5, but it differs in several aspects: the rostrum is shorter and wider in the new species than in *M. leura*; the cervical groove is distinct in *M. takeda* vs. indistinct in *M. leura*; the lower margin of the palm of the first pereopod (cheliped) is distinctly convex in *M. takeda* vs. straight in *M. leura*; the telson bears proximal lateral lobes in the new species vs. absent in *M. leura*. The new species is also generally similar in rostrum and telson shape to *Micheleopsis orlik* Sakai, 2010, also found in the Beibu Gulf. It can be distinguished from the latter by the carapace being devoid of dorsolateral carinae running from the lateral rostral margins to the gastric region and the lower margin of the merus in pereopod 1 with a single tooth (vs. with 2 small teeth in *Micheleopsis orlik*).

Distribution. — Known only from the type locality.

ACKNOWLEDGMENTS

This work was supported by the Knowledge Innovation Program of Chinese Academy of Sciences under contract No. KSCX2-EW-Z-8. We thank Dr. Peter Kee Lin Ng for inviting us to contribute to the special volume for Takeda in Crustaceana Monographs.

REFERENCES

GORE, R. H., 1987. *Callianidea vandoverae* species nova (Decapoda, Thalassinidea, Callianideidae) from off the central eastern Florida coast, U. S. A. Crustaceana, **53**: 186-194.

KENSLEY, B. & R. W. HEARD, 1991. An examination of the shrimp family Callianideidae (Crustacea: Decapoda: Thalassinidea). Proceedings of the Biological Society of Washington, **104**: 493-537.

POORE, G. C. B., 1997. A review of the thalassinidean families Callianideidae Kossman, Micheleidae Sakai, and Thomassiniidae de Saint Laurent (Crustacea, Decapoda) with descriptions of fifteen new species. Zoosystema, **19**: 345-420.

— —, 2008. Thalassinidean shrimps (Crustacea: Decapoda) from north-western Australia, including five new species. Records of the Western Australian Museum, (Supplement) **73**: 161-179.

POORE, G. C. B. & D. J. G. GRIFFIN, 1979. The Thalassinidea (Crustacea: Decapoda) of Australia. Records of the Australian Museum, **32**: 217-321.

SAKAI, K., 1987. Two new Thalassinidea (Crustacea: Decapoda) from Japan, with the biogeographical distribution of the Japanese Thalassinidea. Bulletin of Marine Science, **41**: 296-308.

— —, 1992. The families Callianideidae and Thalassinidae with the description of two new subfamilies, one new genus and two new species (Decapoda: Thalassinidea). Naturalists, Publications of Tokushima Biological Laboratory, Shikoku University, **4**: 1-33.

— —, 2010. Callianassoidea from the Gulf of Tonkin and the Red Sea, in the Zoological Museum of Moscow University (Decapoda, Thalassinidea). Crustaceana, **83**: 1431-1467.

First received 3 July 2011.
Final version accepted 3 February 2012.

ANOTHER SHELL-CARRYING DROMIID CRAB, *DESMODROMIA TRANTERAE* McLAY, 2001, FROM THE DAMPIER ARCHIPELAGO, WESTERN AUSTRALIA AND OBSERVATIONS OF SHELL-ACQUISITION BEHAVIOUR OF *CONCHOECETES ARTIFICIOSUS* (FABRICIUS, 1798) (DECAPODA, BRACHYURA, DROMIIDAE)

BY

COLIN L. McLAY[1,3]) and ANDREW HOSIE[2,4])
[1]) Biological Sciences, Canterbury University, Christchurch, 8004, New Zealand
[2]) Western Australian Museum, Welshpool DC, WA 6106, Australia

ABSTRACT

Recent field surveys have yielded more specimens of *Desmodromia tranterae* McLay, 2001, from Western Australia confirm that it carries bivalve shells as a means of camouflage. We also provide the first account of shell-acquisition behaviour by *Conchoecetes artificiosus* (Fabricius, 1798). We compare the morphological adaptations of the three known shell-carrying dromiid genera *Conchoecetes*, *Desmodromia* and *Hypoconcha*. The species in these genera each have different ways of grasping bivalve shells, although *Desmodromia* is most similar to *Hypoconcha* from tropical America. We discuss the evolution of shell-carrying in the Dromiidae.

INTRODUCTION

Shell-carrying by crabs from the family Dromiidae has been reported in two genera: *Hypoconcha* Guérin-Méneville, 1854 (six species) from tropical America and *Conchoecetes* Stimpson, 1858 (three species) from the Indo-Pacific (Wicksten, 1986a; Guinot et al., 1995; Guinot & Tavares, 2000). The discovery of two new species of unusual dromiid crabs in the genus *Desmodromia* McLay, 2001, from Western Australia and Darwin, Australia lead to speculation that these crabs may also carry bivalve shells (McLay, 2001).

[3]) Corresponding author; e-mail: Colin.mclay@canterbury.ac.nz
[4]) e-mail: Andrew.hosie@museum.wa.gov.au

New material from Western Australia (WA), reported here, confirms that *Desmodromia tranterae* McLay, 2001, does indeed carry bivalve shells. The way in which this crab carries shells for camouflage is illustrated. We also describe how *Conchoecetes artificiosus* (Fabricius, 1798) from the same habitat carries and acquires new shells. There are some similarities in the way shells are carried by *Desmodromia*, *Conchoecetes* and *Hypoconcha*, but also some significant differences that may imply that shell-carrying in the dromiids has evolved more than once. We provide a summary of relevant morphological evidence and a hypothesis about the evolution of camouflage in dromiid crabs. The morphological and behavioural characters will need to be eventually mapped on to a molecular phylogeny.

Specimens that were used for this study came from the Western Australia Museum (WAM), Queensland Museum, Brisbane (QM) and the Allan Hancock Collection, now held by the LA County Museum, Los Angeles (AHF). The following abbreviations are used: CW is maximum carapace width, CL is carapace length (dimensions given as CW × CL); P1 is the cheliped, P2–P3 are the walking legs and P4–P5 are the shell-carrying legs; G1 and G2 are the first and second male gonopods. Live *Conchoecetes artificiosus* were dredged from Moreton Bay, Brisbane.

Desmodromia tranterae McLay, 2001
(fig. 1)

Material examined. — QM-W16119: Northwest Shelf, WA, epibenthic sledge, 19°59.0′S 117°51.4′E, collected by R. V. 'Soela', CSIRO, 42 m, 18 February 1983; 2 ovig. females, damaged, no pereopods, 8.0 × 7.9 mm (12 eggs, diameter = 0.9 mm), and 7.9 × 8.0 mm (20 eggs, diameter = 0.85 mm). QM-W16153: Northwest Shelf, WA, station, trawled, 19°03.5′S 119°03.6′E, collected by R. V. 'Soela', CSIRO, 81 m, 28 March 1983; 1 immature female, damaged, 6.3 × 6.0 mm. WAM C25544: Dampier Archipelago, Angel Island, dive 8.5 m, 20°27.96′S 116°49.69′E, collected by M. Hewitt et al., 29 October 1998; habitat: many sponges & soft corals on pavement, coarse shelly sediment, some hard coral (faviids & *Turbinaria* Oken, 1815); 1 male 7.9 × 8.5 mm, no accompanying shell, several limbs missing. WAM C25771: Dampier Archipelago, North Coast of Goodwyn Is., 20°32′S 116°32.7′E, collected by hand in intertidal by M. Hewitt, 31 August 1999; habitat: wide intertidal limestone platform with algal turf, large pools, some loose rocks, many red and green algae; 1 male 10.2 × 9.8 mm, accompanying shell is 20.5 mm wide × 26.2 mm long; shell carried is *Vasticardium vertebratum* (Jonas, 1844) (Cardiidae).

Remarks. — The first specimen of *Desmodromia tranterae* reported by McLay (2001), was an immature female, but the present material includes ovigerous females and the first male so we first augment the original description herein. The sternal sutures 7/8 end apart on tubercles between the

Fig. 1. *Desmodromia tranterae* McLay, 2001, WAMC25771, male 10.2 × 9.8 mm: A, dorsal view without shell; B, ventral view without shell; C, frontal view without shell; D, ventral view showing crab grasping hinge line of *Vasticardium vertebratum* (Jonas, 1844) with P4's. Shell size is 20.5 mm wide × 26.2 mm long. The crab is sharing the space inside the shell with a sponge growing near the ventral edge of the shell covered with sand grains, micro-gastropods and coral colonies.

P2 bases and the P1, as is found in *D. griffini* McLay, 2001. The first male gonopods (= G1) are stout and flattened, tipped with long setae, while the second gonopods (= G2) are long, needle-like and approximately as long as the G1. Most dromiids have this kind of gonopods. Vestigial male pleopods on abdominal somites 3-5 are absent. The ovigerous females had only small broods, 12-20 eggs, of relatively large eggs, diameter 0.8-0.9 mm. This species may have abbreviated development with fewer zoeal stages. Previously known depth range of 9-15 m is considerably extended to intertidal — 81 m. The shell, *Vasticardium vertebratum*, carried by the larger male, was complete but showed some wear and tear and had been colonized near its ventral edge by a small (7.7 × 7.0 mm) mud-coloured sponge that had itself been colonized by three colonies of a red branching coral (see fig. 1D). The sponge provided a matrix in which were embedded coral and coralline algae fragments, micro-gastropods, and sand grains. The dimensions of the live red corals were 2.4 × 2.2 mm, 4.1 × 1.5 mm, and 5.5 × 2.4 mm. The sponge occupied ~10% of the shell ventral surface and the corals covered ~46% of the sponge surface. The remaining space under the shell, between the hinge-line and the sponge, was neatly filled by the dimensions of the crab, suggesting that both had been resident on the shell for a long period of time.

Desmodromia tranterae carries its shell using the last two pairs of pereopods. The P4 is much shorter than the P5 with all of the articles reduced except for the coxae. The ridged merus of the P4 is directed laterally, followed by the carpus which is rotated so as to be directed ventrally and then the propodus is rotated to point postero-laterally, with the result that the short hooked dactylus is directed upwards. The P5 is inserted postero-dorsally and all the articles are flattened; the limb lies on top of the carapace, with the merus directed anteriorly and the meral-carpal joint almost forms a right angle, with the carpus and propodus directed laterally, and the short hooked dactylus is directed antero-dorsally depending upon whether it is extended out of its socket. The P4 and P5 in this species do not have a sub-chelate mechanism for shell carrying.

The occurrence of both a sponge + coral and a crab, which comes from a crab family in which sponge-carrying is common, occupying the same shell, is curious indeed. However, it seems unlikely that the crab is responsible for the presence of the sponge + coral and much more likely that it is all a curious coincidence, although whether the crab or the sponge was first to arrive is impossible to know.

McLay (2001) recognized two species of *Desmodromia*: *D. griffini* and *D. tranterae*, mainly because of differences in the shape of the carapace anterolateral margin. However, it is possible that these differences only reflect individual variation resulting from the crabs moulting under bivalve shells whose inner surface sculpturing influence the carapace shape as it hardens.

Conchoecetes artificiosus (Fabricius, 1798)
(fig. 2)

Material examined. — WAM C26222: Dampier Archipelago, Cape Legendre, 36-33 m, rake box dredge, 20°21.69′S 116°52.40′E–20°21.23′S 116°52.36′E, 14 July1999; habitat: coarse muddy sand and limestone rocks; gorgonians, sponges; male 13.5 × 13.7 mm accompanying shell is 39 mm wide × 23 mm long; shell carried is *Cardita marmorea* Reeve, 1843 (Carditidae).

Remarks. — The last two pairs of pereopods in *Conchoecetes artificiosus* are reduced compared to the P2 and P3, but the P4 is much stronger than the P5. The P4 merus is directed laterally while the carpus is directed ventrally and the propodus is rotated posteriorly resulting in the stout hook-like (or talon-like) dactylus curving upwards. When closed the tip of the dactylus meets a projection of the distal propodal margin forming a sub-chelate mechanism to grasp the shell hinge line. The P5 is inserted postero-dorsally and lies on the dorsal surface, extending to about half the carapace length; the merus is directed anteriorly, while the rest of the limb is directed at about 45° with the carpus and propodus becoming progressively shorter and terminated by the small hooked, curved dactyl. As in *D. tranterae*, the P5 coxae are not reduced. This is directed laterally although the claw points antero-vertically. The abdomen cannot be involved in control of the shell because the talons on the P4 extend posteriorly beyond the abdominal margin.

Observations were made on shell use by *C. artificiosus* collected from Moreton Bay, Brisbane, Australia. Four of the crabs collected (2 males and 2 females, CW 8.2-21 mm) were kept in aquaria or white plastic trays 35 × 30 cm with sand and fed pieces of squid and mussels weekly. These crabs moulted and grew during a period of ~6 months. Bivalve shells used by *C. artificiosus* include *Anadara trapezia* (Deshayes, 1840), *Circe plicatum* (Lamarck, 1816), *Dosinia sculpta* (Hanley, 1845), *Lunulicardia hemicardium* (Linnaeus, 1758), *Mactra abbreviata* Lamarck, 1819, *Mimachlamys gloriosa* (Reeve, 1853), *Paphia undulata* (Born, 1780), *Pitar affinis* (Gmelin, 1791), *Plagiocardium setosum* (Redfield, 1848), *Tapes dorsatus* (Lamarck, 1818) and *Vasticardium vertebratum* (Jonas, 1844). The crabs were given a random variety of labelled shells to use, which came from the same habitat, and the average number of days that a crab resided under the same shell was 11.4 d (range 9-14 d). All shells were larger than the crabs themselves, thereby offering complete cover. Shell-change was not the result of deterioration in

Fig. 2. *Conchoecetes artificiosus* (Fabricius, 1798), WAMC26222, male 13.5 × 13.7 mm: A, dorsal view without shell; B, ventral view without shell; C, frontal view without shell; D, ventral view showing crab grasping the hinge line of *Cardita marmorea* Reeve, 1843, with P4's. Shell size is 39.0 mm wide × 23.0 mm long.

a shell or eviction by a competitor, thereby forcing crabs to move on to a new shelter, because sometimes crabs returned to the same shells that they had previously vacated. As hazardous as it might appear to be, shell-change seemed to be a customary habit. Having discarded their old exoskeleton, post-moult crabs remained under the same shell until the new exoskeleton hardened.

The behaviour used when changing shells depended upon whether the new shell was concave side down or concave side up. In the former case, crabs sometimes did so by crawling under a new shell (usually larger), still carrying the old one, so the manner in which the change was affected was not apparent. Sometimes the crab crawled over a new shell for 30-40 min beforehand and then, lifted up the edge of the shell, before quickly discarding the old one and slipping underneath. However, other crabs who encountered an upside-down shell (concave side up) first of all investigated it by stroking the chelae and walking legs over the surface, then discarded their old shell, turned themselves upside-down and grabbed the shell margin with their P4's. Now both the crab and the shell were lying exposed on the bottom. From this position, the crab then elevated itself, reaching out with the chelae and pushing with its walking legs in the hope that its body weight would be enough to tip the shell over on top. If the crab had first grasped the shell margin opposite to the hinge line then it was not normally able to overturn the shell despite repeated attempts. So during the course of what might take an hour or more, the crab would move around the edge of the shell and try again, sometimes pausing for a 'bit of a lie down'. Of course the only part of the shell that can be used to tip the shell over is the hinge line because on the outside of the shell this is where the curvature is great enough that the crab can roll the shell over using its own weight to get the shell past the vertical and thus making it over-balance and fall on top! In other cases the crab would spend more time investigating the shell to find the hinge line before turning itself over and grabbing it correctly first time. It may be that some shells are more suitable than others and that the crabs can recognize where the hinge line is located.

Hypoconcha lowei Rathbun, 1933

(fig. 3)

Material examined. — AHF 1972-17: Gulf of California, 16 km east of San Felipe, approx. 31°1.39′N 113°50.07′W, 13-24 m, collected by E. Norse, 3 November 1971, identified by Mary Wicksten; 1 female 27.6 × 24.9 mm, 1 male 29.0 × 27.0 mm; no accompanying shells.

Remarks. — *Hypoconcha lowei* carries its bivalve shell using the P4–P5 and also uses the angular posterior part of the abdomen to fit under the shell hinge-line. The P4 is ventral and shorter (~0.5 × CL), than P5 (~0.6 × CL). All articles are reduced in length and the limb turns a right angle at the meral-carpal joint so that the rest is directed posteriorly with the lunate dactylus rotated so as to lie horizontal. The claw of the dactylus points vertically and it can swivel laterally so as to catch the hinge-line of the shell. The flattened

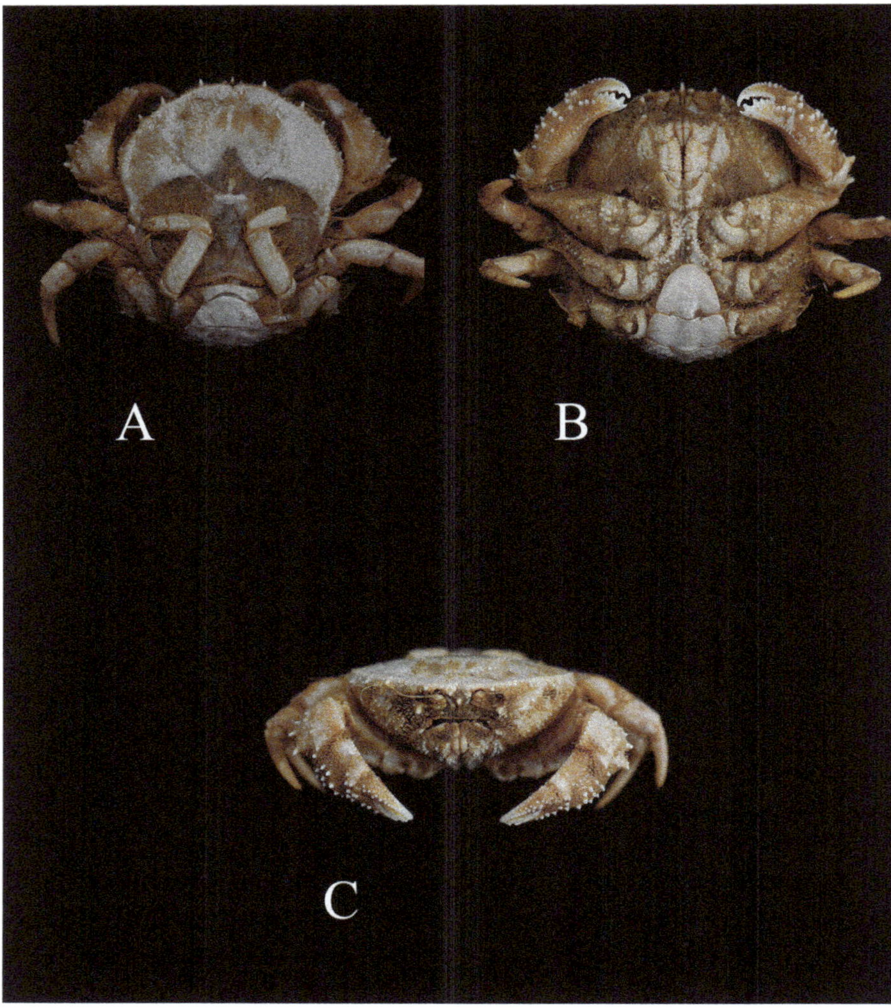

Fig. 3. *Hypoconcha lowei* Rathbun, 1933, AHF1972-17, female 27.6 × 24.9 mm: A, dorsal view without shell (note propodus and dactyl of right P5 are missing); B, ventral view without shell; C, frontal view without shell.

P5 is dorsal, forms a right angle at the meral-carpal joint so that the rest of the limb is directed laterally, with the lunate dactylus lying in the horizontal plane and its claw directed vertically. A slot in the propodus houses the dactylus which can be rotated distally to grasp the edge of the shell. As with the other shell-carriers the P5 coxae are not reduced. The P4 and P5 in this species do not have a sub-chelate mechanism for shell carrying. An unusual feature of all *Hypoconcha* species is that part of the dorsal carapace surface is membranous. In *H. lowei* this makes up ~40% of the surface, although this includes a small calcified cardiac area that protects the heart. The whole area of the carapace is densely covered with long setae. Rathbun (1937: pl. 8 figs. 5, 6) illustrates *H. lowei* carrying a bivalve shell. Off the Californian coast, this species has been recorded as carrying an abalone shell (*Haliotis* sp.) (Wicksten, 1986b) thus providing the only example of a brachyuran crab making use of a gastropod for camouflage.

Comparison of shell-carrying in *Desmodromia*, *Conchoecetes* and *Hypoconcha*

A comparison of shell-carrying adaptations is presented in table I. Additional information about shell-carrying by *Conchoecetes artificiosus*, *Hypoconcha arcuata* Stimpson, 1858, and *H. panamensis* Smith, in Verrill, 1869, is presented by Guinot & Tavares (2000). The P5 is very similar for all three genera and the P4 is also similar in *Hypoconcha* and *Desmodromia* so that the major point of difference is the P4 of *Conchoecetes*. The sub-chelate P4 of this species is constructed differently from the sub-chelate limbs in other dromiids. Most dromiids have grasping limbs that are based on a ground plan of a short curved dactylus that is opposed by one or more sharp spines on the distal propodal margin. Camouflage is held by inserting the sharp spines into a piece of sponge or ascidian and gripping it by closing dactylus. However, in *Conchoecetes*, the stout talon-like dactyl is not closed against structures on the distal propodal margin, but against a projection on the proximal margin, which itself carries a short mobile process (see Guinot & Tavares, 2000). This mobile process is a unique structure that acts like a car shock-absorber that can absorb sudden movements of the shell thereby helping to maintain the grip on the hinge-line by the talon. *Hypoconcha* does not have sub-chelate grasping P4 or P5, but instead it relies on a close fit of the carapace to the shell and a suction-mechanism to hold the shell. *Desmodromia* has similar limbs to *Hypoconcha*, but does not use its abdomen, although this could depend upon the shape of the space available in different shells. The main difference is that the P1-3 in *Desmodromia* are moulded such that they all fit closely together and can be folded away tightly when at rest. In that way the crab can fit efficiently into the shelter afforded by the shell.

A feature of *Hypoconcha*, not shared by the other genera, is the presence of a decalcified area on the carapace: in *H. lowei* around 40% of the surface is membranous and flexible. As first suggested by Wicksten (1986b: 319), the membranous surface, along with the marginal setae, may make a tight seal between the shell and the crab, by acting like a suction device. This may

TABLE I
Comparison of the adaptations of the three shell-carrying dromiid genera

Character	*Hypoconcha*	*Conchoecetes*	*Desmodromia*
Carapace	As wide as long, posterior part membranous	Sub-pentagonal, strongly calcified throughout	Sub-pentagonal, weakly calcified
Supra-orbital margin	Beneath edge of carapace, eyes not visible dorsally	Not eave-like, eyes visible dorsally	Eave-like, covering eyes in dorsal view
Antero-lateral carapace margin	Not eave-like, evenly curved, may be interrupted by small spines	Not eave-like, teeth not well developed, usually absent	Eave-like, divided into irregular teeth
Carapace surface	Smooth, densely covered with long setae	Smooth evenly covered in a tomentum of minute brown setae	Smooth, not setose
Abdomen	Involved in carrying or holding shell; angular abdomen inserted under hinge of shell	Not involved in carrying or holding shell	Fits closely to hinge line of shell, but not involved in carrying or holding shell
Abdominal locking mechanism	None, uropods vestigial, abdomen only loosely folded ventrally	Uropods well developed, visible externally, fitting in front of serrated flange on coxae of P2	Uropods well developed, visible externally, fitting in front of small tubercles on coxae of P2
Walking legs P2 & P3	Well developed, longer than P1	Well developed, almost as long as P1	Well developed, almost as long as P1, margins crest-like
Shell carrying leg, P4	Directed posteriorly, length $\sim 0.5 \times$ CL, dactylus short, lunate and mobile, held against inner margin of shell hinge-line	Large talon-shaped, length $\sim 0.64 \times$ CL, dactylus directed posteriorly that grasps hinge-line of shell	Directed posteriorly, length $\sim 0.4 \times$ CL, dactylus short, curved and mobile, held against inner margin of shell hinge-line
Shell carrying leg, P5	Dorsal, directed anteriorly, length $\sim 0.6 \times$ CL, dactylus lunate and mobile, holds shell margin	Dorsal, directed anteriorly, length $\sim 0.5 \times$ CL, dactylus notched and mobile, holds shell margin	Dorsal, directed anteriorly, length $\sim 0.8 \times$ CL, dactylus notched and mobile, holds shell margin

explain how *Hypoconcha* clings so tightly to its shell such that removal of the shell without damaging the crab is almost impossible (Williams, 1984: 257). This is reminiscent of how hermit crabs make use of gastropod shells by their soft flexible abdomen. The shell also provides protection when the crab moults: captive *H. sabulosa* remain under the shell while moulting (Wicksten, 1986b: 320). The same is probably true for *C. artificiosus* because newly moulted crabs are found under the same shell as they were pre-moult.

DISCUSSION

Shell-carrying by crabs is a form of camouflage whose adaptive advantage is to reduce mortality caused by predators. Use of camouflage allows crabs to live in places where they would not normally be found and at times when they would normally be inactive. For example, dromiids can feed on sponges during the day as long as they are concealed under a matching "sponge-umbrella", which may also have toxic properties (McLay, 1983). Shell-carrying dromiids can also be active during the day, feeding on substrates where bivalve shells are abundant, without being attacked by predators.

McLay (1993: 228-229) has presented an argument, based on morphological features, that the use of camouflage is ancestral for dromiids: in brief the crabs that have the most gills, and cleaning epipods, have the most elaborate development of spines on their P4 and P5 (e.g. *Sphaerodromia* spp.) which are used to carry soft-bodied camouflage such as sponges and ascidians. Other characters including vestigial male pleopods and female sternal sutures 7/8, analysed by Guinot & Tavares (2003), also support the hypothesis that *Sphaerodromia* retains many ancestral features. Polarity within the family is provided by species of *Takedromia* McLay, 1993, which lack prehensile P4 and P5, never carry camouflage and have fewer gills. Species of this genus have a body that has become more heavily calcified and the carapace adorned with decorative granules and areolae resembling pieces of broken coral fragments (McLay, 1993: 229).

The argument for the evolution of camouflage in these podotreme crabs clearly assumes that the common ancestor of the dromiids was not a crab that had both P4 and P5 reduced, but may have carried pieces of sponge broken off by other damage. It seems likely that the sub-chelate limbs evolved from ordinary walking legs by the development of propodal spines that could be opposed to the dactylus thereby creating a grasping mechanism that facilitated the holding of camouflage. The P2 and P3 of extant dromiids have all the precursors needed for such limbs so no new structures had to be evolved. The next evolutionary step could have been the behaviour that was used to choose and cut out the "umbrella" and fashion pieces of soft animals to fit their requirements. The polarity of camouflage characters obviously depends upon where the boundary between dromiids and non-dromiid ancestors is drawn.

Therefore we are postulating that the limbs now used to carry shells evolved from limbs formerly used to carry soft camouflage. This hypothesis will need to be tested by mapping the characters on to a cladistic phylogeny of molecular characters that may provide confirmation/rejection by using independent evidence.

The proposed analysis using molecular characters may allow us to answer the question as to how many times shell-carrying evolved amongst dromiids. The comparison that we have made between the dromiid shell-carrying genera shows that the P4 and P5 of *Hypoconcha* and *Desmodromia* hold the shell using lunate-shaped dactyli that are braced against the shell hinge or margins. Only *Conchoecetes* still uses an accentuated sub-chelate mechanism on the P4 (reminiscent of the sponge-carriers), which almost solely hold the shell, while the P5 only has a minor supporting or stabilizing role. The structure of the P4 is a stout claw-like dactyl that is opposed to a unique proximal propodal structure. The P4 of *Conchoecetes* looks remarkably like the P4 and P5 of the dorippid *Paradorippe* spp. which also carries bivalve shells (Holthuis & Manning, 1990). In most respects *Desmodromia* closely resembles other sponge crabs like *Cryptodromia* for example, except for the P4 and P5. Finally, although *Hypoconcha* has very similar P4 and P5 limbs it has evolved a semi-membranous carapace, which probably enhances its ability to hold on to its shell. No other dromiids have a carapace like this.

While we know something about the morphological adaptations used for shell-carrying habit, very little is known about how the behaviour. *Conchoecetes* is the only genus in which shell use and shell change has been studied and the behaviour shows some similarity to that seen in sponge-carriers like *Cryptodromia hilgendorfi* (McLay, 1983). For example, the pattern of use of pereopods in cutting out a sponge umbrella and then donning of the new camouflage bears some resemblance to the shell-investigation of *C. artificiosus* that precedes it backing into the new shell and grasping the hinge line. In both *Cryptodromia* and *Conchoecetes*, camouflage changes occur during the intermoult period. However, while the average residence time under a shell for *Conchoecetes artificiosus* was 11.4 d, the sponge crab *Cryptodromia hilgendorfi* used the same sponge cap on average for 33.1 d (McLay, 1983), perhaps reflecting the difference in effort required to fashion a new cap matching the background compared to just picking up any available shell that provided concealment. Unfortunately the shell-acquisition behaviour of the other two genera has not been investigated so it is not possible to say whether it is a shared character. In dorippids (Eubrachyura), we also see prehensile carrying-limbs that are very similar to the P4 of *Conchoecetes*, clearly an example of convergent evolution.

Use of bivalves by crustaceans has also evolved in hermit crabs that normally use gastropod shells (Borradaile, 1916). Perhaps the best known example is *Porcellanopagurus* Filhol, 1885 (13 species), as well as *Solitariopagurus* Türkay, 1986 (four species) and *Alainopagurus* Lemaitre & McLaughlin, 1995 (one species). Adoption of this habit is linked to carcinization in anomurans that involves reduction of the abdomen under the cephalothorax (McLaughlin & Lemaitre, 1997). As in brachyurans, the abdomen is inserted under the hinge line but hermit crabs also use their rasp-like uropods to grasp the shell. Instead of reduced sub-chelate P4 and P5, hermit crabs have reduced P4 and P5 with propodal rasps. In hermit crabs, the shell does not usually cover the entire crab, but at least in females it protects her brood. Shell-carrying and carcinization are adaptive responses to predation and the shortage of places to hide and find shelter.

ACKNOWLEDGEMENTS

We offer this paper as a token of our gratitude and admiration for the contribution that Professor Masatsune Takeda has made to our knowledge about Crustacea of the Indo-Pacific. We are grateful to Shirley Slack-Smith for identifying the bivalve shells that were carried by *Desmodromia tranterae*. The first author wishes to acknowledge the help of the Bill Stephenson and the staff of the Zoology Department for laboratory space and the use of facilities in 1977. Crabs from Dampier Archipelago were collected during the WAM Dampier Archipelago Surveys, funded by Woodside Energy Pty Ltd. Moreton Bay crabs were collected using the research trawler "Wanderer II" and I wish to thank Greg Campbell, Dylan Webb and Stephen Cook for their assistance. Merrill Williams also provided valuable help in finding crabs. Also Bruce Campbell of the Queensland Museum kindly provided help with identifying animals collected from Moreton Bay.

REFERENCES

BORRADAILE, L. A., 1916. Crustacea. Part II. — *Porcellanopagurus*: an instance of carcinization. British Antarctic ("Terra Nova") Expedition, 1910. Natural History Report, **3**(3): 111-126.

GUINOT, D., D. DOUMENC & C. C. CHINTIROGLOU, 1995. A review of the carrying behaviour in brachyuran crabs, with additional information on the symbioses with sea anemones. Raffles Bulletin of Zoology, **43**: 377-416.

GUINOT, D. & M. S. TAVARES, 2000. *Conchoeodromia alcocki* Chopra, 1934: megalopa of *Conchoecetes artificiosus* (Fabricius, 1798) (Decapoda, Brachyura, Dromiidae). Journal of Crustacean Biology, **20**(special number 2): 301-309.

— — & — —, 2003. A new subfamilial arrangement for the Dromiidae de Haan, 1833, with diagnoses and descriptions of new genera and species (Crustacea, Decapoda, Brachyura). Zoosystema, **25**: 43-129.

HOLTHUIS, L. B. & R. B. MANNING, 1990. Crabs of the subfamily Dorippinae MacLeay, 1838, from the Indo-West Pacific region (Crustacea: Decapoda: Dorippidae). Researches on Crustacea, (Special number) **3**: 1-151.

MCLAUGHLIN, P. A. & R. LEMAITRE, 1997. Carcinization in the Anomura — fact or fiction? I. Evidence from adult morphology. Contributions to Zoology, **67**: 79-123.

MCLAY, C. L., 1983. Dispersal and use of sponges and ascidians as camouflage by *Cryptodromia hilgendorfi* (Brachyura: Dromiacea). Marine Biology, **76**: 17-32.

— —, 1993. Crustacea Decapoda: the sponge crabs (Dromiidae) of New Caledonia and the Philippines with a review of the genera. In: A. CROSNIER (ed.), Résultats des Campagnes MUSORSTOM, Volume 10. Mémoires du Muséum national d'Histoire naturelle, Paris, **156**: 111-251.

— —, 2001. A new genus and two new species of unusual dromiid crabs (Brachyura: Dromiidae) from Northern Australia. Records of the Australia Museum, **53**: 1-8.

RATHBUN, M. J., 1937. The oxystomatous and allied crabs of America. United States National Museum Bulletin, **166**: 1-278.

WICKSTEN, M. K., 1986a. Carrying behavior in brachyuran crabs. Journal of Crustacean Biology, **6**: 364-369.

— —, 1986b. Shell-carrying in *Hypoconcha sabulosa* (Herbst, 1799) (Decapoda, Brachyura). Crustaceana, **50**: 19-20.

WILLIAMS, A. B., 1984. Shrimps, lobsters, and crabs of the Atlantic coast of the eastern United States, Maine to Florida: 1-566, figs. 1-380. (Smithsonian Institution Press, Washington).

First received 16 September 2011.
Final version accepted 25 January 2012.

A NEW GENUS AND SPECIES OF DEEP-WATER XANTHID CRAB (DECAPODA, BRACHYURA, XANTHIDAE) FROM THE PHILIPPINES

BY

JOSE C. E. MENDOZA[1,3]) and PETER K. L. NG[1,2,4])

[1]) Department of Biological Sciences, National University of Singapore, 14 Science Drive 4, 117543 Singapore

[2]) Tropical Marine Science Institute, National University of Singapore, Blk. S2S, 18 Kent Ridge Road, 119227 Singapore

ABSTRACT

A new genus and species of deep-water xanthid crab is described from the Philippines. *Takedax gabrielae* gen. et sp. nov. resembles *Medaeops* Guinot, 1967, in the cristate anterior margins of the ambulatory meri. It is also similar to *Monodaeus* Guinot, 1967, in the long cheliped fingers, the shape of the ambulatory legs, and the form of the merus of the third maxilliped. *Takedax* gen. nov. is separated from both known genera by a suite of characters related to the carapace, epistome, pereiopods, male thoracic sternum, and male abdomen.

INTRODUCTION

In the period from 2000 to 2008, several crabs of an unknown species were collected from deep waters around the islands of Balicasag and Panglao, in the Bohol Sea; off the eastern coast of Quezon Province, in the Philippine Sea; and off the western coast of Batangas Province, near the Verde Island Passage. These crabs were initially believed to be close to *Medaeops granulosus* (Haswell, 1882) due to similarities in the general form of the carapace and the cristate anterior margins of the ambulatory meri. Subsequent comparisons with *M. granulosus* showed that the specimens represented an undescribed species, with the long fingers of the chelae and relatively longer ambulatory legs suggesting an affiliation with *Monodaeus* Guinot, 1967, a genus known

[3]) Corresponding author; e-mail: jcmendoza@nus.edu.sg

[4]) e-mail: peterng@nus.edu.sg

only from the Atlantic Ocean, Mediterranean Sea and Western Indian Ocean (Guinot & Macpherson, 1988). However, although the new species exhibited characters shared by *Medaeops* and *Monodaeus*, making assignment to either genus difficult, a recent and comprehensive multi-gene phylogenetic analysis of the Indo-West Pacific Xanthidae indicated that it belongs in neither genus (see Lai et al., 2011, as "New genus 2, sp. nov."). This observation, together with its unusual suite of morphological characters, make it necessary to establish a new genus for this new species.

Measurements of the present material are expressed as carapace width by carapace length in millimeters. The nomenclature of the carapace regions follows that of Dana (1851) and Serène (1984). The following abbreviations are used: P1–P5 = pereopods 1-5, respectively (P1 = chelipeds, P2–P5 = ambulatory legs), G1 = male first gonopod, G2 = male second gonopod, coll. = collected by, is. = island, juv. = juvenile, ovig. = ovigerous, purch. = purchased by, stn = station. The thoracic somites and sternites are numbered from 1 to 8. The thoracic sternal sutures are referred to by the numbers of the two sternites involved, e.g., 1/2 to 7/8. The specimens used in this study are deposited in the Muséum national d'Histoire naturelle, Paris (MNHN), the Natural History Museum, London (NHM), the Crustacean Reference Collection, National Museum of the Philippines, Manila (NMCR), the Florida Museum of Natural History, University of Florida, Gainesville (UF), the National Museum of Natural History, Smithsonian Institution, Washington, D.C. (USNM), and the Zoological Reference Collection of the Raffles Museum of Biodiversity Research, Singapore (ZRC).

SYSTEMATIC ACCOUNT

Family XANTHIDAE MacLeay, 1838
Subfamily ANTROCARCININAE Ng & Chia, 1994
Takedax gen. nov.

Type species. — *Takedax gabrielae* sp. nov., by present designation. Gender masculine.

Diagnosis. — Carapace transversely subhexagonal; dorsal surface granulate, rugose, larger granules laterally; regions well defined, 2P with 2 distinct, dorso-laterally projecting, pyramidal tubercles. Front bilobed, moderately advanced beyond orbits; each lobe slightly convex, lateral angle slightly produced; separated laterally from preorbital tooth by deep notch that appears as small fossa in anterior view. Anterolateral margin convex, with 4 teeth, third tooth projecting furthest laterally, last tooth projecting posterolaterally; anterior margin indistinctly descending obliquely toward buccal cavity, marked by low, subhepatic, granulate tubercle. Posterolateral margins strongly converging posteriorly. Posterior carapace margin almost straight, raised, granulate. Subhepatic and pterygostomian regions with round granules. Basal article of antenna subrectangular, slender; flagellum much longer than width of orbit. Epistome thick posterior border projecting gently at median region. Endostome with incomplete ridges. Lacinia of first maxilliped moderately broad distally. Third maxillipeds not completely closing buccal cavity anteriorly; antero-external angle of merus slightly projecting. Male thoracic sternum with small portion of sternite 8 exposed when abdomen closed. Chelipeds unequal or subequal; fingers long, tips pointed, incurved; dactylus of major chela with or without modified proximal tooth. Ambulatory legs moderately long, slender, flattened; anterior margin of meri distinctly cristate. Male abdomen short, tip of telson posterior to level of condyles of cheliped coxae; disto-lateral angles of somite 6 produced; telson subtriangular. G1 moderately long; slightly carinate along lateral margin along central third; with several long plumose subterminal setae; apical lobe narrow, spatuliform; G2 about half length of G1.

Etymology. — This genus is named in honor of Professor Masatsune Takeda, in acknowledgement of his significant and wide-ranging works on the Brachyura. The name is an arbitrary combination of "Takeda", and "-x", from *Xantho*, the type genus of the family. Gender masculine.

Remarks. — *Takedax* gen. nov., is perhaps most similar to *Monodaeus* Guinot, 1967, in the long fingers of the chelipeds, the produced antero-external angle of the merus of the third maxilliped and the slightly exposed male thoracic sternite 8 when the abdomen is closed. However, it can be clearly distinguished from *Monodaeus* by: 1) the relatively longer, narrower male abdomen, with the tip of the telson just reaching a level immediately posterior to the coxosternal condyles of the P1 (fig. 1C) (vs. a shorter, broader male abdomen, tip of telson more posteriorly located, in *Monodaeus*); 2) the proportionately longer telson, which is generally as long as, or longer, than it is wide (fig. 1C) (vs. a shorter telson, broader than long, in *Monodaeus*, cf. Guinot & Macpherson, 1988: figs. 16, 18, 20); 3) the nearly straight or slightly convex lateral portions of the posterior border of the epistome (fig. 1B) (vs. prominently convex in *Monodaeus*, cf. Guinot, 1967: fig. 22); and 4) the presence of a short notch on the anterior and posterior borders of the vulva (fig. 2H) (vs. absent in *Monodaeus*). Several other characters

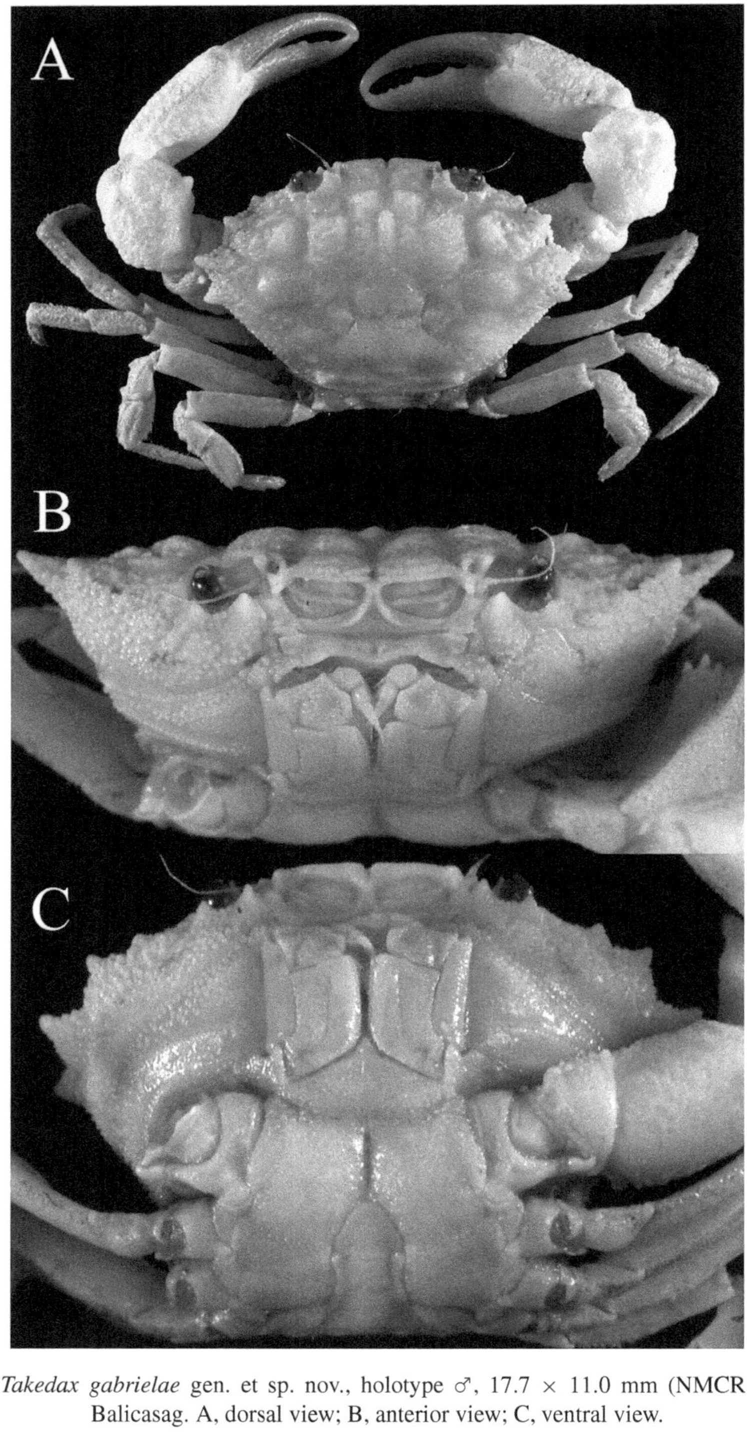

Fig. 1. *Takedax gabrielae* gen. et sp. nov., holotype ♂, 17.7 × 11.0 mm (NMCR 30932), Balicasag. A, dorsal view; B, anterior view; C, ventral view.

also distinguish *Takedax* from the known species of *Monodaeus*, but these may be significant only at the species level, viz. 1) the distinct, pyramidal tubercle on the lateral 2P region of the carapace (fig. 1A) (vs. indistinct or absent in *Monodaeus*, cf. Guinot & Macpherson, 1988: pls. 1-3); 2) the more strongly convergent posterolateral margins of the carapace (fig. 1A) (vs. less convergent in *Monodaeus*, cf. Guinot & Macpherson, 1988: pls. 1-3); 3) the relatively more slender ambulatory legs (figs. 1A, 2D) (vs. stouter in *Monodaeus*, particularly in *M. couchii* (Couch, 1851) and *M. cristulatus* Guinot & Macpherson, 1988, cf. Guinot & Macpherson, 1988: figs. 1-9, pls. 1-3); and 4) the relatively shorter distal, tapering portion of the G1 (fig. 3A, B) (vs. relatively longer in *Monodaeus*, cf. Guinot & Macpherson, 1988: fig. 21). For the moment at least, there is a clear separation in their known distributions, with *Monodaeus* restricted to the Eastern Atlantic, including the Mediterranean, and the Western Indian Ocean (Serène, 1984; Guinot & Macpherson, 1988) while *Takedax* is present in the western Pacific.

Takedax gen. nov. also resembles *Medaeops* Guinot, 1967, particularly *M. granulosus* (Haswell, 1882) (type species) in the cristate anterior margins of the ambulatory meri, the form of the G1 and the general form of the carapace; and the two have been mistaken for each other (e.g. Ho et al., 2004). However, *Takedax* can be distinguished from *Medaeops* s. str. by 1) the relatively longer and more slender ambulatory legs (figs. 1A, 2D) (vs. much shorter and stouter in *Medaeops*, cf. Mendoza et al., 2009: figs. 1A, 2A, B, 3B, D, F); 2) the longer and more slender cheliped fingers (fig. 2E, F) (vs. shorter and stouter in *Medaeops*, cf. Mendoza et al., 2009: fig. 3A, C, E); 3) produced antero-external angle of the merus of the third maxilliped (fig. 2A) (vs. not produced in *Medaeops*, cf. Guinot, 1967: fig. 21; Mendoza et al., 2009: fig. 2C, F); 4) the distinct, pyramidal tubercle on the lateral 2P region of the carapace (fig. 1A) (vs. absent in *Medaeops*, cf. Mendoza et al., 2009: figs. 1A, 2A, D); 5) the partially exposed male thoracic sternite 8 when the abdomen is closed (fig. 2C) (vs. not exposed in *Medaeops*); and 6) the relatively broader abdominal somite 6 (fig. 1C) (vs. narrower in *Medaeops*, cf. Mendoza et al., 2009: figs. 1C, 2C, F). Furthermore, although *Takedax* is found within the range of *Medaeops*, species such as *M. granulosus* and *M. neglectus* (Balss, 1922) are generally found in the lower intertidal and shallow subtidal zones (Serène, 1984; Davie, 2002), while *Takedax* is found at greater depths (from at least 50 to 356 m or more).

There is also molecular support for *Takedax* as a separate genus. Recently, a phylogeny of the Xanthidae (Lai et al., 2011: fig. 1), using five molecular markers and supported by morphological characters, showed that *Takedax* gen. nov. (as "New genus 2, sp. nov.") was immediately basal to a subclade (within clade "Eux 3") containing the genera *Monodaeus*, *Medaeops*, *Paramedaeus* Guinot, 1967, *Cranaothus* Ng, 1993, and *Danielea* Ng & Clark, 2003. While there was only moderate statistical support at this node, with a Bayesian posterior probability (BI) and Maximum Likelihood (ML) values at 0.84 and 52, respectively, the support for two of the nested clades, representing the genera *Medaeops* and *Monodaeus*, was very high (both with BI = 1, ML = 100). *Takedax* was consistently excluded from both these well-supported clades, corroborating our hypothesis that it is a distinct genus from *Monodaeus* and *Medaeops*. *Takedax* is currently represented only by the type species, *T. gabrielae* sp. nov.

Lai et al. (2011: 430, 431) commented that the subfamily Euxanthinae Alcock, 1898, as presently recognized (see Ng et al., 2008) was polyphyletic, being split into two major clades. Morphological diagnoses for the two major clades, "Eux 1" (which they considered Euxanthinae s. str.) and "Eux 3", were provided, including the respective synapomorphies noted in the thoracic sternum, male abdomen and G1. Furthermore, a clade, "Eux 2", composed of only *Glyptoxanthus* A. Milne-Edwards, 1879, has recently been recognized as a new subfamily, Glyptoxanthinae Mendoza & Guinot, 2011. Lai et al. (2011: 431) also noted that there were five subclades within "Eux 3", one of which included genera presently classified under Antrocarcininae Ng & Chia, 1994 (e.g. *Cyrtocarcinus* Ng & Chia, 1994, and *Glyptocarcinus* Takeda, 1973) (cf. Ng et al., 2008). While Lai et al. (2011) did not formally recognize or provide a diagnosis for an expanded Antrocarcininae, they nevertheless recognized this taxon in the discussion. Following this, *Takedax* is here assigned to the Antrocarcininae. A complete systematic revision of the Euxanthinae Alcock, 1898, and Antrocarcininae Ng & Chia, 1994, is currently being undertaken by the authors.

Comparative material. — *Medaeops granulosus* (Haswell, 1882): same as material examined in Mendoza et al. (2009).

Medaeops neglectus (Balss, 1922): 5 ♂, 11.5 × 7.7 mm–18.4 × 12.4 mm (MNHN-B6533), intertidal, Nosy Be, Madagascar, coll. A. Crosnier, 20 May 1958; 2 ♂, 14.1 × 9.5 mm, 16.1 × 10.8 mm, 2 ♀, 9.7 × 6.5 mm, 12.6 × 8.6 mm (MNHN-B7958), intertidal, Nosy Be, Madagascar, coll. A. Crosnier, Sept. 1958; 1 ♂, 13.0 × 8.8 mm (UF 14004), 1 ♂, 8.3 × 5.6 mm (UF 17972), under rocks, 0-3 m, sand flat with seagrass & patch reefs, W of Hellville, Nosy Be, Madagascar, coll. A. Anker et al., 25 May 2008; 1 ♂, 15.7 × 10.8 mm (UF 14198), 0-12 m, NW side of Nosy Komba, Madagascar, coll. G. Paulay, 26 May 2008; 1 ♂, 17.1 × 11.6 mm (UF 14599), lagoonal fringing reef slope with sand & rubble, 1-2 m, Trois Frères islet, near Nosy Be, Madagascar, coll. A. Anker et al., 20 May 2008.

Monodaeus couchii (Couch, 1851): 1 ♂, 20.4 × 13.8 mm (NHM 1956.5.29.4), Cumbrae Deep, Firth of Clyde, Scotland, coll. 23 Nov. 1948; 1 ♂, 21.7 × 14.3 mm, 1 ♀, 28.8 × 17.6 mm (NHM 1956.5.29.3-4), 106 m, Cumbrae Deep, Firth of Clyde, Scotland, coll. R. B. Pike, Jul. 1949; 2 ♂, 16.9 × 11.6 mm, 31.5 × 20.4 mm, 3 ovig. ♀, 20.9 × 13.1 mm–24.4 × 15.9 mm, 1 juv. ♀, 11.9 × 8.2 mm (NHM 1960.70.25.134-136), stn 115, Port Erin Area, Scotland, coll. R. Hartnoll, no other data; 5 ♂, 9.5 × 6.6 mm–14.9 × 10.1 mm, 6 ♀, 13.0 × 8.9 mm–23.0 × 15.4 mm (NHM 1962.7.5.26-30), 20 m, Farland Point, Millport, Firth of Clyde, Scotland, coll. R. B. Pike, 11 Jan. 1957; 2 ovig. ♀, 18.9 × 13.2 mm, 27.0 × 18.2 mm (NHM 1981: 566), Tiree Passage, Scotland, coll. S. Smith, Jun. 1981; 8 ♂, 18.3 × 12.5 mm–30.5 × 20.3 mm, 10 ♀, 20.3 × 13.3 mm–28.4 × 17.7 mm (NHM 1979: 283), 55 m, 4 miles south of Spanish Head, Isle of Man, Irish Sea, coll. R. V. "Cumia", 12 Jul., 1979; 1 ♂, 19.2 × 13.0 mm (NHM 1981: 106), stn 47, 588-562 m, 51°50.2′N 11°51.5′W–51°47.3′N 11°46.1′W, Dingle, Ireland, coll. A. Wheeler, "Cirolana", 21 Jul. 1973; 1 ♂, 18.7 × 13.0 mm (NHM 1981: 94), stn 37, 520-568 m, 55°37.5′N 09°30.5′W–55°33.5′N 09°33.5′W, Tory Is., Donegal, Ireland, coll. A. Wheeler "Cirolana", 19 Jul. 1973.

Monodaeus cristulatus Guinot & Macpherson, 1988: holotype ♂, 23.9 × 16.4 mm (MNHN-B17498), stn 57, 232 m, 29°21′S 14°49′E, coll. E. Macpherson, "Benguela VI", 1984. Other material: 5 ♂, 9.6 × 6.6 mm–18.0 × 12.5 mm, 7 ♀, 10.5 × 7.3 mm–18.2 × 12.0 mm (USNM 221965), stn P42, 185 m, 28°57′S 14°56′E, coll. "Benguela XII", 24 Jan. 1988, det. E. Macpherson.

Takedax gabrielae sp. nov.
(figs. 1-3)

Medaeops granulosus — Ho et al., 2004 (in part): 657, fig. 5H (not *Leptodius granulosus* Haswell, 1882).

Material examined. — Holotype ♂, 17.7 × 11.0 mm (NMCR 39032), tangle net, ca. 50-500 m, Balicasag Is., Bohol Sea, coll. local fishermen, purch. P. Ng et al., Dec. 2000. Paratypes: 1 ♂, 15.6 × 10.4 mm (ZRC 2001.0464), tangle net, ca. 50-500 m, Balicasag Is., Bohol Sea, coll. local fishermen, purch. P. Ng et al., Dec. 2000; 2 juv. ♀, 7.1 × 4.9 mm, 7.5 × 5.0 mm (ZRC 2012.0001), tangle net, ca. 100-300 m, Maribohoc Bay, north of Panglao Is., coll. J. Arbasto, Nov. 2003–Apr. 2004; 2 ♂, 13.1 × 8.7 mm, 14.2 × 9.4 mm, 1 ♀, 7.2 × 5.0 mm (ZRC 2010.0164), stn CP2712, beam trawl, 140-139 m, 15°20.56′N 121°29.76′E, off Quezon Province, eastern Luzon Is., coll. R/V "DA-BFAR", AURORA 2007, Apr. 2007; 1 ♂, 14.6 × 9.8 mm, 1 ♀, 14.8 × 10.1 mm (ZRC 2012.0002), 1 ♂, 13.5 × 8.7 mm (MNHN-IU-2009-704), 1 ovig. ♀, 10.2 × 7.5 mm (MNHN-IU-2009-705), 1 ♂, 12.5 × 8.2 mm, 1 ♀, 10.2 × 7.5 mm (USNM), stn CP2716, beam trawl, 335-356 m, 14°31.20′N 121°60.77′E, off Quezon Province, eastern Luzon Is., coll. RV "DA-BFAR", AURORA 2007, Apr. 2007; 1 ♂, 8.5 × 5.5 mm (NMCR 39033), stn CP2780, beam trawl, 183-188 m, off western coast of Batangas, southwestern Luzon Is., coll. RV "DA-BFAR", 24 Mar. 2008.

Description. — Carapace transversely ovate, width 1.5-1.6 times length; most of dorsal surface granulate, rugose, with larger granules laterally; regions well defined; 1F, 2F confluent; 1M distinct from 2F, fused to inner branch of 2M; 2M partially divided longitudinally, 3M prominent, 4M indistinct; 2L, 3L confluent, 4L fused to third anterolateral tooth, 5L, 6L distinct; 1P prominent, 2P with 2 distinct, dorsolaterally projecting, pyramidal tubercles. Front moderately advanced beyond orbits, bilobed, median notch moderately deep; anterior margin of lobes granulate, slightly convex, lateral angles slightly produced; separated laterally from preorbital tooth by deep notch appearing as small fossa in anterior view. Anterolateral margin convex, with 4 coarsely granulate and broadly triangular teeth, third tooth projecting furthest laterally, last tooth projecting posterolaterally; anterior part vaguely descending toward buccal cavity, marked by low, subhepatic, granulate tubercle. Posterolateral margins straight, strongly converging posteriorly, subequal in length to anterolateral. Posterior margin straight, raised, granulate. Subhepatic and pterygostomian regions regularly covered with round granules. Antennules folding slightly obliquely. Basal article of antenna subrectangular, slender; flagellum much longer than width of orbit. Posterior region of epistome extensive, posterior border with median and lateral projections. Endostome with incomplete ridges. Lacinia of first maxilliped moderately broad distally.

Third maxillipeds granulate, not completely closing anterior portion of buccal cavity; antero-external angle of merus slightly projecting, anterior margin oblique, slightly notched mesially.

Male thoracic sternum granulate; sternites 1, 2 fused completely; complete suture between sternites 2, 3; deep transverse groove between sternites 3, 4, suture 3/4 visible only laterally; median longitudinal line present on sternite 4, interrupted within sternoabdominal cavity, but continued on posterior portion of sternite 4, absent on sternites 5, 6, complete on sternites 7, 8; sutures 4/5, 5/6 joining medially inside sternoabdominal cavity, continuing medially from that point as one suture, but interrupted medially; sternal press-button on posterior half of sternite 5; small decalcified region visible between sternites 6, 7 in exposed parts of thoracic sternum and in central portion of sternoabdominal cavity; small portion of sternite 8 exposed between abdominal somite 2 and coxa of P5 when abdomen closed; sutures 6/7 and 7/8 complete, clearly demarcated.

Chelipeds unequal or subequal; supero-external surfaces rugose, coarsely granulate; posterior margin of merus with large, conical tubercles, postero-ventral surface with transverse, granulate ridges of variable length; carpus with 2 teeth on internal angle; palm finely granulate on internal surface and on ventral half of external surface, with distinct groove along supero-external surface, most prominent proximally near articulation with carpus; fingers long, tips pointed, incurved; dactylus of major chela with or without modified proximal tooth.

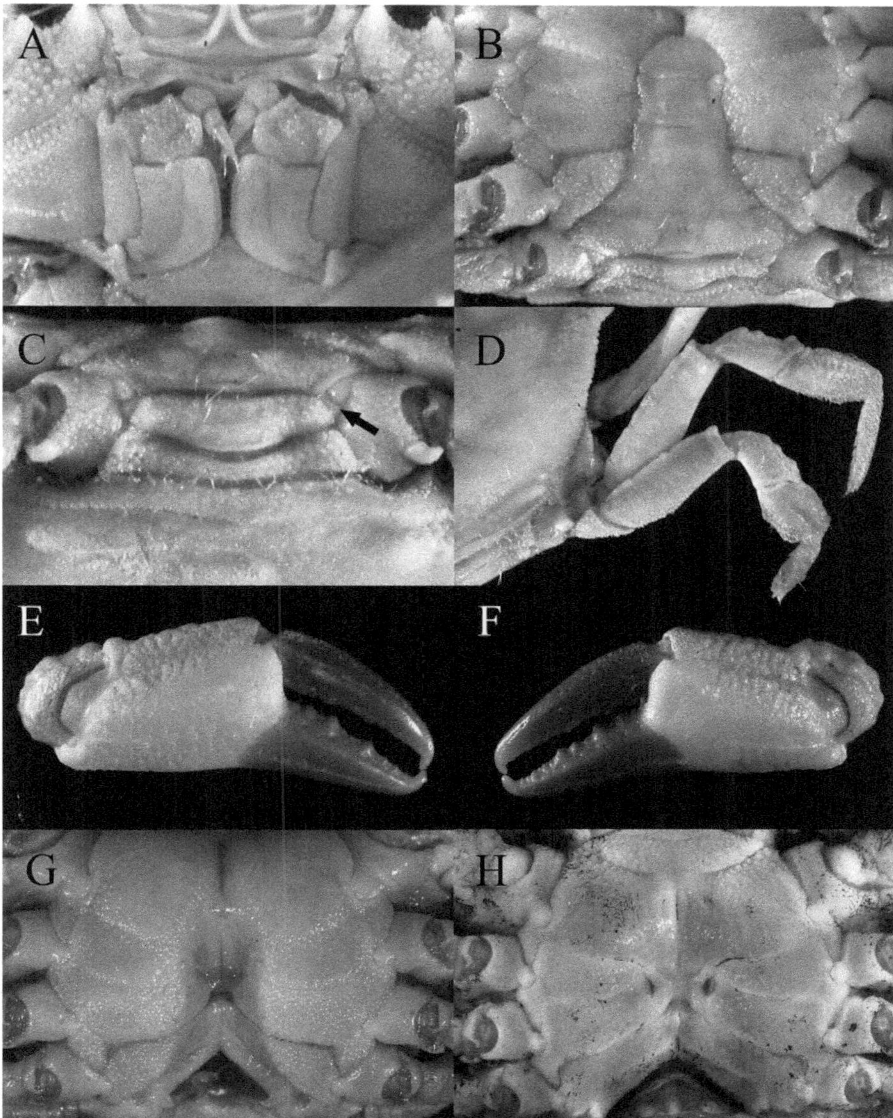

Fig. 2. *Takedax gabrielae* gen. et sp. nov., (A–G) holotype ♂, 17.7 × 11.0 mm (NMCR 30932), Balicasag; (H) paratype ♀, 14.8 × 10.1 mm (ZRC 2012.0002), Quezon Province. A, buccal field, antero-ventral view; B, thoracic sternum and abdomen, posteroventral view; C, thoracic sternite 8 (arrow), posterior view; D, left P4 and P5, dorsal view; E, major chela, external view; F, minor chela, external view; G, sternoabdominal cavity of male, ventral view; H, same of female.

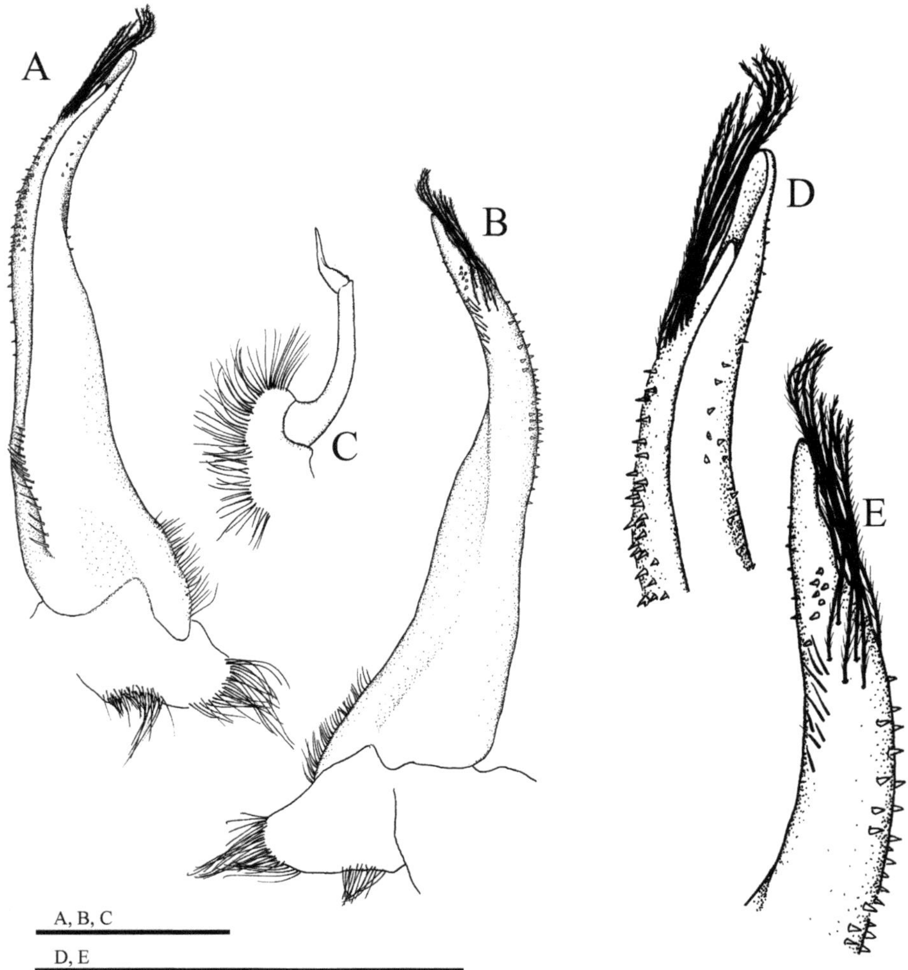

Fig. 3. *Takedax gabrielae* gen. et sp. nov., holotype (NMCR 30932) gonopods. A, left G1, external view; B, same, internal view; C, left G2; D, distal tip of left G1, external view; E, same, internal view. Scale bars = 1.0 mm.

Ambulatory legs moderately long, slender, flattened; anterior margin of meri distinctly cristate, straight to crenulate, sometimes cut into low teeth proximally; anterior margin of carpi cristate, trilobate.

Male abdomen short, tip of telson just short of level of condyles of cheliped coxae; somites 3-5 fused, other somites and telson free; somite 6 broad, disto-lateral angles produced; telson subtriangular. G1 moderately long; stout proximally, tapering distally; slightly carinate along lateral margin over central third; distal half moderately spinose, with several long plumose subterminal setae; apical lobe narrow, spatuliform; G2 about half length of G1.

Female morphology. — The morphology of the female is similar to that of the male, except in sexually dimorphic characters, i.e., thoracic sternum, abdomen and vulvae (fig. 2H). The thoracic sternum has a wider sternoabdominal cavity, with the median line of sternite 4 fully included within; sutures 2/3, 6/7 and 7/8 are complete and clearly demarcated; sutures 4/5 and 5/6 join medially at a point directly anterior to the vulvae, from which point only one suture continues medially, but does not quite reach the median longitude of the thoracic sternum; suture 6/7 has a centrally located, decalcified, subtriangular area. The vulvae are vertically oval, anterior and posterior tips somewhat notched; and they are positioned on the anterior portion of sternite 6, near sternite 5. The abdomen is much wider and longer, the tip of telson is at a level anterior to the coxo-sternal condyles of P1; and all the abdominal somites and telson are freely articulated.

Variation. — Some morphological variation among the type material was noted. Unlike the holotype, the male and female paratypes had distinctly unequal chelipeds, with most possessing a more coarsely granulate palm, and some have a well-differentiated dactylar cutting tooth. With regard to the basal dactylar cutting tooth, its absence has also been observed by the authors in individuals of species in other related genera, e.g., *Monodaeus* (pers. obs.) and *Guitonia* Garth & Illiffe, 1992 (cf. Lasley et al., 2010: 2), particularly in larger individuals, and this appears to be independent of sex (pers. obs.). This may be due to the breaking or wearing out of the tooth due to vigorous and/or prolonged use between moult cycles.

Life colour. — Background colour of carapace and pereopods white, with extensive mottling of reddish-orange, particularly on the anterior part of the carapace, and on the pereopods; the fingers of the chelipeds are dark brown over their entire length, but this does not extend onto the palm.

Etymology. — This species is named in honor of a Philippine heroine, Gabriela Silang, a military leader during one of the early revolutions against the Spanish colonizers in the northern Philippines. She became a martyr to her cause in 1763.

Remarks. — We re-examined the recent record of *Medaeops granulosus* from Taiwan (Ho et al., 2004: 657, fig. 5H), and find that the figure they provided agrees well in all respects with the present new species, and we are confident they are conspecific. *Takedax gabrielae* sp. nov. is thus known from the Bohol Sea, from the eastern and southwestern coasts of Luzon, and from

Taiwan. It has been caught by tangle net or beam trawl from depths of at least 50 m. The depth records from tangle net fishermen are not reliable (e.g. 50-500 m), but other specimens that have been collected by beam trawling indicate a depth range of 83-356 m for the species. The misidentified *M. granulosus* from Taiwan (Ho et al., 2004: 657) was collected from 83-132 m, and this is also much deeper than what was previously known for *M. granulosus*, which is otherwise a shallow-water species (see Mendoza et al., 2009).

ACKNOWLEDGEMENTS

The authors would like to thank the tangle net fishermen of Balicasag and Panglao, particularly Jose Arbasto; the organizers, participants and sponsors of the AURORA 2007 expedition — principal investigators Philippe Bouchet (MNHN) and Marivene Manuel-Santos (National Museum of the Philippines), the Philippine government through its Bureau of Fisheries & Aquatic Resources (BFAR), TOTAL (Philippines) Corporation, Lounsbery Foundation, CoMarges component of Census of Marine Life, Aurora State College of Technology (ASCOT), University of San Carlos (USC), National Taiwan Ocean University (NTOU), MNHN & NUS; and the following, for their kind and generous assistance during museum visits and facilitation of specimen loans: Régis Cleva & Danièle Guinot (MNHN), Paul Clark (NHM), Marivene Manuel-Santos (NMCR), Gustav Paulay, Arthur Anker, John Slapcinsky & Sandy Beemis (UF), Rafael Lemaitre, Karen Reed, Chad Walter & the staff of the Department of Invertebrate Zoology (USNM), as well as Tan Swee Hee, Tan Siong Kiat and Martin Low (ZRC). This study was supported by the National University of Singapore (ARF Grant # R-154-000-334-112), and also in part by a Singapore–France Merlion travel grant to the first author. The authors are most grateful to Peter Davie, Paul Clark & Sammy De Grave for their comments and suggestions for the improvement of this paper, and to Hironori Komatsu for the invitation to contribute to this volume.

REFERENCES

ALCOCK, A., 1898. Materials for a carcinological fauna of India. No. 3. The Brachyura Cyclometopa. Part I. The family Xanthidae. Journal of the Asiatic Society of Bengal, Calcutta, **67**(2): 67-233.

BALSS, H., 1922. Diagnosen neuer japanischer Decapoden. Zoologischer Anzeiger, **54**(1-2): 1-6.

COUCH, R. Q., 1851. Notice of a crustacean new to Cornwall. Transactions of the Penzance Natural History & Antiquities Society, **2**: 13-14.

DANA, J. D., 1851. On the classification of the Cancroidea. American Journal of Science and Arts, **12**(34): 121-131.

DAVIE, P. J. F., 2002. Crustacea : Malacostraca : Eucarida (Part 2) : Decapoda–Anomura, Brachyura. In: A. WELLS & W. W. K. HOUSTON (eds.), Zoological Catalogue of Australia, **19.3B**: i–xiv, 1-641. (CSIRO Publishing, Melbourne).

GARTH, J. S. & T. M. ILIFFE, 1992. *Guitonia troglophila*, a new genus and species of anchialine crab from the Galápagos, Isla Santa Cruz, Grieta de Caleta la Torta (Crustacea: Decapoda: Brachyura). Proceedings of the Biological Society of Washington, **105**(2): 310-316.

GUINOT, D., 1967. Recherches préliminaires sur les groupements naturels chez les Crustacés Décapodes Brachyoures. II. Les anciens genres *Micropanope* Stimpson et *Medaeus* Dana. Bulletin du Muséum national d'Histoire naurelle, Paris, (2) **39**(2): 345-374.

GUINOT, D. & E. MACPHERSON, 1988. Remarques sur le genre *Monodaeus* Guinot, 1967, avec la description de deux espèces nouvelles (Crustacea Decapoda Brachyura). Bulletin du Muséum national d'Histoire naurelle, Paris, (4) **10**(4): 731-757.

HASWELL, W. A., 1882. Catalogue of the Australian stalk- and sessile-eyed Crustacea: iii–xxiv, 1-324, pls. 1-4. (The Australian Museum, Sydney).

HO, P.-H., P. K. L. NG, T.-Y. CHAN & D.-A. LEE, 2004. New records of 31 species of brachyuran crabs from the joint Taiwan-France expeditions, "TAIWAN 2000" and "TAIWAN 2001", off deep waters in Taiwan. Crustaceana, **77**(6): 641-668.

LAI, J. C. Y., J. C. E. MENDOZA, D. GUINOT, P. F. CLARK & P. K. L. NG, 2011. Xanthidae MacLeay, 1838 (Crustacea : Decapoda : Brachyura) systematics: a multi-gene approach with support from adult and larval morphology. Zoologischer Anzeiger, **250**(4): 407-448.

LASLEY, R. M. JR., J. C. E. MENDOZA & P. K. L. NG, 2010. Two new species of *Guitonia* Garth & Iliffe, 1992 (Crustacea: Decapoda: Brachyura: Xanthidae) from the central and western Pacific Ocean. Zootaxa, **2684**: 1-13.

MACLEAY, W. S., 1838. Illustrations of the Annulosa of South Africa; being a portion of the objects of natural history chiefly collected during an expedition into the interior of South Africa, under the direction of Dr. Andrew Smith, in the years 1834, 1835, and 1836; fitted out by the "Cape of Good Hope Association for Exploring Central Africa." In: A. SMITH (ed.), Illustrations of the zoology of South Africa investigations: 1-75, pls. 1-4. (Smith, Elder and Co., London).

MENDOZA, J. C. E., V.-C. CHONG & P. K. L. NG, 2009. A new xanthid crab of the genus *Medaeops* Guinot, 1967, from Peninsular Malaysia, with a note on *Leptodius granulosus* Haswell, 1882 (Crustacea: Decapoda: Brachyura: Xanthidae). Zootaxa, **2297**: 44-54.

MENDOZA, J. C. E. & D. GUINOT, 2011. Revision of the genus *Glyptoxanthus* A. Milne-Edwards, 1879, and establishment of Glyptoxanthinae nov. subfam. (Crustacea: Decapoda: Brachyura: Xanthidae). Zootaxa, **3015**: 29-51.

MILNE-EDWARDS, A., 1873-1880. Études sur les xiphosures et les crustacés de la région mexicaine. Mission scientifique au Mexique et dans l'Amérique centrale, ouvrage publié par ordre du Ministre de l'Instruction publique. Rechercheszoologiques pour servir à l'histoire de la faune de l'Amérique central et du Mexique, publiées sous la direction de M.H. Milne Edwards, membre de l'Institut. Cinquième partie. Tome premier: 8 [unnumbered], 1-368. (Imprimerie Nationale, Paris).

NG, P. K. L., 1993. On a new genus and species of xanthid crab (Crustacea: Decapoda: Brachyura) from Chesterfield Island, Coral Sea. Proceedings of the Biological Society of Washington, **106**(4): 705-713.

NG, P. K. L. & D. G. B. CHIA, 1994. The genus *Glyptocarcinus* Takeda, 1973, with descriptions of a new subfamily, two new genera and two new species from New Caledonia (Crustacea: Decapoda: Brachyura: Xanthidae). Raffles Bulletin of Zoology, **42**(3): 701-731.

NG, P. K. L. & P. F. CLARK, 2003. Three new genera of Indo-West Pacific Xanthidae (Crustacea, Decapoda, Brachyura, Xanthoidea). Zoosystema, **25**(1): 131-147.

NG, P. K. L., D. GUINOT & P. J. F. DAVIE, 2008. Systema Brachyurorum: Part I. An annotated checklist of extant brachyuran crabs of the world. Raffles Bulletin of Zoology, (Supplement) **17**: 1-286.

SERÈNE, R., 1984. Crustacés Décapodes Brachyoures de l'océan Indien occidental et de la mer Rouge. Xanthoidea: Xanthidae et Trapeziidae. Avec un addendum par A. Crosnier: Carpiliidae et Menippidae. Faune tropicale, **24**: 1-400, pls. 1-48.

TAKEDA, M., 1973. A new genus and a new species of the Parthenopidae from the sea off the Ogasawara lslands (Crustacea, Brachyura). Bulletin of the National Science Museum, Tokyo, **16**(1): 31-36.

First received 8 October 2011.
Final version accepted 12 Januray 2012.

ADDITIONAL DEEP-SEA PONTONIINE SHRIMPS (DECAPODA, PALAEMONIDAE) FROM TAIWAN, WITH DESCRIPTION OF ONE NEW SPECIES

BY

MASAKO MITSUHASHI[1]), XINZHENG LI[2]) and TIN-YAM CHAN[3,4])

[1]) Laboratory of Biology, Osaka Institute of Technology, Ohmiya, Asahi-ku, Osaka 535-8585, Japan

[2]) Institute of Oceanology, Chinese Academy of Sciences, 7 Nanhai Road, Qingdao 266071, China

[3]) Institute of Marine Biology and Center of Excellence for Marine Bioenvironment and Biotechnology, National Taiwan Ocean University, 2 Pei-Ning Road, Keelung 202, Taiwan, R.O.C.

ABSTRACT

Two additional deep-sea pontoniine shrimps are reported from Taiwan and Dongsha (= Pratas). *Periclimenes ngi* Li, Mitsuhashi & Chan, 2008, originally only known from the Philippines, is a new record for Taiwan. The other species collected from Dongsha is new to science and although very similar to *P. ngi*, differs in the first pereiopod with the carpus being as long as the merus, the third pereiopod having a stouter propodus and a proportionately shorter accessory tooth of the dactylus, as well as its distinctive colouration in life.

RÉSUMÉ

Deux nouvelles crevettes pontoniines de profondeur sont signalées de Taiwan et Dongsha (= Pratas). *Periclimenes ngi* Li, Mitsuhashi & Chan, 2008, seulement connue des Philippines à ce jour, représente un nouveau signalement pour Taiwan. L'autre espèce récoltée à Dongsha est nouvelle pour la science et, bien que très proche de *P. ngi*, elle diffère par le premier péréiopode dont le carpe est aussi long que le mérus, par le troisième péréiopode avec un propode plus fort et l'épine accessoire du dactyle proportionnellement plus courte, ainsi que par sa coloration sur le vivant bien distincte.

[4]) Corresponding author; e-mail: tychan@mail.ntou.edu.tw

INTRODUCTION

Recent deep-sea biodiversity surveys in Taiwan have yielded two new genera and three new species of pontoniine shrimp: *Patonia mclaughlinae* Mitsuhashi & Chan, 2006; *Periclimenes sandybrucei* Mitsuhashi & Chan, 2009; and *Neoclimenes holthuisi* Mitsuhashi, Li & Chan, 2010. During these surveys more deep-sea pontoniine specimens were obtained and this work reports upon two additional species. One of them is *Periclimenes ngi* Li, Mitsuhashi & Chan, 2008, which was previously only known from the Philippines, however two specimens of *P. ngi* were collected off the northeastern coast of Taiwan. The other species, of which only one specimen was obtained from Dongsha (= Pratas), a territory under the jurisdiction of Taiwan in the South China Sea, is a new species close to *P. ngi*.

The carapace length (cl, dorsally from the postorbital margin to the posterior border of the carapace) and body length (bl, dorsally from the postorbital margin to the end of the telson) are given as standard measurements. The material examined is deposited in the National Taiwan Ocean University, Keelung (NTOU).

TAXONOMY

Periclimenes ngi Li, Mitsuhashi & Chan, 2008
(figs. 1A, 2)

Periclimenes ngi Li, Mitsuhashi & Chan, 2008: 397, figs. 9-11, 17C, D [type locality: Bohol Sea, the Philippines].

Material examined. — One ovigerous female (cl 12.1 mm, bl 43.4 mm), Dasi fishing port, northeastern coast of Taiwan, commercial trawler, 4 Aug. 1996 (NTOU M01217); 1 ovigerous female (cl 9.5 mm, bl 43.3 mm), TAIWAN 2004, station CP268, 24°30.46′N 122°6.28′E, beam trawl, 421-531 m, 2 Sep. 2004 (NTOU M01222).

Remarks. — Both specimens have the left second pereiopods missing and the posterior pereiopods partly damaged. Nevertheless, they generally agree quite well with the original description of *P. ngi*, except for the rostrum of the Dasi specimen which is deeper and more straight (fig. 2B). On the other hand, the specimen from station CP268 has four ventral rostral teeth (fig. 2A) vs. three ventral teeth in all other known specimens (Li et al., 2008). The colouration was recorded only for the specimen from station CP268 (fig. 1A) and it is almost identical with the type material from the Philippines (Li et al., 2008: fig. 17C, D).

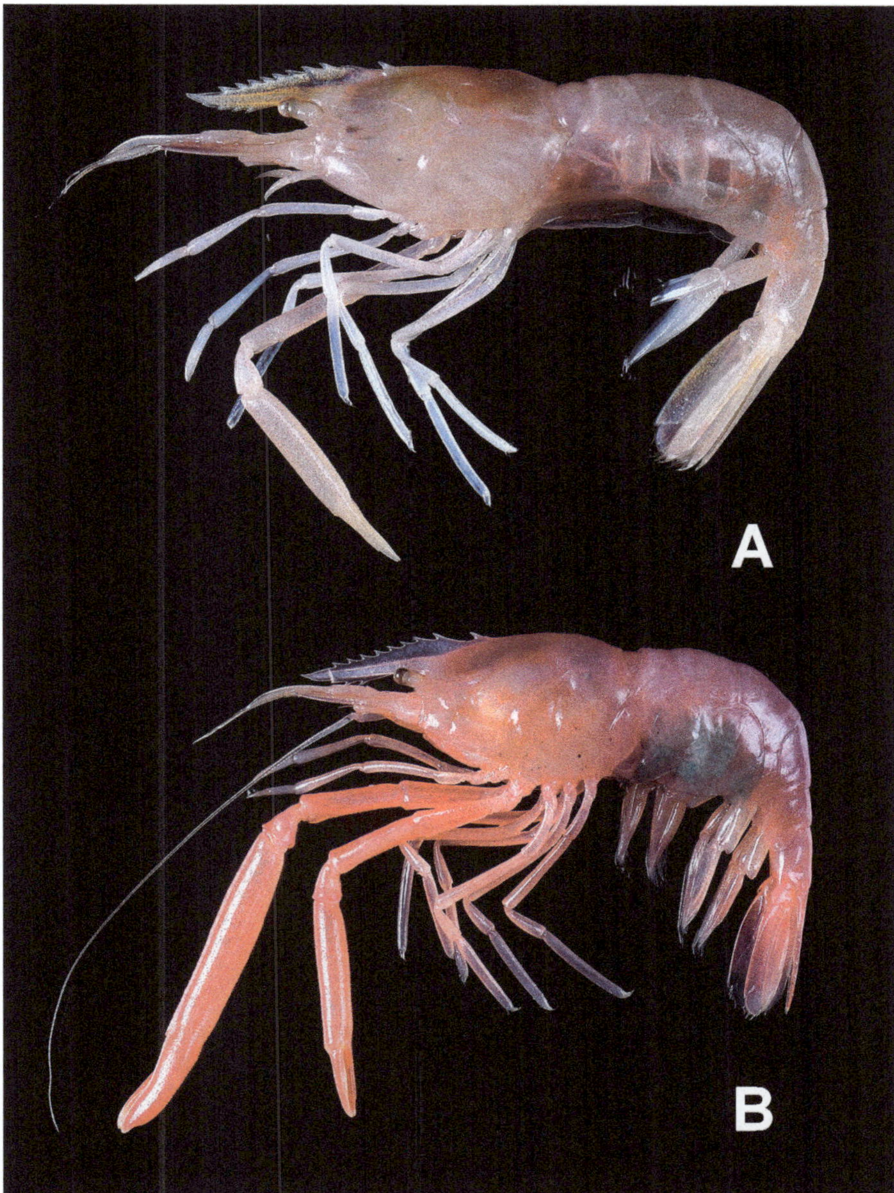

Fig. 1. A, *Periclimenes ngi* Li, Mitsuhashi & Chan, 2008, ovigerous female (cl 9.5 mm), station CP268. B, *P. takedai* n. sp., holotype, ovigerous female (cl 15.9 mm).

Distribution. — Originally reported from the Bohol Sea, the Philippines, 273-302 m. The present record extends its distribution northwards to north-eastern Taiwan and its depth range to 421-531 m.

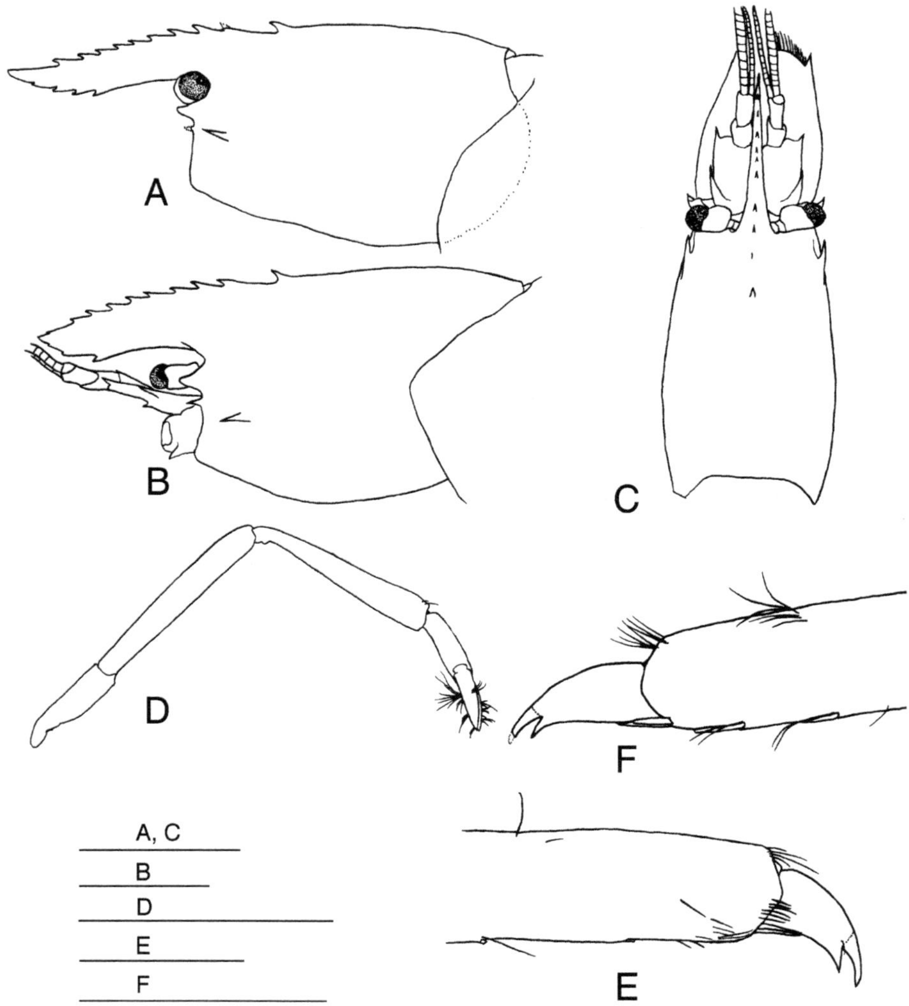

Fig. 2. *Periclimenes ngi* Li, Mitsuhashi & Chan, 2008: A, C, D, F, ovigerous female (cl 9.5 mm), station CP268; B, E, ovigerous female (cl 12.1 mm), Dasi fishing port. A, B, carapace, lateral; C, carapace and cephalic appendages, dorsal; D, first pereiopod, ventral; E, dactylus and distal part of propodus of right third pereiopod, lateral; F, same, left fourth pereiopod, medial. Scales: A–D, 5 mm; E–F, 1 mm.

Periclimenes takedai sp. nov.

(figs. 1B, 3-5)

Material examined. — Holotype, ovigerous female (cl 15.9 mm, bl 54.8 mm), TAIWAN 2005, station CP 315, 21°40.2′N 117°43.17′E, beam trawl, 509-506 m, 17 Aug. 2005 (NTOU M01223).

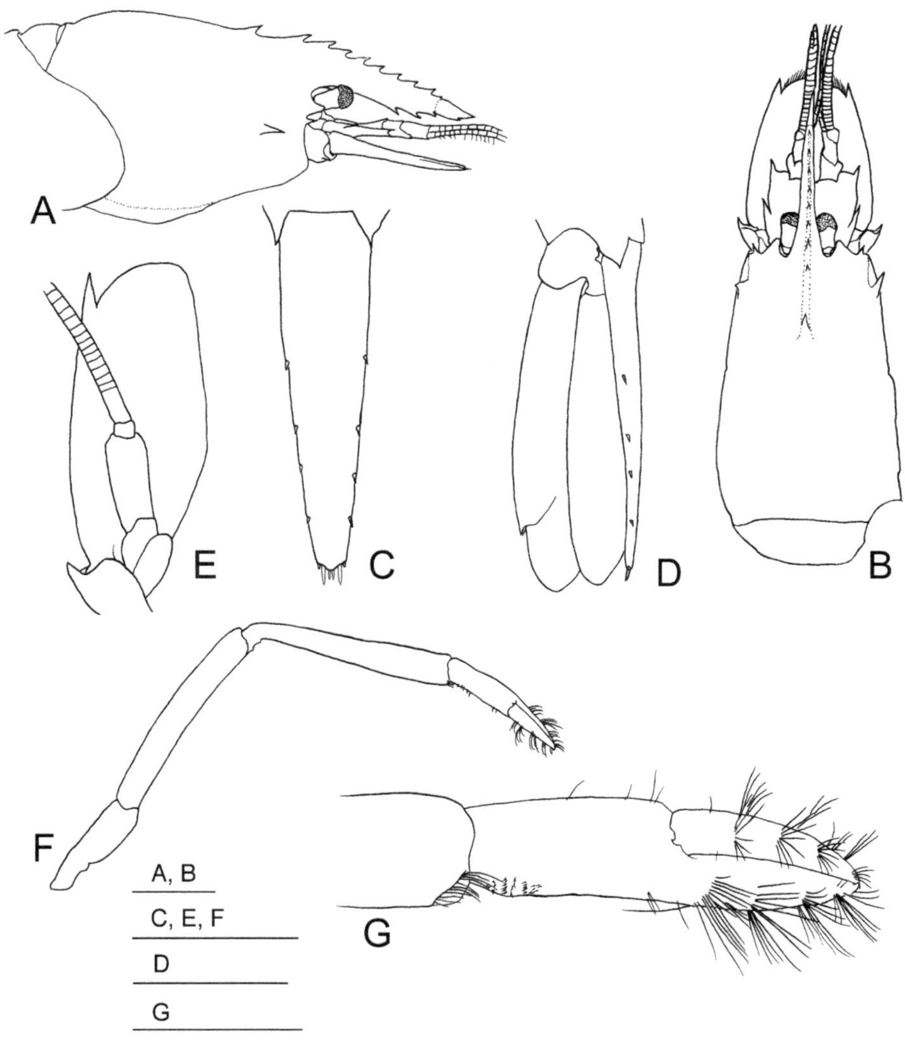

Fig. 3. *Periclimenes takedai* sp. nov., holotype, ovigerous female (cl 15.9 mm). A, carapace and cephalic appendages, lateral; B, same, dorsal; C, telson, dorsal; D, tailfan, lateral, setae omitted; E, right antenna, ventral; F, first pereiopod, ventral; G, same, chela and distal part of carpus, medial. Scales: A–F, 5 mm; G, 1 mm.

Etymology. — The species is named after Dr. Masatsune Takeda in honor of his many important contributions in the taxonomy of decapod crustaceans.

Description. — Large pontoniine shrimp with subcylindrical body form (fig. 1B).

Rostrum (fig. 3A, B) well developed, deep, straight, directed slightly anteroventrally, 0.6 of carapace length, extending distinctly beyond antennular peduncle but falling slightly short of distal margin of scaphocerite; dorsal margin with 8 acute, similar, evenly distributed teeth, distal-most tooth far distant from tip of rostrum, posterior-most tooth slightly posterior to level of orbital margin, second posterior tooth just above orbital margin; ventral margin with 4 acute and evenly distributed teeth; posterior-most tooth slightly anterior to midlength of rostrum, 2 anterior teeth smaller than remaining teeth.

Carapace (fig. 3A, B) smooth, glabrous; epigastric spine acute, located at about anterior 1/3 of carapace and posterior to level of hepatic spine, distinctly separated from dorsal rostral teeth; supraorbital spine absent; antennal spine acute, marginal; inferior orbital angle produced, blunt, overreaching antennal spine; hepatic spine acute, distinctly erect and outwardly directed, at level slightly lower than antennal spine but twice as long as latter; pterygostomial margin blunt, not distinctly produced.

Abdomen smooth; pleura of anterior 5 somites broadly rounded; sixth somite 1.6 times as long as fifth, 1.5 times as long as proximal depth, posteroventral angle blunt. Telson (fig. 3C, D) 1.9 times as long as sixth abdominal somite, 0.6 times of carapace length, 3.6 times as long as proximal width, lateral margins straight; bearing 4 pairs of dorsolateral spines located at anterior 0.4, 0.6, 0.7 and 0.8; these dorsolateral spines small, similar in size and about 0.02 of telson length; posterior margin medially produced with 3 pairs of spines, lateral and submedian spines similar in size to dorsolateral spines, intermediate spines larger and twice as long as dorsolateral spines.

Antennule (fig. 3A, B) with peduncle reaching to distal 1/3 of scaphocerite length. Basal segment 1.6 times longer than wide measured along medial margin, ventromedial margin with small acute tooth located at midlength; stylocerite acute, slender, extending to midlength of basal segment; distolateral tooth slender, acute, reaching distolateral margin of intermediate segment, anterior margin of basal segment rounded; intermediate segment about as long as wide; distal segment 1.7 times as long as wide; upper flagellum biramous, with proximal 17 (left) or 18 (right) articles fused.

Antenna (fig. 3E) with basicerite armed with acute distoventral tooth. Carpocerite slightly shorter than half scaphocerite length, 2.8 times as long as distal width. Scaphocerite twice as long as wide, lateral margin slightly convex, distolateral tooth acute, slightly shorter than rounded distal margin of lamina.

Eye (fig. 3A, B) with cornea well developed, cornea globular, diameter 0.08 carapace length, ocellus absent. Stalk short, slightly longer than broad, narrower than cornea.

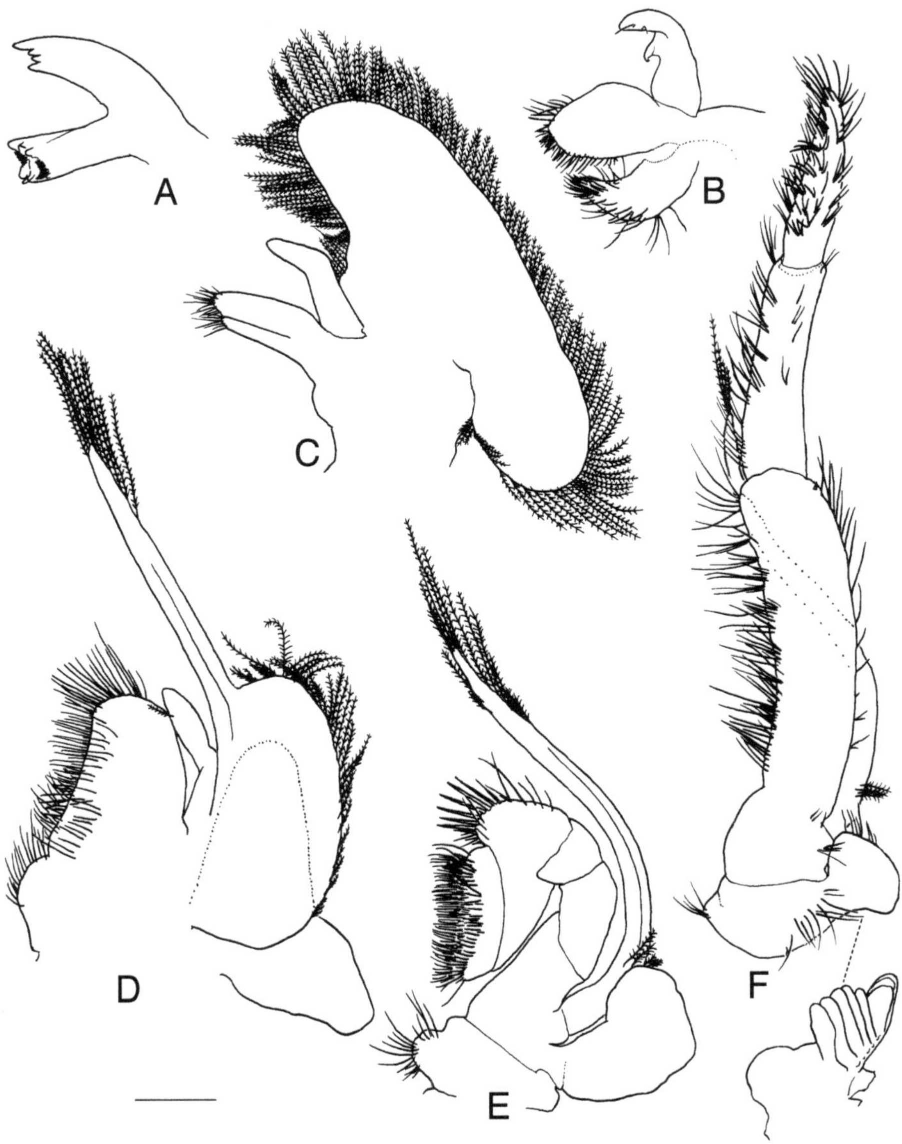

Fig. 4. *Periclimenes takedai* sp. nov., holotype, ovigerous female (cl 15.9 mm), mouthparts, left side. A, mandible; B, maxillula; C, maxilla; D, first maxilliped; E, second maxilliped; F, third maxilliped (arthrobranch shown). Scale: 1 mm.

Left mouthparts dissected. Mandible (fig. 4A) as usual for genus, molar process with 5 blunt marginal teeth and 2 groups of short, dense marginal setae; incisor process bearing 4 triangular distal teeth, distal most tooth largest. Maxillule (fig. 4B) with bilobed palp, upper lobe well developed, bearing few setae on posterior margin, lower lobe with very short apical seta; upper lacinia with several stout spines and setae distally; lower lacinia slender, tapering distally, with numerous distoventral setae. Maxilla (fig. 4C) with simple palp lacking setae; basal endite well developed, bilobed, distal lobe 3 times as wide as proximal lobe, both with distal setae; coxal endite obsolete; scaphognathite large, 2.6 times as long as wide. First maxilliped (fig. 4D) with simple palp, slender, curved along distolateral margin of basal endite, with subapical plumose seta; basal endite large and broad, densely fringed with long setae along entire medial margin; coxal endite medially convex, with long mediodistal setae; caridean lobe large, elongate; exopod well developed, flagellum with long plumose setae distally; epipod large, elongate, deeply bilobed. Second maxilliped (fig. 4E) with normal endopod; dactylar segment narrow, 4 times broader than long, medial margin densely fringed with long finely serrulate setae; propodal segment with several long setae along disto- and dorsomedial margins; carpal segment short, broader than long, unarmed; exopod with long plumose setae on distal margin and proximal part; coxal segment medially produced, with long simple setae medially; epipod large. Third maxilliped (fig. 4F) slender, extending to midlength of distal segment of antennular peduncle; ischiomerus and basis incompletely fused, 5 times longer than wide, medial and lateral margins sparsely setose, distolateral part with single small spine, basal part convex medially, unarmed; coxa with medial process bearing few setae on medial margin and on basal part of lateral plate; penultimate segment 3 times as long as proximal width, slightly longer than ultimate segment, fringed with numerous setae laterally, bearing several long setae medially; ultimate segment tapering distally, 4 times as long as proximal width; exopod well developed, reaching base of penultimate segment, with several long plumose setae on distal margin and proximal part; arthrobranch well developed with 10 lamellae.

First pereiopod (fig. 3F, G) slender, extending beyond scaphocerite by distal 1/5 of carpus; chela (fig. 3G) 4.0 times as long as deep, subcylindrical; fingers slender, as long as palm, bearing tufts of long setae, with cutting edges entire; palm with short grooming setae medioproximally; carpus 1.5 times as long as chela, 6.7 times as long as distal width, slightly tapering proximally; merus as long as carpus, 6.7 times as long as medial width; ischium half length of merus, 3.2 times as long as distal width.

Second pereiopods well developed, unequal, finely granulated. Major (right) second pereiopod (figs. 1B, 5A, B) exceeding scaphocerite by length of carpus and chela when fully extended anteriorly; chela 1.6 times as long as carapace;

palm twice as long as fingers, 5.0 times as long as distal width; dactylus slightly curved, 4.3 times as long as proximal width, laminar, with lateral expansion; cutting edge with 1 large and 1small triangular teeth at proximal fourth and mid-length respectively, small notch present between teeth, distal part entire, tip with strongly hooked unguis; fixed finger swollen; 4.5 times as long as deep; ventral surface convex; cutting edge with broad irregular tooth near base, large acute triangular tooth at proximal 0.4, deep notch present between teeth, distal part of cutting edge entire, tip with strongly hooked unguis; carpus short, cup-like, 0.2 of palm length, slightly longer than distal width; merus half as long as palm, 0.3 of medial width; ischium as long as merus, 3.2 times as long as distal width; basis and coxa without special features. Minor (left) second pereiopod (fig. 5C) generally similar to major second pereiopod, exceeding scaphocerite by chela when fully extended anteriorly; chela 1.1 times as long as carapace; palm twice as long as fingers, 4.3 times as long as distal width; dactylus 4.8 times as long as proximal width; cutting edge with 1 large triangular tooth and 1 small tooth at proximal fourth and third respectively, distal part entire, tip with strongly hooked unguis; fixed finger 4.4 times as long as deep, proximally with broad irregular tooth and large acute triangular tooth at proximal third of cutting edge, distal part entire; carpus short, 0.3 of palm length, 1.3 times as long as distal width; merus 0.6 of palm length, 0.3 times as long as medial width; ischium slightly longer than merus, 3.4 times as long as distal width.

Third pereiopod (fig. 5D–F) moderately slender, exceeding scaphocerite by lengths of dactylus and distal 0.2 of propodus; dactylus biunguiculate, 2.4 times as long as proximal depth; unguis feebly demarcated, 3.2 times longer than basal depth, 0.4 of corpus length; corpus 2.5 times as long as basal depth, dorsal margin convex, ventral margin concave, with acute accessory tooth extending to 0.3 of unguis length; propodus about half of carapace length, 6.7 times dactylus length, 7.5 times as long as wide, with pair of distoventral and subdistoventral spines, 3 more ventral spines present at distal 0.1, 0.2, 0.4 of propodal length; tufts of simple setae present at distal margin and near spines; carpus 0.5 times as long as propodus, 3.6 times distal width; merus 1.1 times longer than propodus, 7.6 times medial width; ischium half merus length, 3.5 times longer than distal width; basis and coxa without special features.

Fourth and fifth pereiopods generally similar to third, more slender, with propodi 8.4 and 10.6 times as long as dactyli, respectively.

First pleopod with exopod 2.5 times as long as endopod, 4.7 times longer than maximum width; endopod 4.9 times as long as proximal width, covered

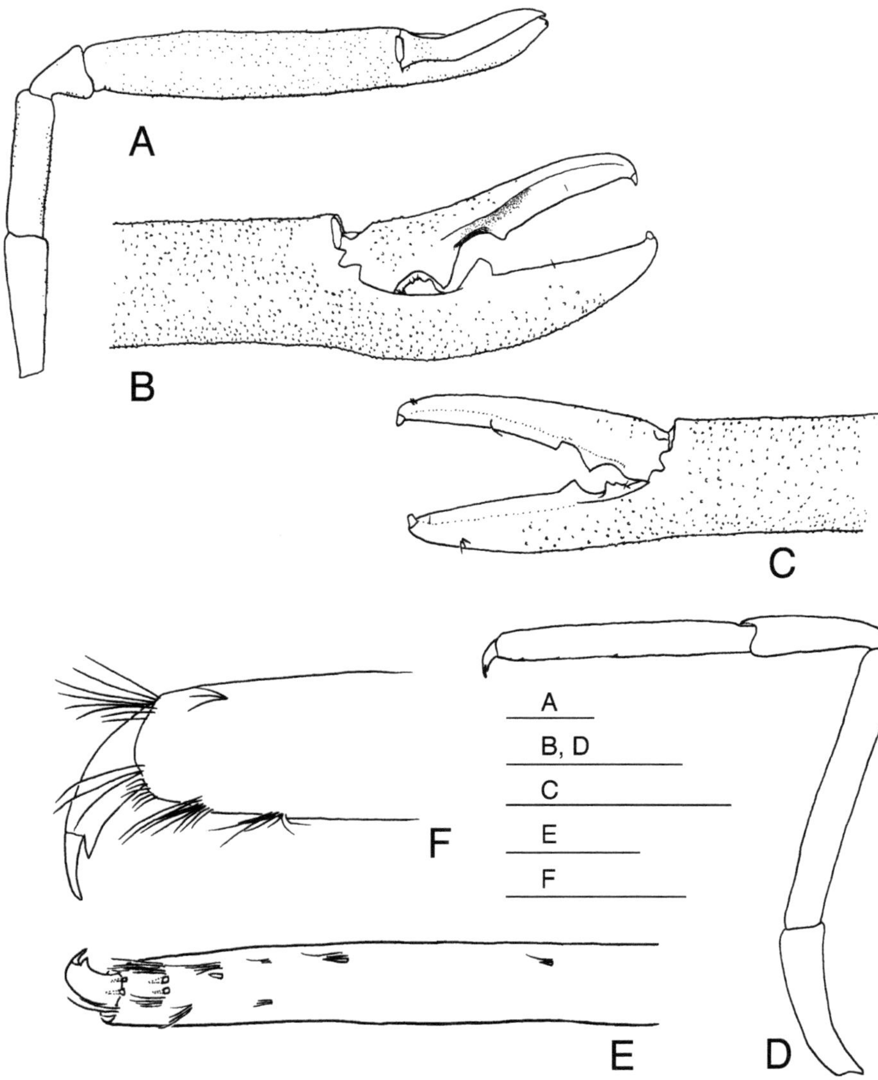

Fig. 5. *Periclimenes takedai* sp. nov., holotype, ovigerous female (cl 15.9 mm). A, major right second pereiopod, lateral; B, same, distal part of chela, medial; C, distal part of chela of minor left second pereiopod, medial; D, left third pereiopod, lateral; E, same, dactylus and distal part of propodus, ventral; F, same, lateral. Scales: A–D, 5 mm; E–F, 1 mm.

with long setae along distal and lateral margins. Second pleopod with exopod slightly longer than endopod, 3.1 times longer than maximum width; endopod 4.0 times as long as maximum width; appendix interna 0.3 times of endopod length.

Uropods (fig. 3D) exceeding telson, with protopodite rounded posterolaterally; exopod slightly longer than endopod, 2.5 times as long as maximum width, lateral margin convex, with acute distolateral tooth and small movable spine medially; endopod 2.9 times as long as maximum width.

Ova in early developmental stage, numerous, small, about 0.5-0.6 mm in diameter.

Colouration (fig. 1B). — Body uniformly rose-pink but with rostrum, distal parts of first and ambulatory pereiopods semi-translucent; eyes dark brown; eggs olive-coloured.

Remarks. — The general body form, the granulate second pereiopods (fig. 5A–C) and the biunguiculate dactylus of the ambulatory pereiopods (fig. 5F) align the present new species with the "*P. alcocki* Kemp, 1922 species group". Currently, this group contains 14 species from deep water of the Indo-West Pacific (Li et al., 2008). Of these, *P. takedai* sp. nov. is most similar to *P. alcocki*, *P ngi* and *P. tangeroa* Bruce, 2005 in having the rostrum not overreaching the scaphocerite and bearing about seven dorsal and two postorbital teeth, four minute dorsolateral spines on the telson, and the cornea markedly reduced and about 0.08 times carapace length (fig. 3A-D).

Nevertheless, *P. takedai* sp. nov. differs from these latter three species in the different proportion of the accessory tooth on the dactyli of the ambulatory pereiopods. Each accessory tooth is 0.30 times as long as the unguis in *P. takedai* (fig. 5F), 0.45 in *P. ngi* (fig. 2E, F,), 0.20 in *P. alcocki* and 0.07 in *P. tangeroa*. Moreover, *P. tangeroa* has only one ventral rostral tooth (vs. four in the new species) and the dactylus of the third pereiopod considerably more slender (0.28 propodal length vs. 0.15 in *P. takedai* sp. nov.). The epigastric spine in *P. alcocki* is situated at the same level as the hepatic spine, while it is distinctly behind the hepatic spine in the new species. Other than the length of the accessory tooth on the dactyli of the ambulatory pereiopods, the new species also differs from *P. ngi* in the more erect and outwardly directed hepatic spine (cf. fig. 3B and fig. 2C), the carpus of the first pereiopod being as long as merus (fig. 3F) rather than slightly shorter than merus in *P. ngi* (fig. 2D), the propodus of the third pereiopod being stouter (7.5 times as long as width, vs. 10 times as long in *P. ngi*). Furthermore, the body is uniformly rose-pink with the rostrum semi-translucent in *P. takedai* sp. nov. (fig. 1B) whereas generally pale pink with the yellowish rostrum and whitish teeth in *P. ngi* (fig. 1A; also see Li et al., 2008: fig. 17C, D). Although these distinguishing characters between *P. takedi* sp. nov. and *P. ngi* are rather subtle, these characters seems to be constant. For example, the shape of the hepatic spine, the size of the accessory tooth on the dactyli of the ambulatory pereiopods, and the relative length between the carpus and merus of the first pereiopod are almost identical amongst the Taiwanese and Philippines material of *P. ngi*. Colouration in life is now known to be a reliable character in distinguishing morphologically similar species of shallow water pontoniines (e.g., Bruce, 1974; Okuno, 2004; Mitsuhashi & Takeda, 2008; Okuno & Bruce, 2010). Although deep-sea pontoniines usually do not have striking colour patterns, the difference in the colouration between *P. takedai* sp. nov. and *P. ngi* appears to be specific.

An identification key to the species of the "*P. alcocki* species group" has been given by Li et al. (2008). The following amended key for the couplet "6" in Li et al. (2008: 407) can be used to accommodate the present new species:

6. Accessory tooth of dactylus of third pereiopod 0.45 time as long as unguis; R. 2 + 7-9/3-4 . . . . *P. ngi*
– Accessory tooth of dactylus of third pereiopod 0.08-0.30 times as long as unguis 7
7. Cornea diameter less than 0.08 carapace length; hepatic spine at same level as epigastric spine; dactylus of third pereiopod 0.17 of propodal length, accessory tooth 0.20 times as long as unguis; R. 1-2 + 6-9/2-4 . . . . *P. alcocki*
– Hepatic spine anterior to level of epigastric spine 7a
7a. Cornea diameter 0.10 of carapace length; dactylus of third pereiopod 0.28 of propodal length, accessory tooth 0.07 length of unguis; dorsal spines of telson 0.04 length of telson; R. 1 + 7/1 . . . . *P. tangeroa*
– Cornea diameter 0.08 of carapace length; dactylus of third pereiopod 0.15 of propodal length, accessory tooth 0.30 length of unguis; dorsal spines of telson 0.02 length of telson; R. 2 + 7/4 . . . . *P. takedai* sp. nov.

ACKNOWLEDGEMENTS

Sincerely thanks are extended to Dr. A. J. Bruce for his useful comments and information on the studied material. We also thank Dr. Charles H. J. M. Fransen of Netherlands Centre for Biodiversity Naturalis, Leiden, Dr. Sammy De Grave of the Oxford University Museum of Natural History, and an anonymous reviewer for their critical reviewing and valuable comments. This work was supported by the National Science Council and Academia Sinica, Taiwan, R.O.C., and the National Natural Science Foundation of China (31071889).

REFERENCES

BRUCE, A. J., 1974. *Coralliocaris viridis* sp. nov., a preliminary note (Decapoda, Natantia, Pontoniinae). Crustaceana, **26**(2): 222-224.

— —, 2005. Pontoniine shrimps from the 2003 NORFANZ Expedition, 10-May–16 June (Crustacea: Decapoda: Palaemonidae). Zootaxa, **981**: 1-20.

KEMP, S., 1922. Notes on Crustacea Decapoda in the Indian Museum, XV: Pontoniinae. Records of the Indian Museum, **24**: 113-288.

LI, X., M. MITSUHASHI & T.-Y. CHAN, 2008. Deep-sea pontoniines (Decapoda: Palaemonidae) from the Philippine "PANGLAO 2005" Expedition, with descriptions of four new species. Journal of Crustacean Biology, **28**(2): 385-411.

MITSUHASHI, M. & T.-Y. CHAN, 2006. A new genus and species of deep-water pontoniine shrimp (Decapoda: Caridea: Palaemonidae) from Taiwan. Zoosystema, **28**(2): 389-398.

— — & — —, 2009. A new deep-sea pontoniine shrimp (Decapoda: Palaemonidae) of the "*Periclimenes foresti* Bruce, 1981" species group from Taiwan. In: S. DE GRAVE, S. T. AHYONG & N. L. BRUCE (eds.), Special Issue on the Occasion of the 80th Birthday of A. J. (Sandy) Bruce — 50 Years of Shrimping in the Indo-West Pacific. Crustaceana, **82**(7): 919-929.

MITSUHASHI, M., X. LI & T.-Y. CHAN, 2010. *Neoclimenes holthuisi* n. gen., n. sp. a new deep-sea pontoniine shrimp from Taiwan (Crustacea: Decapoda: Palaemonidae). In: C. H. J. M. FRANSEN, S. DE GRAVE & P. K. L. NG (eds.), Studies on Malacostraca: Lipke Bijdeley Holthuis Memorial Volume. Crustaceana Monographs, **14**: 515-527. (Brill, Leiden).

MITSUHASHI, M. & M. TAKEDA, 2008. Identity of the coral-associated pontoniine shrimp species, *Coralliocaris nudirostris* (Heller, 1861) and *C. venusta* Kemp, 1922 (Crustacea: Decapoda: Palaemonidae), with descriptions of two new species. Zootaxa, **1703**: 1-24.

OKUNO, J., 2004. *Periclimenes speciosus*, a new species of anthozoan associated shrimp (Crustacea: Decapoda: Palaemonidae) from Southern Japan. Zoological Science, **21**: 865-875.

OKUNO, J. & A. J. BRUCE, 2010. Designation of *Ancylomenes* gen. nov., for the '*Periclimenes aesopius* species group' (Crustacea: Decapoda: Palaemonidae), with the description of a new species and a checklist of congeneric species. Zootaxa, **2372**: 85-105.

First received 7 October 2011.
Final version accepted 17 November 2011.

A NEW GENUS AND SPECIES OF APHANODACTYLIDAE AHYONG & NG, 2009 (DECAPODA, BRACHYURA, PINNOTHEROIDEA) FROM THE RYUKYU ISLANDS, JAPAN

BY

TOHRU NARUSE[1,3]) and TADAFUMI MAENOSONO[2])

[1]) Iriomote Station, Tropical Biosphere Research Center, University of the Ryukyus, 870 Uehara, Taketomi, Okinawa 907-1541, Japan

[2]) Kankyosha, 1-4-5-102 Kyozuka, Urasoe, Okinawa 901-2111, Japan

ABSTRACT

A new genus and new species of pinnotheroid crab of the family Aphanodactylidae, *Takedactylus masatsunei*, is described from Sesoko Island, adjacent to Okinawa Island, Ryukyu Islands, Japan. The new species differs markedly from known aphanodactylid genera by the characters of the carapace, epistome, third maxilliped and ambulatory legs, and as such, must be placed in its own new genus. This study brings the number of taxa in the family to nine species in five genera.

INTRODUCTION

The Aphanodactylidae Ahyong & Ng, 2009, is a small family in the Pinnotheroidea De Haan, 1833, and comprises four genera, *Aphanodactylus* Tesch, 1918 (four species), *Gandoa* Kammerer, 2006 (two species), *Gustavus* Ahyong & Ng, 2009 (monotypic), and *Uruma* Naruse, Fujita & Ng, 2009 (monotypic) (Ahyong & Ng, 2009; Naruse et al., 2009; Ng & Naruse, 2009). Members of these genera (when known) are associated with terebellid polychaete worms and morphologically united by several distinct features: proportionately short, stout, ambulatory legs with one or more spines lining the flexor margins of the ischiomerus of at least some of the legs; a row of distal, flexor spines on the ambulatory propodus opposing a claw-like dactylus (to be confirmed in *Gandoa*); a very short ambulatory dactylus that is less than

[3]) Corresponding author; e-mail: naruse@lab.u-ryukyu.ac.jp

half the respective propodus length; and marked sexual dimorphism in which the female is distinctly wider than the male (to be confirmed in *Gandoa* and *Uruma*) (Ahyong & Ng, 2009: 34).

Recently the second author collected an unusual small crab at Sesoko Island that is adjacent to Okinawa Island, Ryukyu Islands, Japan. The crab is clearly referable to Aphanodactylidae, and is here described as a new species in a new genus.

The measurements provided, in millimetres, are carapace length (CL) and width (CW) respectively. The abbreviations P2–P5 are used for pereopods 2-5, respectively. The type specimen is deposited in the Ryukyu University Museum, Fujukan (RUMF), University of the Ryukyus, Okinawa, Japan.

TAXONOMY

Family APHANODACTYLIDAE Ahyong & Ng, 2009

Takedactylus n. gen.

Type species. — *Takedactylus masatsunei* n. sp., by present designation.

Diagnosis. — Carapace rectangular, with proportionally wide fronto-orbital width; epibranchial angle located at level of orbit in dorsal view; lateral margins of carapace subparallel. Orbit entire, mesial end of infraorbital margin not produced, rounded. Posterior margin of epistome strongly produced medially. Thoracic sternites 1 and 2 fused; sternites 2/3 demarcated by shallow groove, lateral margins of sternites 2 and 3 convergent anteriorly; sternites 3/4 fused, laterally unarmed; sternites 4-8 demarcated by narrow grooves, those demarcating sternites 4-7 incomplete medially; without longitudinal median groove. Third maxillipeds covering entire buccal cavern; ischium small, mid-length slightly longer than half-length that of merus; carpus to dactylus articulated end to end. Chelipeds subequal; chela strong, palm high. Ambulatory legs robust, P2–P4 similar in shape, P5 smallest; meri of P2-4 stout, distal corner of lower-flexor margin of P2 and P3 with tuft of setae; propodus of P2-4 relatively short, slightly longer than respective carpus; distoflexor ends of propodus with a strong spine on upper and lower angle each; dactylus of P2-4 strong, with strong median spine on flexor-lower margin.

Etymology. — We are honoured to name the new genus after Prof. Masatsune Takeda, who has greatly enhanced our knowledge of the decapod fauna; not only through his own studies but also through the mentoring of many colleagues. The name is derived from an arbitrary combination of "Takeda" and

"-dactylus", the suffix of the type genus of Aphanodactylidae. Gender masculine.

Remarks. — *Takedactylus* n. gen. possesses the diagnostic features of the Aphanodactylidae, such as a non-enlarged palp of the third maxilliped and the presence of distoflexor spines of ambulatory propodi and opposing claw-like dactyli. However, it has several distinct characters that are not observed in the other genera. *Takedactylus* can also be characterized as a aphanodactylid by its rectangular shape of the carapace with subparallel lateral margins, short merus and exopod of the third maxilliped, setose distal ends of the lower-flexor margin of the meri of P2 and 3, and the presence of a median spine on the flexor-lower margin of the ambulatory dactyli. *Takedactylus* appears to be closer to *Gandoa* and *Uruma* than other genera in the rectangular shape of the carapace. However, in addition to the above-mentioned characters, *Takedactylus* can be distinguished from *Gandoa* and *Uruma* by its relatively wider fronto-orbital width to the carapace width (0.76 times CW, figs. 1a, 2a) [vs. about 0.35 in *Uruma ourana* Naruse, Fujita & Ng, 2009 (Naruse et al., 2009, figs. 1, 2a, 3a); 0.59 in *Gandoa zanzibarensis* (Lenz, 1905) (Lenz, 1905, pl. 47 fig. 9b); and about 0.51 in *G. brevipes* (H. Milne Edwards, 1853) (Ng & Naruse, 2009, fig. 6)] and strongly producing triangular projection on the posterior margin of the epistome (fig. 1b) [vs. gently producing entirely in *U. ourana* (Naruse et al., 2009, fig. 3a); gently produced medially in *G. zanzibarensis* (Lenz, 1905, pl. 47 fig. 9b) and in *G. brevipes* (Ng & Naruse, 2009, fig. 6b)].

Takedactylus masatsunei n. sp.
(figs. 1, 2)

Material examined. — Holotype: female, 1.8 × 2.9 mm, RUMF-ZC-1716, Sesoko Island, adjacent to Okinawa Island, Ryukyu Islands, Japan, coll. T. Maenosono, 22 Jul. 2009.

Description of holotype. — Carapace rectangular, CW 1.58 times CL; dorsal surface gently convex longitudinally and transversely, smooth, glabrous, regions poorly defined, H-shaped gastric groove and cervical groove slightly discernible. Front 0.24 times fronto-orbital width, fronto-orbital width 0.79 times external orbital width, 0.76 times CW, front gently sloping anteroventrally, converging distally. Orbital margin entire, rimmed; lateral corner not produced; mesial end of infraorbital margin not produced, rounded, leaving gap to front. Epibranchial angle rounded, projected anteriorly, reaching level of orbit in dorsal view. Epibranchial angle to posterolateral corner connected by entire rim, lateral margins of carapace subparallel.

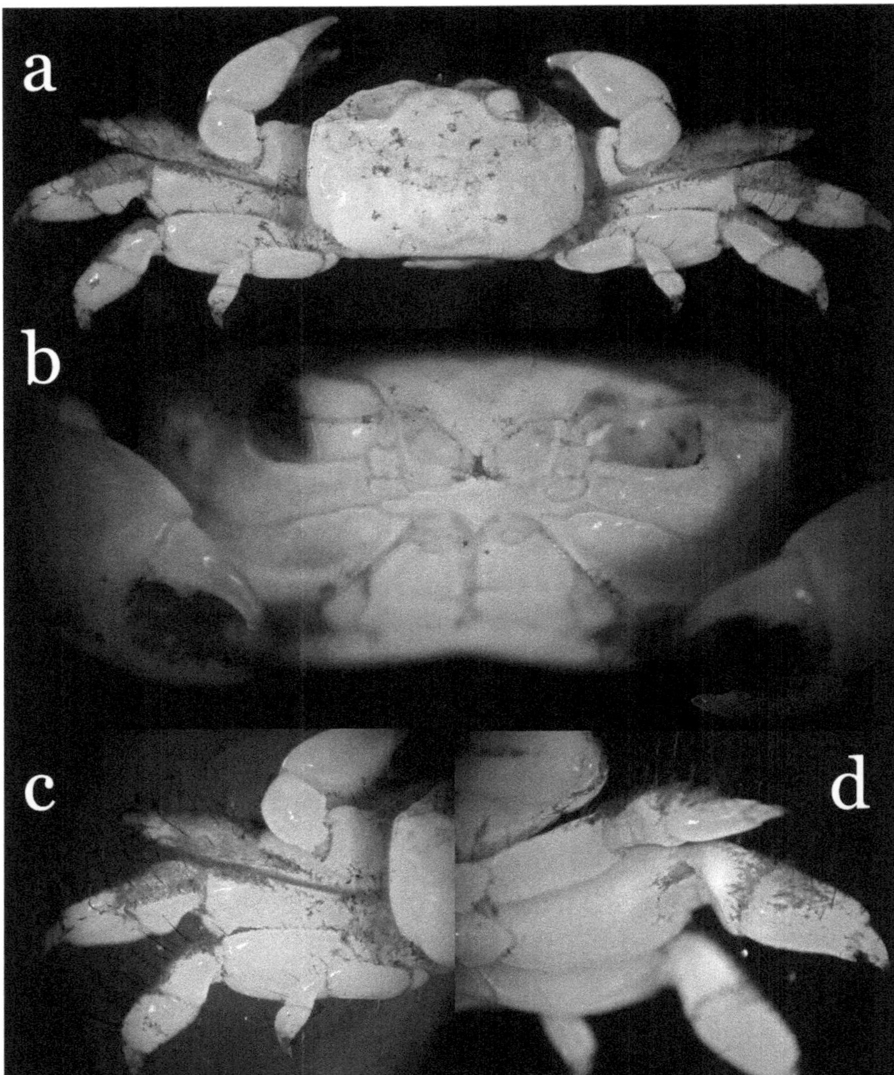

Fig. 1. *Takedactylus masatsunei* n. gen., n. sp. Holotype female (RUMF-ZC-1716). a, habitus, dorsal view; b, cephalothorax, anterior view; c, left pereiopods, upper view; d, left P2 and P3, lower view.

Eye large, with constant width from eyestalk to cornea. Left eye lost.

Basal antennal segments entering to orbit, antennal flagellum reaching distal end of eye.

Posterior margin of epistome sharply produced medially.

Third maxillipeds covering entire buccal cavern. Ischium small, mid-length slightly longer than half-length of that of merus. Carpus to dactylus articulated end to end, tip of dactylus reaching proximolateral end of merus when folded. Exopod slightly exceeding distolateral end of ischium, distally covered by lateral margin of merus in outer view, with long flagellum.

Chelipeds subequal. Merus triangular in cross-section, with distinct subdistal projection on dorsal margin; ventrolateral margin slightly produced, lined with granules. Carpus smooth, with rounded inner angle. Chela strong; palm high, length equal to height, inner surface setose around mid-line, outer surface glabrous; groove running from proximolower part of palm to base of occlusal margin of immovable finger. Immovable finger about 0.66 times palm, occlusal margin lined with small teeth. Movable finger curved distally, shorter than immovable finger, with a large subproximal tooth and several small teeth.

Ambulatory legs robust, similar in shape in P2–P4, small in P5; P3 longest, followed by P4, P2, P5. Meri of P2-4 stout, distal end of lower-posterior margin of P2 and P3 setose; distoanterior part of merus to extensor surface of carpus of P2 and P3 granulated, setose. Carpus of P2-4 granulated on outer surface. Propodus of P2-4 relatively short, slightly longer than respective carpus; distoflexor end of propodus with strong spine on upper and lower angle each, followed laterally by 1 small spine (2 spines on distal-upper surface in right P4). Dactylus of P2-4 strong, short, claw-like, with a strong median spine on flexor-lower margin.

Etymology. — The species is named after Prof. Masatsune Takeda.

Ecological note. — The holotype of *Takedactylus masatsunei* n. sp. was collected from a subtidal (about 1 m depth) silty layer under a buried rock. A new species of pinnotherid of the genus *Indopinnixa* Manning & Morton, 1987, was also collected from the same habitat (Naruse & Maenosono, in press), together with sipunculan (peanut) worms. However, when *T. masatsunei* n. sp. was collected, these possible host animals were not observed in the sieved net.

Remarks. — See discussion for genus.

ACKNOWLEDGMENTS

We thank Peter K. L. Ng (National University of Singapore) and Tomoyuki Komai (Natural History Museum and Institute, Chiba) for reviewing our manuscript.

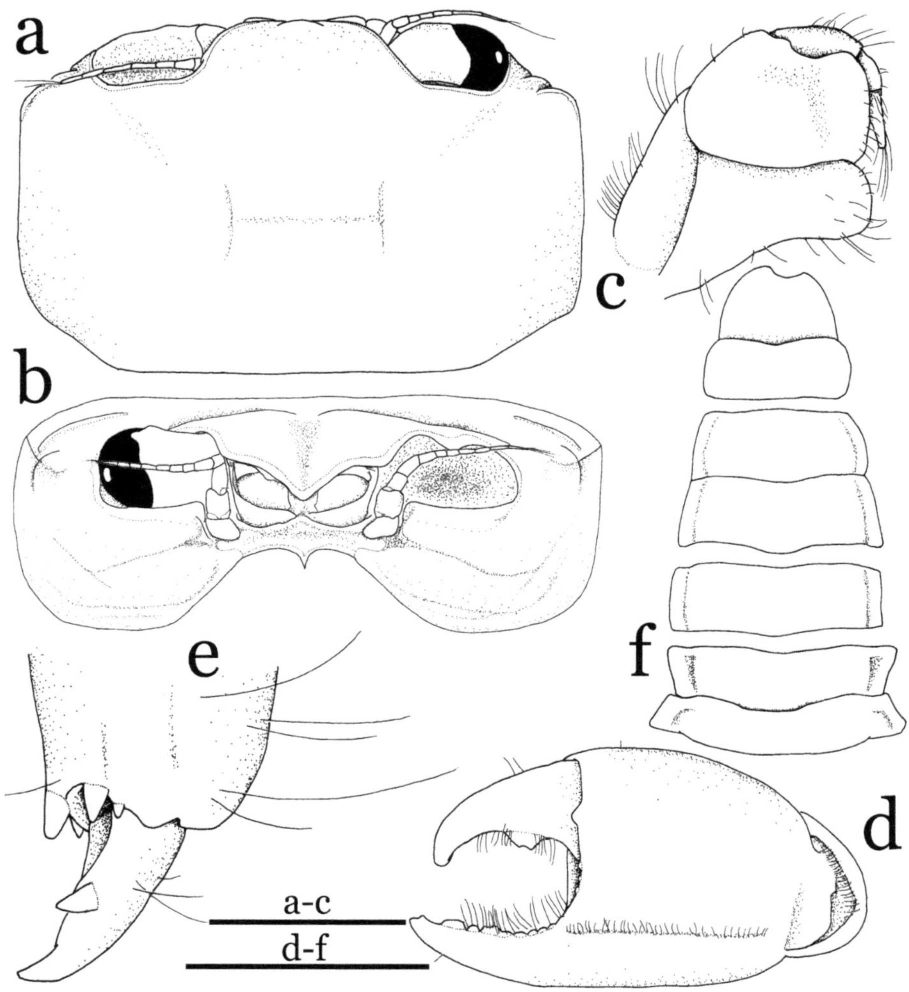

Fig. 2. *Takedactylus masatsunei* n. gen., n. sp. Holotype female (RUMF-ZC-1716). a, cephalothorax, dorsal view; b, right third maxilliped; c, left chela, outer view; d, left third pereopod, distal part of propodus and dactylus, inner-lower view; e, abdomen and telson. Scales, a, c, e = 1.0 mm; b, d = 0.5 mm.

REFERENCES

AHYONG, S. T. & P. K. L. NG, 2009. Aphanodactylidae, a new family of thoracotreme crabs (Crustacea: Brachyura) symbiotic with polychaete worms. Zootaxa, **2289**: 33-47.

HAAN, W. DE, 1833. Crustacea. In: P. F. VON SIEBOLD (ed.), Fauna Japonica, sive Descriptio animalium, quae in itinere per Japoniam, jussu et auspiciis superiorum, qui summum in India Batava imperium tenent, suscepto, annis 1823-1830 collegit, notis, observationibus et adumbrationibus illustravit P.F. de Siebold. Conjunctis studiis C.J. Temminck et H. Schlegel pro Vertebratis atque W. de Haan pro Invertebratis elaborata Regis aupicus edita. Decas II: 25-64, pls. 9-15, 17, C, D. (Lugundi-Batavorum, Leiden).

KAMMERER, C. F., 2006. Notes on some preoccupied names in Arthropoda. Acta Zootaxonomica Sinica, **31**(2): 269-271.

LENZ, H., 1905. Ostafrikanische Dekapoden und Stomatopoden: Gesammelt von Herrn Prof. Dr. A. Voeltzkow. Abhandlungen herausgegeben der Senckenbergischen naturforschenden Gesellschaft, **27**(4): 341-392, pls. 47, 48.

MANNING, R. B. & B. MORTON, 1987. Pinnotherids (Crustacea: Decapoda) and leptonaceans (Mollusca: Bivalvia) associated with sipunculan worms in Hong Kong. Proceedings of the Biological Society of Washington, **100**(3): 543-551.

MILNE EDWARDS, H., 1853. Mémoire sur la famille des Ocypodiens. Suite. Annales des Sciences Naturelles, Zoologie et Biologie Animale, **20**(3): 163-228, pls. 6-11.

NARUSE, T., Y. FUJITA & P. K. L. NG, 2009. A new genus and new species of symbiotic crab (Crustacea: Brachyura: Pinnotheroidea) from Okinawa, Japan. Zootaxa, **2053**: 59-68.

NARUSE, T. & T. MAENOSONO, in press. Two new species of *Indopinnixa* Manning & Morton, 1987 (Decapoda: Brachyura: Pinnotheridae) from the Ryukyu Islands, Japan. In: T. NARUSE, T.-Y. CHAN, H. H. TAN, S. T. AHYONG & J. D. REIMER (eds.), Scientific Results of the Kumejima Marine Biodiversity Expedition — KUMEJIMA 2009. Zootaxa.

NG, P. K. L. & T. NARUSE, 2009. On the identity of *Pinnixa brevipes* H. Milne Edwards, 1853, and a new species of *Aphanodactylus* Tesch, 1918 (Crustacea: Decapoda: Brachyura: Pinnotheroidea) from the Philippines. Raffles Bulletin of Zoology, (Supplement) **20**: 283-290.

TESCH, J. J., 1918. The Decapoda Brachyura of the Siboga-Expedition. II. Goneplacidae and Pinnotheridae. Siboga-Expeditie Monographie, **39c1**: 149-295, 12 pls.

First received 19 September 2011.
Final version accepted 31 January 2012.

A NEW SPECIES OF CNIDARIAN-ASSOCIATED SHRIMP OF THE GENUS *CUAPETES* CLARK, 1919 (DECAPODA, PALAEMONIDAE) FROM SURUGA BAY, JAPAN

BY

JUNJI OKUNO[1])

Coastal Branch of Natural History Museum and Institute, Chiba, 123 Yoshio, Katsuura, Chiba 299-5242, Japan

ABSTRACT

A new species of pontoniine shrimp, *Cuapetes takedai* sp. nov., is described and illustrated on the basis of five specimens from Suruga Bay, Honshu, Japan. The new species is closely related to *C. nilandensis* (Borradaile) on account of the lack of meral and ischial distoventral teeth on the second pereiopod and the presence of elongate distoventral spines on the propodi of the ambulatory pereiopods, but can be readily distinguished from the latter species by the lack of a supraorbital spine and the carpus of the second pereiopod being considerably longer than the palm. The new species is associated with the hydroid, *Lytocarpia niger* (Nutting) and an unidentified gorgonarian.

INTRODUCTION

To date, the pontoniine shrimp genus *Cuapetes* Clark, 1919 contains 25 species, occurring worldwide in the tropics (De Grave & Fransen, 2011). Most members of the genus are free-living, but a few species are known as associates of cnidarians, viz. *C. amymone* (De Man, 1902) and *C. kororensis* (Bruce, 1977) are associated with scleractinian corals (Patton, 1966; Bruce, 1998, 2004), and *C. nilandensis* (Borradaile, 1915) with a wide variety of cnidarians (see below).

Three specimens of an unfamiliar pontoniine shrimp living on a sessile hydroid, *Lytocarpia niger* (Nutting, 1906), from Suruga Bay, Pacific coast of central Honshu, Japan, were available for our taxonomic studies of pontoniine

[1]) e-mail: okuno@chiba-muse.or.jp

shrimps from Japan. In addition, two specimens of the same species were collected from an unidentified gorgonarian coral at the same collecting site. In morphology and ecology, this species is closely allied to *C. nilandensis*, which is the type species of the genus and known as an associate of several cnidarians, such as hydroids, alcyonarians and antipatharians (Bruce, 2004). Based on a direct comparison with specimens of *C. nilandensis*, the species from Suruga Bay is readily distinguished by the lack of a supraorbital spine on the carapace. Therefore, a new species is established for the pontoniine shrimp from Suruga Bay under the name of *Cuapetes takedai* sp. nov.

MATERIAL AND METHODS

Specimens of the new species were collected from the sublittoral zone with SCUBA equipment. The illustrations were made with the aid of a drawing tube mounted on a LEICA MZ12 stereomicroscope. The postorbital carapace length is abbreviated as CL in the text. The specimens examined in this study are deposited in the Coastal Branch of Natural History Museum and Institute, Chiba, Japan (CMNH) and National Museum of Nature and Science, Tokyo, Japan (NSMT).

For comparative purpose, the following specimens were examined in this study: *Cuapetes nilandensis*. CMNH-ZC 01899, 1♀, CL 2.0 mm, Seragaki, Onna Village, Okinawa Island, Ryukyu Islands, Japan, 48-50 m, in association with an unidentified Antipatharia, 4 May 2005, leg. T. Yanagisawa; CMNH-ZC 02412, 1 ovig.♀, CL 2.4 mm, Kyoda, Nago, Okinawa Island, Ryukyu Islands, Japan, 30 m, in association with an unidentified gorgonarian, 19 November 2010, leg. M. Tsuchiya and Y. Yamada.

TAXONOMIC ACCOUNT

Family PALAEMONIDAE Rafinesque, 1815
Subfamily PONTONIINAE Kingsley, 1879
Genus *Cuapetes* Clark, 1919
Cuapetes takedai sp. nov.
(figs. 1-5)

Material examined. — Holotype. NSMT-Cr 21979, ♂, CL 1.6 mm, Ose-zaki, Numazu, Suruga Bay, Honshu, Japan, 35°01.5′N 138°47.6′E, 10 m, 23 November 2003, leg. T. Yanagisawa.

Paratypes. CMNH-ZC 02413, 1♀, CL 2.0 mm, same locality as holotype, 15-19 m, November 1997, leg. R. Minemizu; NSMT-Cr 21980, 1♀, CL 1.2 mm, same locality as

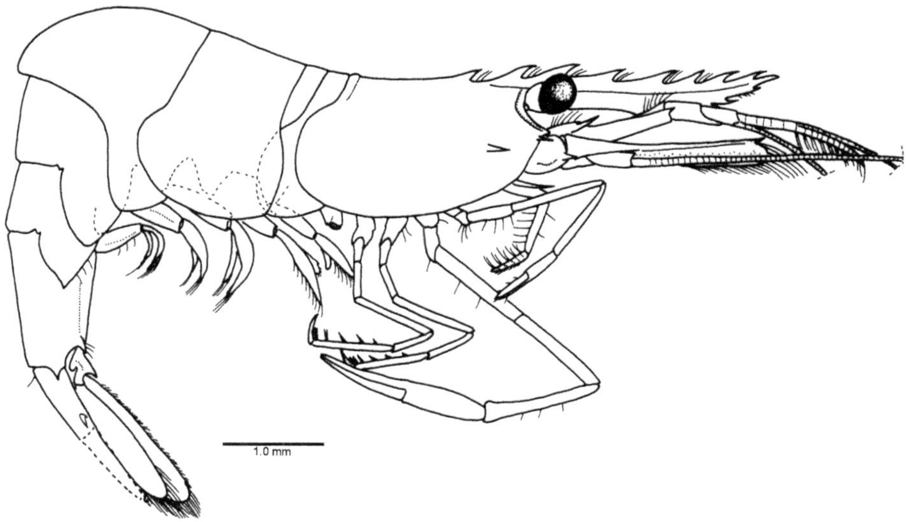

Fig. 1. *Cuapetes takedai* sp. nov. Holotype male (NSMT-Cr 21979). Entire animal in lateral view. Distal half of telson broken.

holotype, 12 October 2003, leg. T. Yanagisawa; CMNH-ZC 02414, 2♂♂, CL 1.8, 1.7 mm, same locality as holotype, 16 m, 24 August 2011, leg. R. Minemizu.

Host. — The holotype and two paratypes (CMNH-ZC 02413 and NSMT-Cr 21980) were collected from *Lytocarpia niger* (Cnidaria: Hydrozoa). Other paratypes (CMNH-ZC 02414) were associated with an unidentified gorgonarian (Cnidaria: Anthozoa).

Description. — Body small size. Carapace (fig. 1) glabrous, supraorbital spine absent; orbit feebly developed, with distinct posterior marginal ridge, inferior orbital angle strongly produced, somewhat angular; antennal spine well developed, slender, submarginal, overreaching level of distal margin of inferior orbital angle; hepatic spine large, arising slightly ventral to level of antennal spine; epigastric spine present; pterygostomial margin rounded.

Rostrum (fig. 2B) slender, mainly horizontal, distal part feebly sinuous dorsally, 1.4-1.7 times as long as median post-orbital carapace length, distinctly overreaching level of distal margin of antennular peduncle; dorsal carina armed with 7-8 equidistant, acute teeth, proximal tooth situated posterior to posterior orbital margin, interspaces with short setae; ventral carina armed with 4 (rarely 3) acute teeth; lateral carina distinct.

Fourth thoracic sternite (fig. 2D) with slender, acute median process; fifth sternite (fig. 2D) with pair of triangular submedian teeth.

Abdomen (fig. 1) glabrous; pleura of first to third somites broad, rounded, those of fourth and fifth somites posteriorly produced, fifth somite terminating

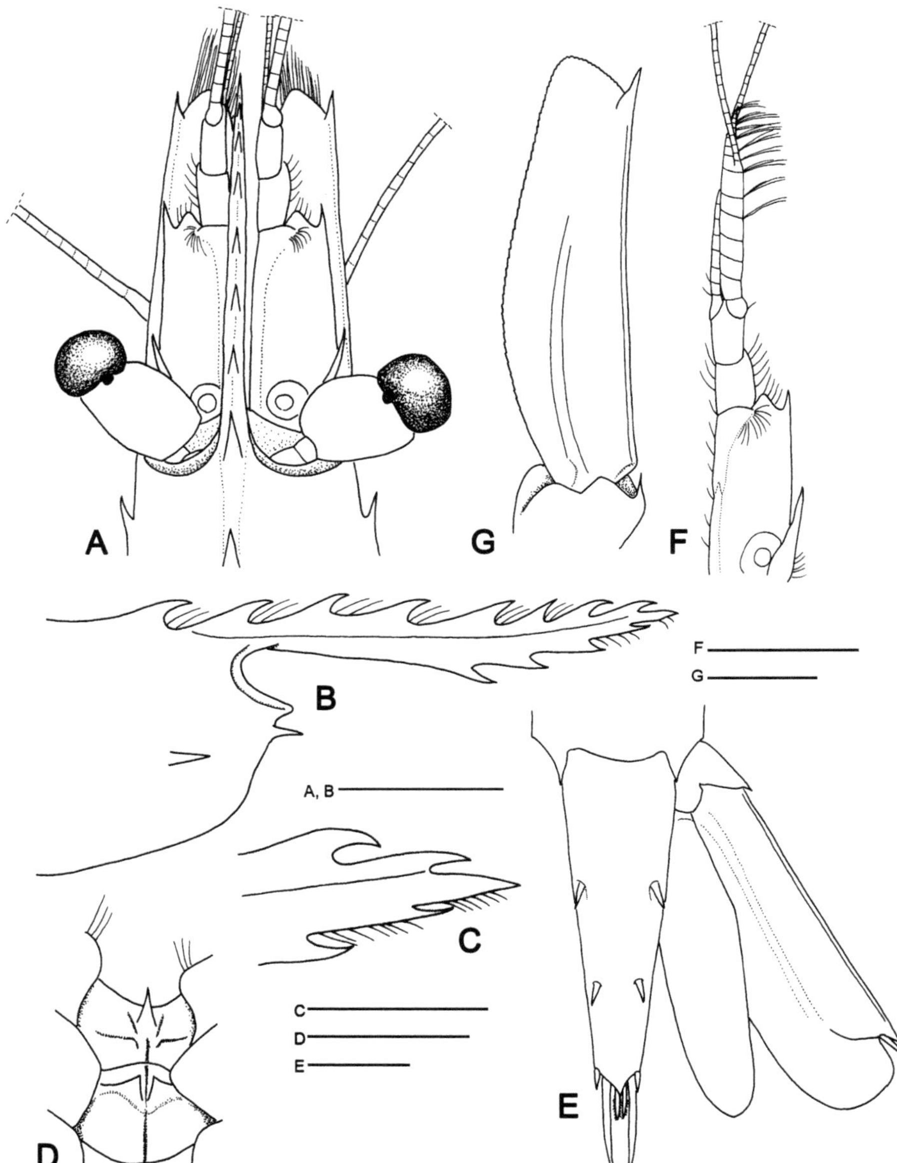

Fig. 2. *Cuapetes takedai* sp. nov. A–D, F–G, holotype male (NSMT-Cr 21979); E, paratype male (CMNH-ZC 02414). A, anterior part of carapace and cephalic appendages, dorsal view; B, anterior half of carapace and rostrum, lateral view; C, tip of rostrum, lateral view; D, fourth and fifth thoracic sternites, ventral view; E, telson and right uropod, dorsal view; F, right antennular peduncle, dorsal view; G, right scaphocerite, dorsal view. E, G, marginal setae omitted. Scales: A, B, F, 1.0 mm; C–E, G, 0.5 mm.

posteroventrally in acute tooth; sixth somite 0.7-0.9 times as long as carapace, 0.8-0.9 times as long as telson, posterolateral process acute, posteroventral margin produced, acute. Telson (fig. 2E) tapering posteriorly, armed dorsolaterally with 2 pairs of robust spines at proximal third and posterior third length respectively; posterior margin triangular, with 3 pairs of spines, lateral spines simple, slightly smaller than dorsolateral spines, subdorsal, intermediate spines simple, longest, submedian spines plumose.

Ophthalmic somite without interocular beak. Eye (fig. 2A) with large, globular cornea, with minute accessory pigment spot, stalk distinctly longer than corneal diameter, becoming slightly narrower distally, maximum width subequal to maximum corneal diameter.

Antennular peduncle (fig. 2F) falling slightly short of level of base of distolateral tooth of scaphocerite; proximal segment distinctly longer than combined length of distal two segments, broad, distolateral margin strongly produced, somewhat triangular, reaching level of proximal third of intermediate segment, with row of setae, lateral margin almost straight, terminating distally in large acute tooth, falling slightly short of midlength of intermediate segment, ventromedial margin armed with small acute tooth, stylocerite short, slender, acute, falling slightly short of level of midlength of proximal segment, statocyst well developed, rounded; intermediate segment feebly lobed laterally, about third of proximal segment length, slightly obliquely articulated with distal segment; distal segment slightly longer than intermediate segment, slender, non-setose. Upper flagellum biramous, proximal 6-7 segments fused, shorter free ramus containing 5-8 segments; lower flagellum slenderer than upper flagellum.

Antenna (fig. 2G) with stout basicerite armed ventrolaterally with acute tooth; scaphocerite falling slightly short of level of rostral apex, 2.6-3.2 times as long as maximum width, lateral margin, feebly concave, terminating distally in strong tooth reaching level of distal margin of obliquely truncated lamella; carpocerite reaching proximal third of scaphocerite.

Epistome unarmed.

Mandible (fig. 3A) robust, without palp; molar process obliquely truncated distally, with few large, blunt teeth; incisor process tapering distally, with 2 or 3 slightly acute distal teeth. Maxillule (fig. 3B) with feebly bilobed palp, internal lobe with small distal protuberance with short, curved inferior seta; upper lacinia broad, distal margin truncate, with 2 rows of simple spines and sparse, short spiniform setae; lower lacinia tapering distal to subquadrate tip, with some serrulate setae distally. Maxilla (fig. 3C) with palp slender,

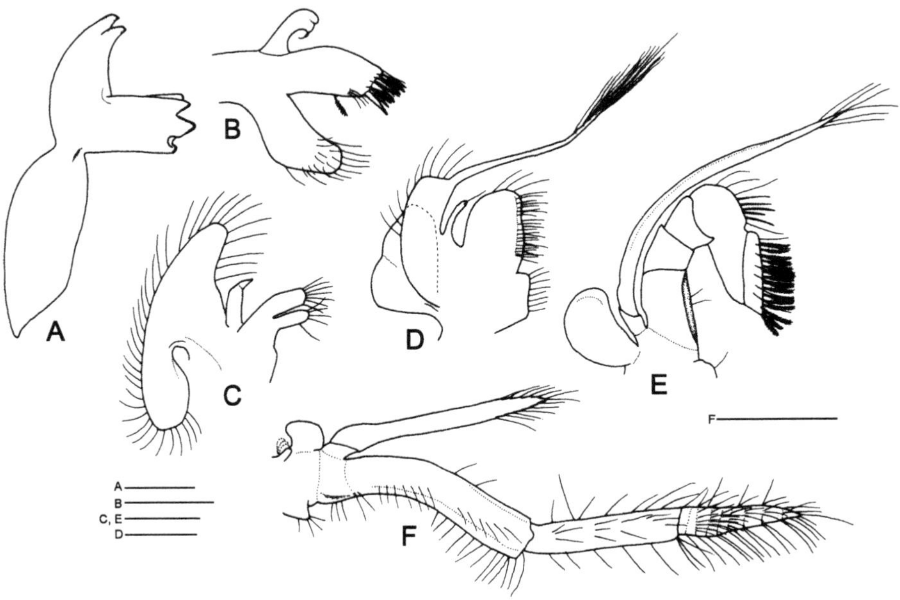

Fig. 3. *Cuapetes takedai* sp. nov. Right mouthparts, external view. A–B, paratype female (CMNH-ZC 02413); C–D, holotype male (NSMT-Cr 21979); E–F, paratype male (CMNH-ZC 02414). A, mandible; B, maxillule; C, maxilla; D, first maxilliped; E, second maxilliped; F, third maxilliped. Scales: A–E, 0.25 mm; F, 0.5 mm.

tapering distally; distal endite developed, narrow, deeply bilobed, with sparse, simple setae distally; proximal endite obsolete, medial margin feebly sinuous; scaphognathite well developed, posterior lobe short, anterior lobe slightly tapering distally. First maxilliped (fig. 3D) with long, slender, simple palp with subterminal spiniform seta; distal endite with medial margin bearing 2 rows of sparse setae, subquadrate distally; proximal endite small, angular, separated from distal endite by shallow notch; caridean lobe broad; exopod with well-developed flagellum with numerous distal setae; epipod large, subtriangular, feebly bilobed. Second maxilliped (fig. 3E) with normal endopod; dactylus narrow, medial margin almost straight; propodus with anteromedial margin broadly rounded; carpus with distinct ventromedial process; merus subequal to carpus in length; ischium and basis feebly demarcated; exopod with well-developed flagellum with numerous dense setae; coxa inflated medially; epipod oval, without podobranch. Third maxilliped (fig. 3F) with endopod slender, falling slightly short of level of distal margin of scaphocerite; ultimate segment tapering distally, with about 8 transverse rows of long, dense setae; penultimate segment 1.4-2.3 times as long as ultimate segment, furnished sparsely with long simple setae, ischiomerus and basis feebly demarcated;

TABLE I
Cuapetes takedai sp. nov.: branchial formula

	Maxillipeds			Pereiopods				
	I	II	III	I	II	III	IV	V
Pleurobranchs	–	–	–	1	1	1	1	1
Arthrobranchs	–	–	1	–	–	–	–	–
Podobranchs	–	–	–	–	–	–	–	–
Epipods	1	1	1	–	–	–	–	–
Exopods	1	1	1	–	–	–	–	–

exopod with well-developed flagellum subequal to ischiomerus, distally with numerous long setae; coxal plate oval; arthrobranch small, consisting of few rounded lamellae.

Branchial formula as in table I.

First pereiopod (fig. 4A) moderately slender, overreaching distal margin of scaphocerite by length of dactylus. Chela (fig. 4B) 0.5-0.6 times as long as carapace, with palm subcylindrical, slightly compressed, shorter than dactylus, with 3 transverse rows of short serrulate grooming setae proximally; fingers each terminating in small, hooked unguis, cutting edges situated laterally, entire. Carpus long, 0.5-0.6 times as long as carapace, 1.8-2.3 times as long as palm, subequal to chela in length, slightly widened distally, with longitudinal row of serrulate grooming setae distally. Merus unarmed, subequal to carpus in length.

Second pereiopods (fig. 4C) well developed, similar, overreaching distal margin of scaphocerite by length of chelae. Chela subcylindrical, 0.8-1.1 times as long as carapace; dactylus (fig. 4D) 1.1-1.5 times as long as palm, terminating in hooked, acutely pointed unguis with dense long setae subapically, cutting edge situated laterally, armed proximally with small, acute tooth, remaining part entire, feebly setose, sharply edged; fixed finger (fig. 4D) generally similar to dactylus, armed proximally with 2 small, acute teeth. Carpus slender, elongate, 1.5-2.0 times as long as palm, 1.1-1.3 times as long as merus, unarmed, slightly widened distally. Merus slender, unarmed. Ischium slender, unarmed, subequal to merus in length.

Third pereiopod (fig. 4E) slender, falling slightly short of distal margin of scaphocerite. Dactylus (fig. 4F) simple, slender, elongate, 0.4 times as long as propodus, compressed laterally, dorsal margin weakly arched, unguis feebly demarcated, flexor margin of corpus feebly concave distally. Propodus 1.7-2.0 times as long as carpus, distoventrally with 2 sets of paired long spines about

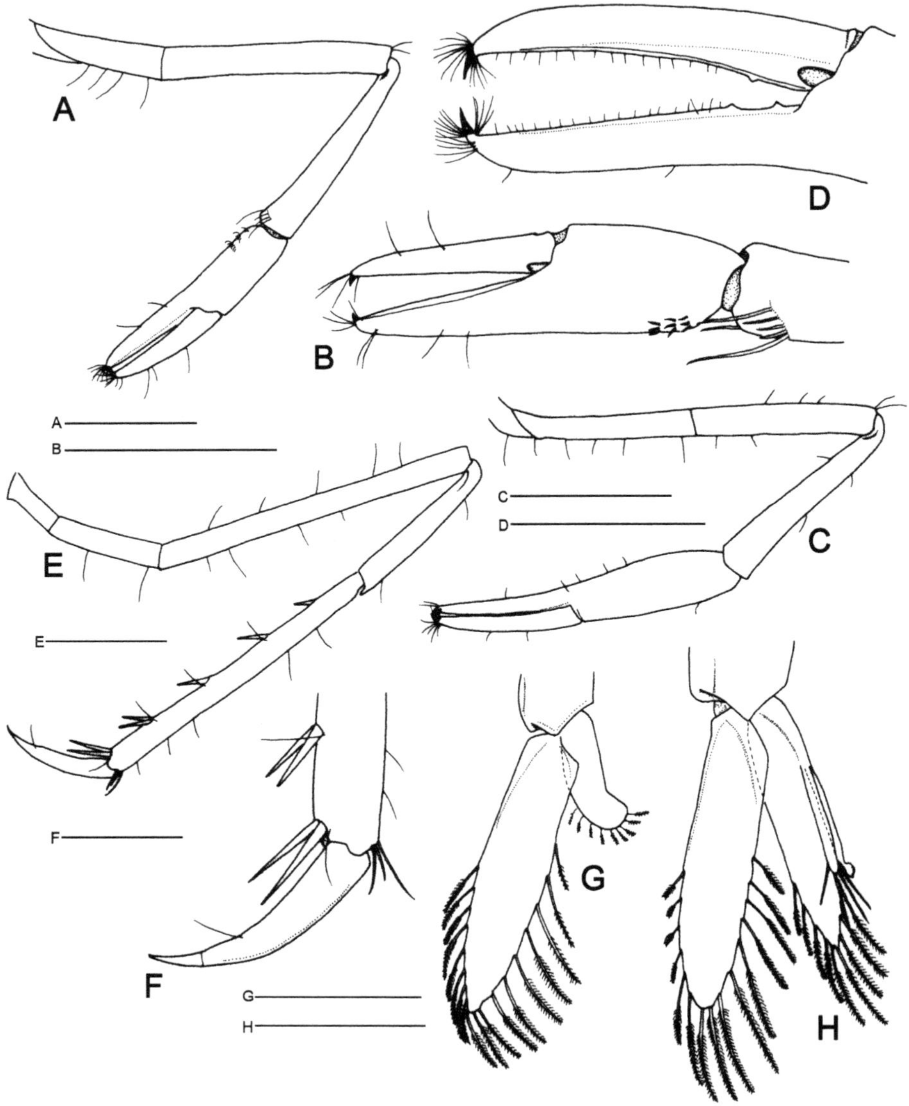

Fig. 4. *Cuapetes takedai* sp. nov. Holotype male (NSMT-Cr 21979). A, right first pereiopod, lateral view; B, same, chela and posterior part of carpus, mesial view; C, right second pereiopod, lateral view; D, same, fingers, mesial view; E, right third pereiopod, lateral view; F, same, dactylus and posterior part of propodus, lateral view; G, right first pleopod, dorsal view; H, right second pleopod, dorsal view. Scales: A, B, D, E, G, H, 0.5 mm; C, 1.0 mm; F, 0.25 mm.

half length of dactylar corpus, with 2-3 single spines on remaining ventral surface, dorsal surface with few short setae. Carpus unarmed. Merus 1.7-2.2 times as long as carpus, unarmed. Armature of fourth and fifth pereiopods similar to that of third. Fourth pereiopod reaching or slightly overreaching level of distal margin of scaphocerite. Fifth pereiopod longer than anterior two ambulatory pereiopods, overreaching level of distal margin of scaphocerite by lengths of dactylus and distal third of propodus.

Endopod of male first pleopod (fig. 4G) elongate, generally oblong, slightly expanded distomedially, distal to distolateral margin furnished with sparse short plumose setae. Endopod of male second pleopod (fig. 4H) with appendices interna and masculina arising from proximal third of medial margin; appendix interna slender, slightly overreaching tip of appendix masculina, with few distal cincinnuli; appendix masculina slender, with 4 long spiniform setae distally.

Uropod (fig. 2E) with protopodite posterolaterally produced, acute; exopod broad, overreaching posterior margin of telson, distal margin broad, subquadrate, lateral margin nearly straight, terminating distally in small acute tooth, with long, mobile spine just medial to distolateral tooth; endopod oblong, falling slightly short of posterior margin of exopod.

Colouration (fig. 5). — Body and thoracic and abdominal appendages generally transparent. Rostrum, antennular peduncle and scaphocerite covered densely with red chromatophores. Pleurobranchs red, visible as five red oblique bands through transparent integument of carapace. Ventral midline of thoracic and abdominal sternites red. Cornea and eyestalk pale gold. Mouthparts yellowish brown. Coxae of thoracic and abdominal appendages white. In individuals associated with Hydrozoa, carapace and abdominal somites covered with fine yellow spots (fig. 5A).

Etymology. — The present new species is named in honor of my respected teacher, Dr. Masatsune Takeda, in recognition of his great contributions of knowledge of the Indo-Pacific carcinology. From "Takeda sensei", I was given warm guidance and several suggestions for taxonomic studies on Crustacea throughout my career.

Common name. — Horoyoi-tenaga-kakure-ebi (new Japanese name).

Distribution. — So far only known from the type locality, Ose-zaki, Numazu, Suruga Bay, Honshu, Japan.

Remarks. — The present new species can be assigned to the genus *Cuapetes* on account of the following diagnostic characters: carapace armed with antennal spine and fixed hepatic spine; fourth thoracic sternite armed with an acute median process; mandible without palp; all maxillipeds with exopodal flagella; first pereiopod with both movable and fixed fingers with cutting borders unarmed, without denticulate or pectinate structure; second pereiopod

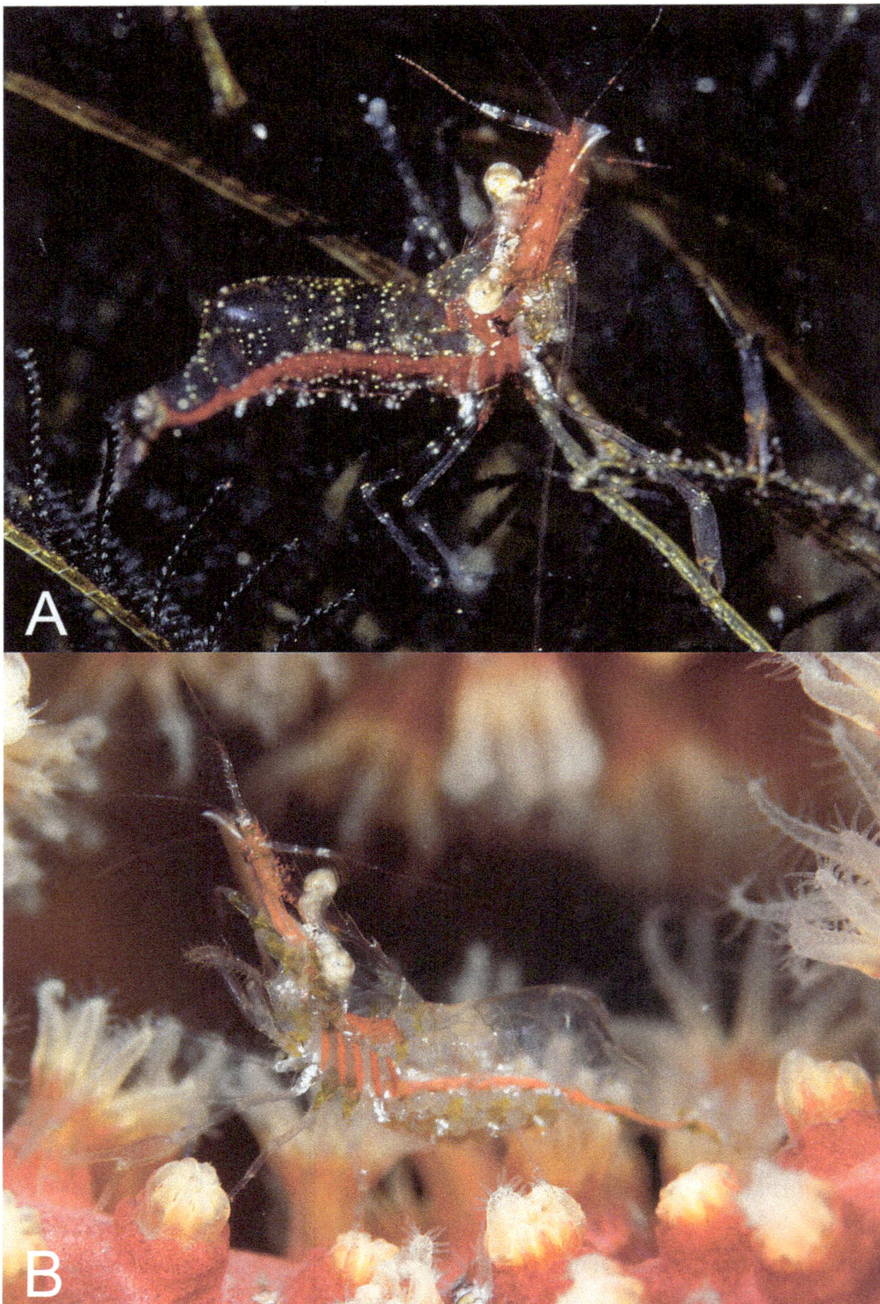

Fig. 5. *Cuapetes takedai* sp. nov., living individuals *in situ* (Ose-zaki, Numazu, Suruga Bay, Honshu, Japan). A, specimen associated with *Lytocarpia niger* (Nutting, 1906); B, specimen associated with an unidentified gorgonarian. Both specimens not collected. Photo R. Minemizu.

elongate, exceeding distal margin of scaphocerite by length of chela and distoventral margin of ambulatory propodi not densely setose but spinose.

Cuapetes takedai is closely related to *C. nilandensis*, the type species of *Cuapetes*, on account of the lack of the ischial and meral distoventral spines on the second pereiopod and the presence of long spines on the distoventral margin of the propodi of the ambulatory pereiopods. The absence of a supraorbital spine on the carapace immediately distinguishes *C. takedai* from *C. nilandensis*, which has a well-developed supraorbital spine. Furthermore, in *C. takedai*, the second pereiopod has the carpus distinctly longer than the palm, while the carpus is distinctly shorter than the palm in *C. nilandensis*.

The hosts of the present new species are included in the cnidarian classes Hydrozoa and Octocorallia. This ecological aspect is also similar to *C. nilandensis*, which is known as an associate of cnidarians, such as hydrozoans, alcyonarians, gorgonarians and antipatharians (Bruce, 2004). As in the case of *Periclimenes laevimanus* Ďuriš, 2010 and species belonging to the genus *Rapipontonia* Marin, 2007, the elongated distoventral spines on the propodi of the ambulatory pereiopods in these two *Cuapetes* species seems to be a functional adapatation for grasping the host cnidarians (Marin, 2007; Ďuriš, 2010).

ACKNOWLEDGEMENTS

I am indebted to a professional underwater photographer, Mr. Ryo Minemizu, and skillful amateur diver, Mr. Tohru Yanagisawa for donating to me the specimens of the new species. Especially, cordial thanks are extended to Mr. Minemizu agreeing to the concept of the memorial volume, and kindly provided his excellent underwater photographs to Dr. Takeda. I am grateful to Dr. Sandy Bruce and two anonymous reviewers for their valuable comments and suggestions to improve the manuscript.

REFERENCES

BORRADAILE, L. A., 1915. Notes on Carides. Annals and Magazine of Natural History, (8) **15**: 205-213.

BRUCE, A. J., 1977. *Periclimenes kororensis* n. sp., an unusual shrimp associate of fungiid coral, *Heliofungia actiniformis*. Micronesica, **13**(1): 33-43.

— —, 1998. New keys for the identification of Indo-West Pacific coral associated pontoniine shrimps, with observations on their ecology (Crustacea: Decapoda: Palaemonidae). Ophelia, **49**: 29-46.

— —, 2004. A partial revision of the genus *Periclimenes* Costa, 1884 (Crustacea: Decapoda: Palaemonidae). Zootaxa, **582**: 1-26.

CLARK, A. H., 1919. Some necessary changes in crustacean nomenclature. Proceedings of the Biological Society of Washington, **32**: 199.

DE GRAVE, S. & C. H. J. M. FRANSEN, 2011. Carideorum catalogus: The Recent species of the Dendrobranchiate, stenopodidean, procarididean and caridean shrimps (Crustacea: Decapoda). Zoologische Mededelingen, Leiden, **85**(9): 195-589.

ĎURIŠ, Z., 2010. *Periclimenes laevimanus* sp. nov. from Vietnam, with a review of the *Periclimenes granulimanus* species group (Crustacea: Decapoda: Palaemonidae: Pontoniinae). In: S. DE GRAVE & C. H. J. M. FRANSEN (eds.), Contributions to shrimp taxonomy. Zootaxa, **2372**: 106-125.

KINGSLEY, J. S., 1879. List of the North American Crustacea belonging to the suborder Caridea. Bulletin of the Essex Institute, **10**(for 1878): 53-71.

MAN, J. G. DE, 1902. Die von Herrn Professor Kükenthal im indischen Archipel gesammelten Decapoden und Stomatopoden. In: W. KÜKENTHAL (ed.), Ergebnisse einer zoologischen Forschungsreise in den Molukken und Borneo. Abhandlungen Senckenbergischen Naturforschenden Gesellschaft, **25**: 467-929, pls. 19-27.

MARIN, I., 2007. A new genus and species of pontoniine shrimp (Crustacea, Decapoda, Palaemonidae, Pontoniinae) associated with plumularid hydroids (Hydroidea, Plumularidae) in Vietnam. Zoosystema, **29**(4): 775-786.

NUTTING, C. C., 1906. Hydroids of the Hawaiian Islands collected by the Steamer Albatross in 1902. Bulletin of the United States Fish Commission, **23**(3): 931-959, pls. 1-13.

PATTON, W. K., 1966. Decapod Crustacea commensal with Queensland branching corals. Crustaceana, **10**(3): 271-295.

RAFINESQUE, C. S., 1815. Analyse de la nature ou tableau de l'univers et des corps organisés: 1-224. (Palerme).

First received 4 November 2011.
Final version accepted 21 December 2011.

NEW RECORDS OF ALBUNEIDAE (DECAPODA, ANOMURA) FROM JAPAN, WITH DESCRIPTION OF A NEW SPECIES OF *PARALBUNEA*

BY

MASAYUKI OSAWA[1,4]) and YOSHIHISA FUJITA[2,3])

[1]) Research Centre for Coastal Lagoon Environments, Shimane University, 1060 Nishikawatsu-cho, Matsue, Shimane 690-8504, Japan

[2]) University Education Centre, University of the Ryukyus, 1 Senbaru, Nishihara, Okinawa 903-0213, Japan

[3]) Marine Learning Centre, 2-95-101 Miyagi, Chatan, Okinawa 904-0113, Japan

ABSTRACT

Three albuneid species are newly reported from Japan on the basis of material from the Ryukyu Islands. *Albunea marquisiana* Boyko, 2000 is recorded for the first time outside of the type locality, Marquises Islands in French Polynesia. A single immature female specimen is provisionally referred to *Albunea* aff. *symmysta* (Linnaeus, 1758), but it is different from the species in the shape of the pereopods II–IV dactyli. *Paralbunea takedai* sp. nov. closely resembles *P. paradoxa* (Gordon, 1938), but the configuration of the distal ocular segment distinguishes the two species.

RÉSUMÉ

Trios espèces albunéides sont signalées pour la première fois du Japon sur la base d'un matériel provenant des îles Ryukyu. *Albunea marquisiana* Boyko, 2000 est trouvée pour la première fois en dehors de la localité type, les îles Marquises en Polynésie française. L'unique spécimen d'une femelle immature est provisoirement rapporté à *Albunea* aff. *symmysta* (Linnaeus, 1758), mais il s'en distingue par la forme du dactyle des péréiopodes II–IV. *Paralbunea takedai* sp. nov. ressemble étroitement à *P. paradoxa* (Gordon, 1938), mais la configuration du segment distal de l'œil distingue les deux espèces.

INTRODUCTION

The anomuran crabs of the family Albuneidae Stimpson, 1858 are specialised forms, living buried in sand and which can burrow deeply. Because of

[4]) Corresponding author; e-mail: osawam@soc.shimane-u.ac.jp

their preferred habitat, they are uncommonly encountered unless specifically targeted for collection, and thus have been poorly studied in general. Worldwide, the family currently includes 48 species belonging to nine genera (Boyko & McLaughlin, 2010). Among them, seven species have been recorded from Japanese waters: *Albunea elioti* Benedict, 1904; *A. groeningi* Boyko, 2002; *A. microps* Miers, 1878; *A. occulta* Boyko, 2002; *A. okinawaensis* Osawa & Fujita, 2007; *A. speciosa* Dana, 1852; and *Paralbunea dayriti* (Serène & Umali, 1965) (Boyko, 2002; Osawa & Fujita, 2007).

We recently obtained some unusual albuneids from the Ryukyu Islands, southwestern Japan. Examination of the specimens revealed that they include three species new to the Japanese fauna. The present paper reports on these three species, including the description of a new species, and increases the Japanese albuneid biodiversity to ten species.

The specimens examined are deposited in the Ryukyu University Museum, Fujukan (RUMF) and the Natural History Museum and Institute, Chiba (CBM). Carapace length (cl), as measured from the tip of the rostrum to the midpoint of the posterior concavity, is provided as an indicator of specimen size. The morphological terminology and numbering system of the carapace grooves (CG) used in the text generally follows those of Boyko & Harvey (1999) and Boyko (2002).

TAXONOMY

Family ALBUNEIDAE Stimpson, 1858

Genus *Albunea* Weber, 1795

Albunea marquisiana Boyko, 2000
(fig. 1)

Albunea marquisiana Boyko, 2000: 109, figs. 2, 3 (type locality: Haava Straits between Île Tahuata and Île Hiva Oa, Marquises Islands, French Polynesia); 2002: 282, figs. 92, 93 (full synonymy). — Boyko & McLaughlin, 2010: 141.

Material examined. — Awa, Nago, Okinawa Island, 12 m, sand, 25 July 2010, coll. Y. Yamada, SCUBA, 3 juveniles (cl 2.9 mm, RUMF-ZC-1523; cl 3.2 mm, RUMF-ZC-1524; cl 3.7 mm, RUMF-ZC-1525).

Distinguishing characters. — Carapace (fig. 1A) slightly wider than long, covered with sparsely setose grooves; anterior margin with 8-10 spines on either side of ocular sinus; posterior margin with submarginal grooves reaching half length of each side of concavity; setal field with narrow anterolateral projections and sinuous anteromedian margin. CG1 with separate posterior lateral

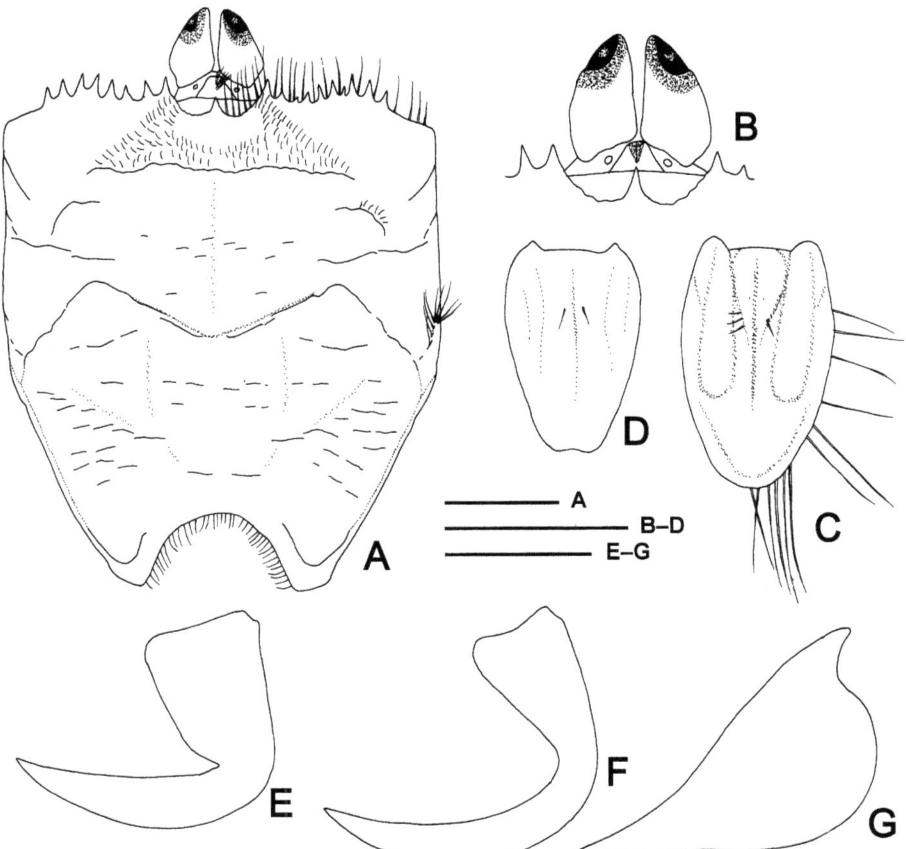

Fig. 1. *Albunea marquisiana* Boyko, 2000. A, B, D–F, juvenile (cl 3.7 mm), RUMF-ZC-1525; C, juvenile (cl 2.9 mm), RUMF-ZC-1523. A, carapace and ocular peduncles, dorsal view, setae omitted from left side; B, rostrum and ocular peduncles, dorsal view, setae omitted; C, telson, dorsal view, setae omitted from left side; D, same, setae omitted; E, left pereopod II dactylus, lateral view, setae omitted; F, left pereopod III dactylus, lateral view, setae omitted; G, left pereopod IV dactylus, lateral view, setae omitted. All plumose setae are simplified. Scales: 1.0 mm.

elements; CG4 with long lateral element terminating above anterior convexity of CG6, and occasionally with some short medial elements; CG5 separated as 2 distant short elements; CG6 and CG7 separate; CG8 with 2-4 median short elements; CG11 present as 2 or 3 short elements. Ocular plate (fig. 1B) subquadrate. Distal peduncular segments dorsoventrally flattened, subtriangular in shape, approximated along mesial margins; lateral margins convex; mesial margins straight; cornea present at distolateral tip. Pereopod II dactylus (fig. 1E) with heel broadly and weakly produced. Pereopod III dactylus (fig. 1F) with heel bluntly produced. Pereopod IV dactylus (fig. 1G) weakly sinuous from base to tip. Telson of male spatulate, distal margin rounded, dorsal surface inflated along median groove bearing dense row of thick short setae; lateral margins weakly calcified (cf. Boyko, 2000, 2002). Telson of female ovate, distal margin rounded (modified from Boyko, 2000, 2002).

Colouration in life. — Body and appendages entirely white. Carapace and abdomen with iridescent tinge. Corneas black.

Distribution. — Previously known only from the Marquises Islands, French Polynesia; 7-130 m depth (Boyko, 2000). The occurrence of *A. marquisiana* in Okinawa Island greatly extends its distribution range to the north and represents the first record from the northern hemisphere.

Remarks. — The specimens examined (cl 2.9-3.7 mm) are on average much smaller than the type material (cl 3.5-10.8 mm, most >5.0 mm), and are considered to be juveniles because they have very short and bud-like pleopods and the gonopores are not present on the coxae of the pereopods III and V.

They agree well with the description of the type material of *A. marquisiana* (Boyko, 2000, 2002) except for some characters. The rostrum reaches or overreaches the proximal margin of the ocular plate in the specimens examined, but it never reaches this margin in the types. The distal segment of the ocular peduncle of the specimens examined does not distinctly taper in the distal corneal part unlike that of the types, and the corneas are comparatively larger in the present specimens than in the types. The heel of the pereopod III dactylus is more broadly rounded and the dorsal margin of the pereopod IV dactylus is more weakly sinuous in the specimens examined than in the types. However, these discrepancies are considered as size-related intraspecific variations. The specimens examined match in size with the juveniles studied by Boyko (2000, 2002; cl 3.1-3.8 mm).

The smallest specimen examined (cl 2.9 mm, RUMF-ZC-1523) also differs from the other two specimens in having only faint CG5, 10 and 11, and a shallow concavity on the distal margin of the telson (fig. 1D). The telson of the two large specimens is rounded (fig. 1C) as in the type specimens.

Albunea aff. **symmysta** (Linnaeus, 1758)
(fig. 2)

Material examined. — Hinai Beach, Iriomote Island, intertidal, sand, 22 June 2005, coll. T. Komai, 1 female (cl 9.6 mm), CBM-ZC 10473.

Distinguishing characters. — Carapace (fig. 2A) wider than long, covered with sparsely setose grooves; anterior margin with 9 or10 spines on either

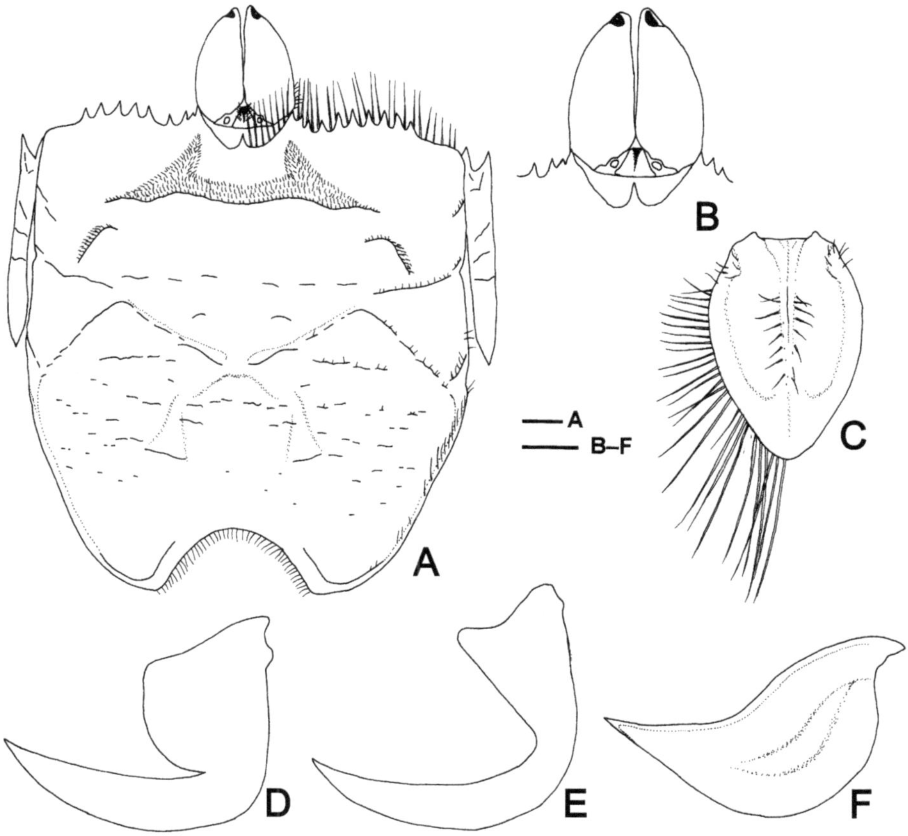

Fig. 2. *Albunea* aff. *symmysta* (Linnaeus, 1758), female (cl 9.6 mm), CBM-ZC 10473. A, carapace and ocular peduncles, dorsal view, setae omitted from left side; B, rostrum and ocular peduncles, dorsal view, setae omitted; C, telson, dorsal view, setae omitted from right side; D, left pereopod II dactylus, lateral view, setae omitted; E, left pereopod III dactylus, lateral view, setae omitted; F, left pereopod IV dactylus, lateral view, setae omitted. All plumose setae are simplified. Scales: 1.0 mm.

side of ocular sinus; posterior margin with submarginal grooves reaching half length of each side of concavity; setal field with narrow anterolateral projections and sinuous anteromedian margin. CG1 with separate posterior lateral elements; CG4 with long lateral element terminating above anterior convexity of CG6, and with some short medial elements; CG5 separated as 2 distant short elements; CG6 and CG7 separate; CG8 with 6 median short elements; CG11 absent. Ocular plate (fig. 2B) subquadrate. Distal peduncular segments dorsoventrally flattened, subtriangular in shape, approximated along mesial margins; lateral margins convex; mesial margins straight; cornea present at distolateral tip. Pereopod II dactylus (fig. 2D) with heel broad, not produced.

Pereopod III dactylus (fig. 2E) with heel bluntly produced. Pereopod IV dactylus (fig. 2F) sinuous from base to tip. Telson of female ovate; distal margin relatively narrowly produced, rounded.

Colouration in life. — Not recorded. Colour in preservative entirely white.

Remarks. — The specimen examined is considered to be an immature female because gonopores are present on the pereopod III coxae, but the pleopods are very slender and fall far short of the distal margin of the telson unlike in mature females.

The specimen is similar to *A. symmysta* in having the following diagnostic characters: CG1 with separate posterior lateral elements, CG11 absent; pereopod III dactylus with produced heel, and female telson with produced distal margin but without distomedian indent. However, the shapes of the heels of the pereopods II–IV dactyli are different between the specimen examined and *A. symmysta* (see Boyko, 2002; Osawa et al., 2010). The heel of the pereopod II dactylus is produced and has a small blunt tip in *A. symmysta*, but it is broadly rounded and not produced in the specimen examined. The heel of the pereopod III dactylus is narrower in *A. symmysta* than in the specimen examined, and its tip is acute in *A. symmysta* unlike blunt in the specimen examined. The heel of the pereopod IV dactylus is blunt but distinct in *A. symmysta*, whereas it is indistinct in the specimen examined. These distinctions probably reflect interspecific variations and the present specimen may perhaps be an undescribed species. However, as the only available specimen is an immature female, a definite assignment is postponed until adult specimens become available for study.

Except for the absence of the CG11, the specimen examined is also generally similar to *A. marquisiana* in the state of the carapace grooves and shapes of the distal segments of the ocular peduncles and pereopods II–IV dactyli. However, the specimen differs from females of *A. marquisiana* in that the telson has more strongly convex lateral margins and a more narrowly produced distal margin.

Genus *Paralbunea* Serène, 1977

Paralbunea takedai sp. nov.

(figs. 3-6)

Material examined. — Oura Bay, Nago, Okinawa Island, St. D5 (Tohkamuri Point), 5 m, muddy sand, 20 June 2009, coll. Y. Fujita, SCUBA, male (cl 4.7 mm), holotype, RUMF-ZC 1526.

Diagnosis. — Carapace wider than long, with grooves bearing sparse short setae; anterior margin unarmed on either side of ocular sinus; posterior margin with submarginal grooves reaching quarter length of each side of concavity; setal field without posterior lateral extensions; anteromedian margin slightly concave. CG1 with separate posterior lateral elements; CG4 present as 2 short oblique lateral elements connected with CG1 lateral elements anteriorly and CG6 posteriorly; no short setose punctae between CG4 lateral elements; CG6 and CG7 fused; CG8 and CG11 absent. Rostrum reaching proximal margin of ocular plate. Ocular plate broadly subtriangular. Distal segment of ocular peduncle dorsoventrally flattened, ovate, separated along mesial margins; lateral margins broadly convex; mesial margins nearly straight; distomesial margin produced anteriorly, terminating in small blunt tip; cornea not delineated. Maxilliped III carpal projection short. Pereopod II dactylus with heel produced and bluntly pointed. Pereopod III dactylus with heel produced, narrow, and acute. Pereopod IV dactylus sinuous, with acute heel and deep indent. Telson of male suboval, narrowed distally; distal margin slightly concave; surface evenly calcified, with pair of rows of short setae along median groove. Telson of female unknown.

Description. — Carapace (figs. 3A, 6A, B) wider than long, with grooves bearing sparse, short plumose setae. Anterior margin concave on either side of ocular sinus, becoming convex laterally, with row of long simple setae but without spines. Rostrum small, acute, reaching proximal margin of ocular plate. Ocular sinus smoothly concave, with row of simple setae but without spines; lateral end subacutely produced. Frontal region smooth; setal field broadened posteriorly, narrowing anteriorly; anterior lateral extensions moderately narrow; posterior lateral extensions absent; anteromedian margin slightly concave. CG1 subparallel to anterior margin of carapace, slightly sinuous and crenulate, with separate lateral elements posteriorly. Mesogastric region smooth, CG2 and CG3 absent; CG4 present as 2 short oblique lateral elements connected with CG1 lateral elements anteriorly and CG6 posteriorly; no short setose punctae between CG4 lateral elements. Hepatic margin unarmed. Brachial region with submarginal groove bearing row of short plumose setae, some anterior setae long and stiff; anterior branchial region generally triangular, smooth; posterior branchial region with some very short, setose and nonsetose punctae. Metagastric region smooth; CG5 absent; CG6 slightly crenulate, not ridged and nonsetose on mesial half, connected with CG7 posteriorly. CG7 slightly oblique, interrupted. Cardiac region nearly smooth, with few very short punctae; CG8-11 absent. Posterior margin deeply and evenly convex, with submarginal grooves reaching to quarter length of each side of concavity. Branchiostegite with short anterolateral submarginal spine; anterior region calcified dorsally, with numerous very short ridges bearing short

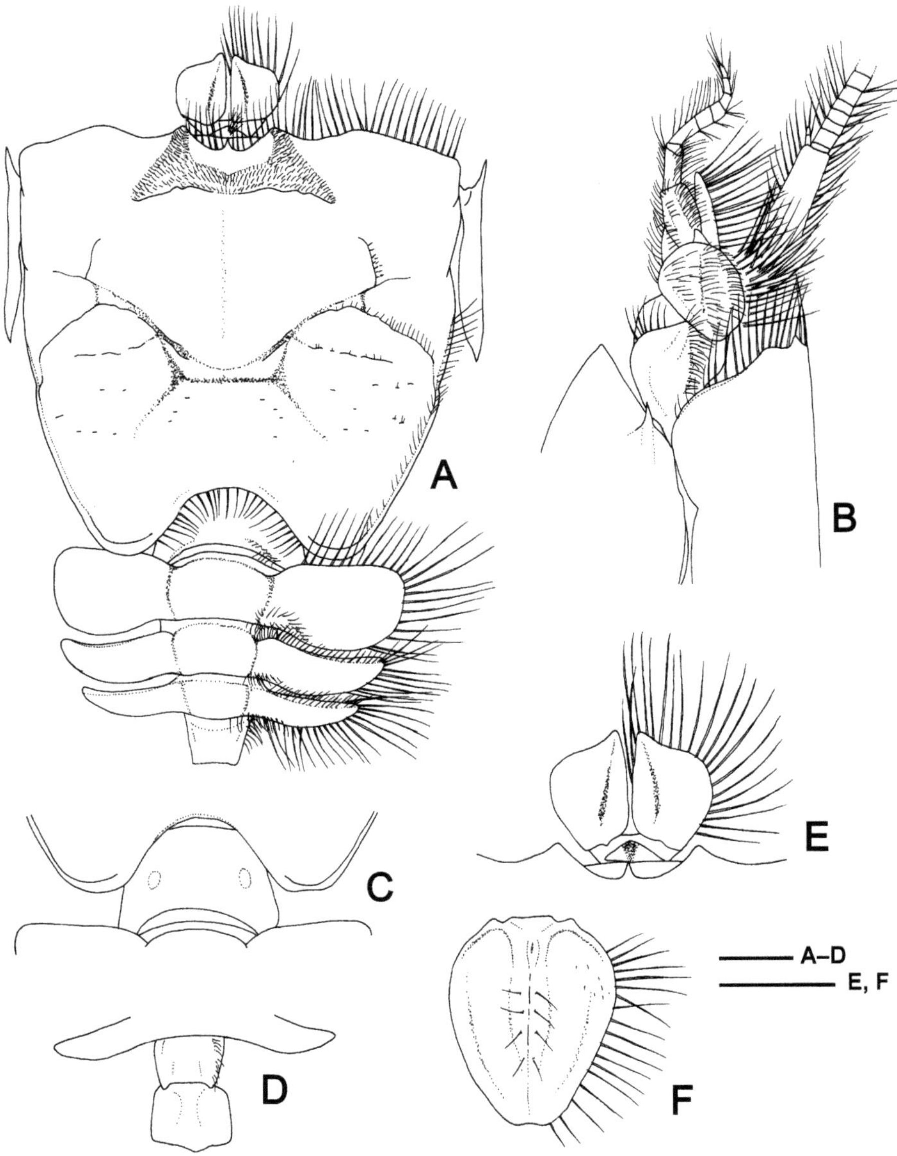

Fig. 3. *Paralbunea takedai* sp. nov., holotype, male (cl 4.7 mm), RUMF-ZC-1526. A, carapace, abdominal somites I–V, and ocular peduncles, dorsal view, setae omitted from left side; B, left anterior parts of carapace and branchiostegite, and antennular and antennal peduncles, lateral view; C, posterior part of carapace and abdominal somites I and II, dorsal view, setae omitted; D, abdominal somites IV–VI, dorsal view, setae omitted; E, rostrum and ocular peduncles, dorsal view, setae omitted from left side; F, telson, dorsal view, setae omitted from left side. All plumose setae are simplified. Scales: 1.0 mm.

plumose setae on membranous ventral surface, anterior margin rounded; posterior region generally membranous, with numerous irregular fragments and oblique rows of short plumose setae.

Ocular peduncles (fig. 3E) apparently lacking median peduncular segment. Ocular plate broadly subtriangular, with shallow median depression. Distal peduncular segment strongly depressed, ovate, with strongly convex lateral margin and nearly straight mesial margin; lateral and mesial margins with row of long plumose setae; mesial margin approximated along entire length, distomesial margin produced anteriorly and terminating in small blunt tip; distolateral margin broadly rounded; dorsal surface weakly convex medially, dorsodistal margin oblique, slightly sinuous; ventral surface slightly concave, with proximal transverse submarginal ridge lined with short plumose setae. Cornea not delineated; narrow, slightly oblique, longitudinal pigmented patch visible from dorsal view near mesial margin of distal peduncular segment.

Antennular peduncle (figs. 3B, 4A) exceeding distal margin of ocular peduncle by half length of penultimate segment. Ultimate segment narrowed distally in lateral view, 4.4 times distal width; dorsal and ventral margins each with row of long plumose setae; lateral surface with few short setae. Penultimate segment medially inflated in lateral view, dorsal and ventral margins with plumose setae, ventral setae longer and denser; lateral surface with some short setae dorsally. Proximal segment broader than long, unarmed; dorsodistal lateral margin with broad projection bearing numerous short plumose setae. Dorsal flagellum (fig. 6A) very long, 7.6 times as long as carapace, consisting of approximately 130 articles each bearing long plumose setae on dorsal and ventral margins; ventral flagellum very short, composed of 2 articles, each with few short plumose setae on dorsal and ventral margins.

Antennal peduncle (figs. 3B, 4B) reaching half length of penultimate segment of antennular peduncle, with supernumerary segmentation. Segment V approximately 1.7 times longer than broad, with plumose setae on margins, ventral setae short. Segment IV ovate, with dense plumose setae on dorsal surface; ventral surface with few short setae. Segment III with row of long simple setae on ventromesial margin; distolateral margin with some short setae. Segment II ovate in dorsal and lateral views; dorsal surface flattened, oblique, with scattered, dense short plumose setae and row of long plumose setae on dorsomesial margin; ventral surface convex, also with scattered short plumose setae. Segment I convex, unarmed on lateral surface; dorsal surface with scattered short plumose setae laterally; ventromesial margin produced in broad rounded lobe; ventral and distal margins each with row of long plumose

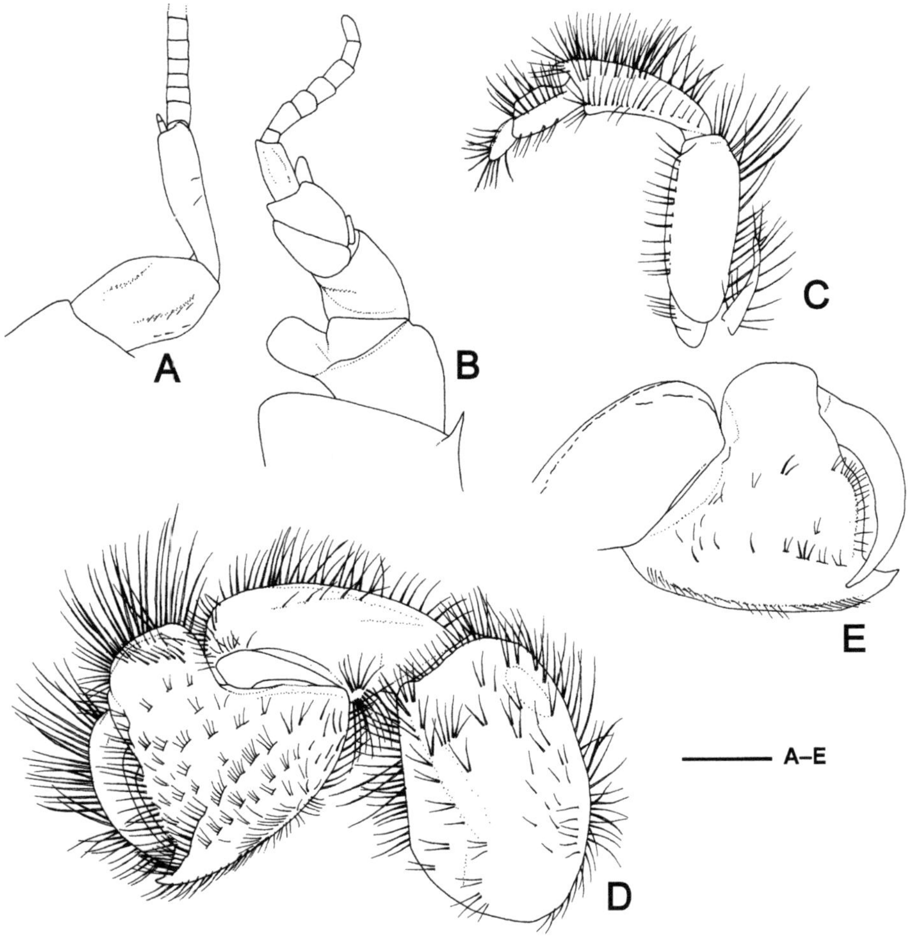

Fig. 4. *Paralbunea takedai* sp. nov., holotype, male (cl 4.7 mm), RUMF-ZC-1526. A, left antennular peduncle, lateral view, setae omitted; B, left antennal peduncle and anterior part of branchiostegite, ventral view, setae omitted; C, left maxilliped III, lateral view; D, left pereopod I, lateral view; E, same, chela and carpus, mesial view, setae omitted. All plumose setae are simplified. Scale: 1.0 mm.

setae. Antennal acicle long, thin, nearly reaching midlength of peduncular segment V, with long plumose setae on dorsal surface; distal margin rounded. Flagellum composed of 7 articles each bearing plumose setae on dorsal, ventral, and distal margins.

Maxilliped III (fig. 4C) slender. Dactylus rounded on distal margin, with short simple setae on margins. Propodus inflated, with row of plumose setae on dorsal margin; lateral surface with blunt crest bearing row of long plumose setae along midline; ventral margin with some short simple setae. Carpus long, with short rounded projection on dorsodistal margin; lateral surface with scattered plumose setae; dorsal and ventral margins each with row of long plumose setae. Merus relatively slender, 2.5 times longer than broad, unarmed on margins; lateral surface with row of short plumose setae along ventral margin; dorsal margin with row of long plumose setae. Basis- ischium fusion incomplete, crista dentata absent. Exopod consisting of 2 segments; proximal segment small (not illustrated); distal segment styliform, tapering, reaching nearly to midlength of merus, with plumose setae on margins; flagellum absent.

Pereopod I (fig. 4D, E) dactylus curved, tapering; dorsal surface unarmed but with plumose setae becoming longer proximally; lateral surface with row of very short setae ventrally; mesial surface smooth. Propodus with numerous short, transverse ridges bearing short simple setae and patch of short plumose setae dorsoproximally on weakly convex lateral surface, long ridge extending onto fixed finger; dorsal margin unarmed but with dense, long plumose setae; ventral margin with row of short simple setae on distal three fourths and long plumose setae on proximal one fourth; mesial surface with some short transverse ridges and short plumose setae on ventral half, submarginal row of short plumose setae present ventrally; fixed finger terminating in small spine; cutting edge lacking teeth, with long plumose setae. Carpus with dorsodistal margin broadly rounded, unarmed; dorsal margin with row of long plumose setae; ventrodistal margin with tuft of long plumose setae; lateral surface with some short plumose setae dorsally and patch of short plumose setae distally; mesial surface with oblique row of interrupted ridges bearing long setae dorsally and row of short setae medially. Merus unarmed on margins; dorsal and ventral margins each with row of long plumose setae; lateral surface with small dorsodistal decalcified region and scattered transverse rows of short plumose setae; mesial surface decalcified on proximal one third, bearing scattered short setae. Basis-ischium incompletely fused, unarmed, bearing few short plumose setae. Coxa unarmed.

Setation on pereopods II–IV generally similar to each other.

Pereopod II (fig. 5A, B) relatively slender. Dactylus base to heel sinuous and shallowly concave, heel with narrowly rounded projection, heel to tip broadly indented, tip subacute, tip to base strongly convex; lateral surface nearly smooth, with row of simple and plumose setae along dorsal margin, proximal setae longer; ventral margin with row of short and long plumose setae, setae on distal half numerous; mesial surface with irregular rows of short plumose setae on heel and blunt transverse ridges bearing short plumose setae near distal base of heel. Propodus longer than broad, with sparse, short plumose setae

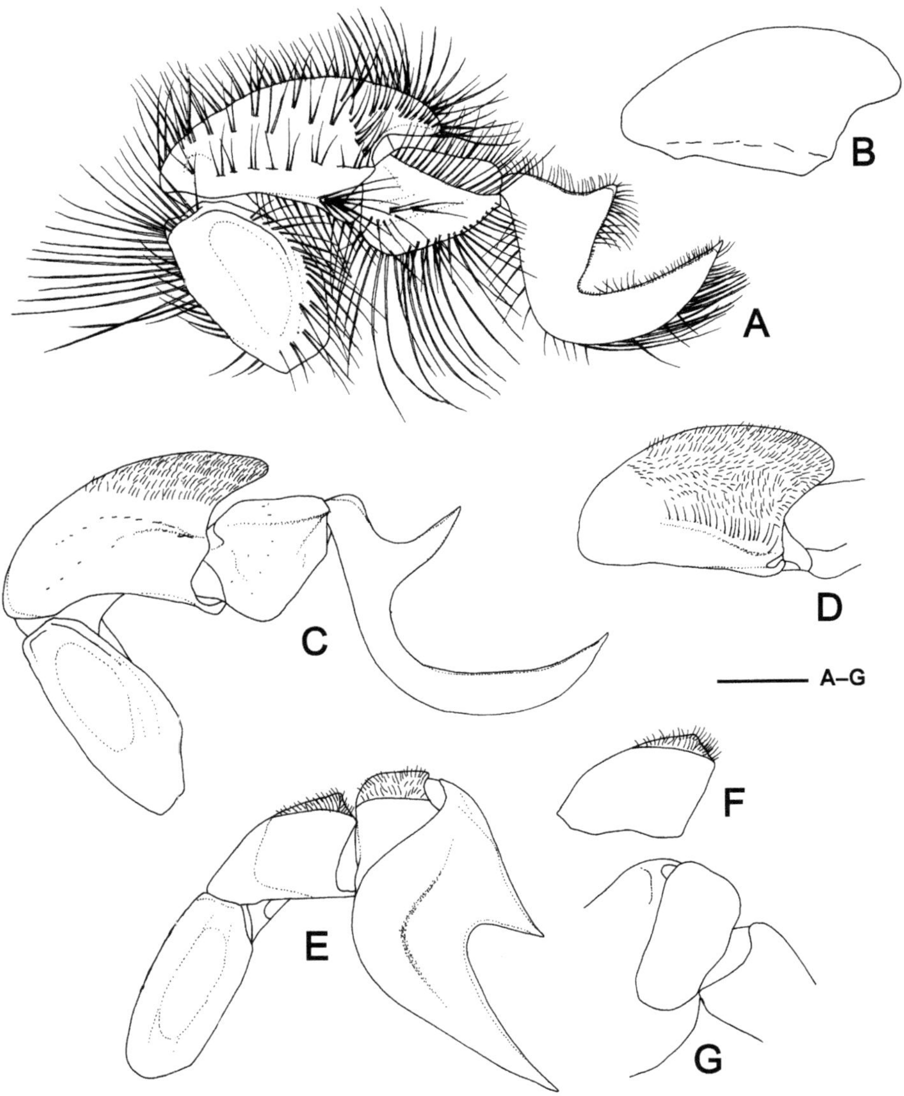

Fig. 5. *Paralbunea takedai* sp. nov., holotype, male (cl 4.7 mm), RUMF-ZC-1526. A, right pereopod II, lateral view; B, same, carpus, perpendicular view, setae omitted; C, right pereopod III, lateral view, most setae omitted; D, same, carpus, perpendicular view, most setae omitted; E, right pereopod IV, lateral view, most setae omitted; F, same, carpus, perpendicular view, most setae omitted; G, same, proximal part of dactylus, distal part of carpus, and propodus, mesial view, setae omitted. All plumose setae are simplified. Scale: 1.0 mm.

on dorsal margin; lateral surface with oblique dorsal facet delimited by blunt ridge bearing row of long plumose setae and with row of short plumose setae on midline; ventral margin inflated proximally, with short and long plumose setae; mesial surface with blunt ridge bearing short plumose setae along ventral margin. Carpus with dorsodistal, broadly rounded projection reaching to one third length of propodus and with dorsodistal mat of short plumose setae; lateral surface with row of short ridges bearing short plumose setae ventrally on midline, dorsal half with scattered short plumose setae and with oblique row of thick plumose setae distally; ventral surface narrow, flat, with row of long setae mesially, ventrodistal margin not produced, bearing tuft of long plumose setae; mesial surface with scattered plumose setae near margins, dorsal setae longer. Merus 1.6 times as long as broad, inflated medially in lateral view, unarmed on margins; lateral surface with median, broad decalcified part and with row of long plumose setae along ventral margin; dorsal margin with row of short and long plumose setae; ventral surface flattened, with scattered long plumose setae; mesial surface with numerous long plumose setae. Basis-ischium incompletely fused, unarmed, bearing scattered short plumose setae. Coxa unarmed.

Pereopod III (fig. 5C, D) slightly longer than pereopod II. Dactylus base to heel broadly and deeply concave, heel narrowly produced, slightly curved, terminating in acute tip, heel to tip very broadly concave, tip acute, tip to base broadly convex; lateral and mesial surfaces nearly smooth. Propodus longer than broad, with oblique dorsal facet delimited by blunt ridge bearing row of short plumose setae and with oblique row of short plumose setae ventromedially on lateral surface; ventral margin inflated proximally; mesial surface nearly smooth. Carpus with dorsodistal broadly rounded projection reaching nearly to half length of propodus and with dense mat of very short plumose setae covering distal half of dorsolateral surface; lateral surface with 2 irregular rows of short ridges bearing short lateral surface setae along midline; ventral surface narrow, flat, ventrodistal margin not produced; mesial surface nearly smooth. Merus 2.0-2.2 times as long as broad, relatively narrow, inflated medially in lateral view, unarmed on margins; lateral surface with median, broad decalcified part; dorsodistal margin with tuft of long plumose setae; ventral surface flattened; mesial surface with scattered plumose setae ventrally. Basis-ischium incompletely fused, unarmed. Coxa unarmed.

Pereopod IV (fig. 5E–G) shorter than pereopod II. Dactylus base to heel slightly concave, heel produced and acute, heel to tip broadly concave, tip acute, tip to base broadly convex; lateral surface weakly concave medially, nearly smooth; mesial surface with median decalcified part demarcated ventrally by longitudinal ridge bearing row of long plumose setae. Propodus broader than long, broader and shorter than carpus, with long plumose setae on dorsal margin; lateral surface with oblique dorsal facet delimited by blunt ridge, facet with mat of very short plumose setae; ventral margin inflated; mesial surface smooth. Carpus with row of long plumose setae on dorsal margin, dorsodistal margin not produced; lateral surface decalcified on proximal two thirds, with oblique dorsodistal facet delimited by blunt ridge, facet with mat of very short plumose setae, elsewhere smooth; ventral surface narrow, flat, ventrodistal margin slightly produced, rounded; mesial surface nearly smooth. Merus 2.2 times as long as broad, narrow, oblong in lateral view, unarmed on margins; lateral surface with median, broad decalcified part; mesial surface smooth, with decalcified part on proximal one third. Basis-ischium incompletely fused, unarmed. Coxa unarmed.

Pereopopod V coxae of male each with gonopore partially obscured by short setae and bearing short protrusion.

Abdomen (fig. 3A, C, D) with somite I broader than long, widened posteriorly; dorsal surface smooth except for short plumose setae along posterior lateral margin, anterior margin nearly straight, posterior margin concave, pair of small decalcified windows present. Somites II–IV weakly convex and generally smooth on dorsal surface, with well developed pleura. Somite II dorsal surface with submarginal blunt ridge along anterior margin, patch of short plumose setae on posterolateral part extending onto pleura; pleura expanded, directed laterally, anterolateral and posterolateral margins rounded, anterior and lateral margins each with row of long plumose setae, posterior margin with short setae. Somite III generally similar to somite II in shape, but narrower and shorter; dorsal surface with patch of short setae on each posterolateral part; pleura narrower and much shorter than those on somite II, directed anterolaterally, anterior and posterior margins each with row of short plumose setae, lateral margin narrowly rounded, with long plumose setae, dorsal surface with row of short plumose setae along anterior margin. Somite IV similar to somite III in shape and setation, but thinner and shorter; pleura also narrower and shorter than those on somite III, directed laterally. Somite V subequal in length and breadth to somite IV; dorsal surface with row of short plumose setae along each lateral margin. Somite VI wider and longer than somite V; dorsal surface smooth.

Male lacking pleopods; female unknown.

Telson of male (fig. 3F) suboval, narrowed distally; lateral margins with short and long plumose setae; distal margin slightly concave; surface evenly calcified, with median groove extending nearly to distal margin and with line of short simple setae along each side of median one third of groove. Telson of female unknown.

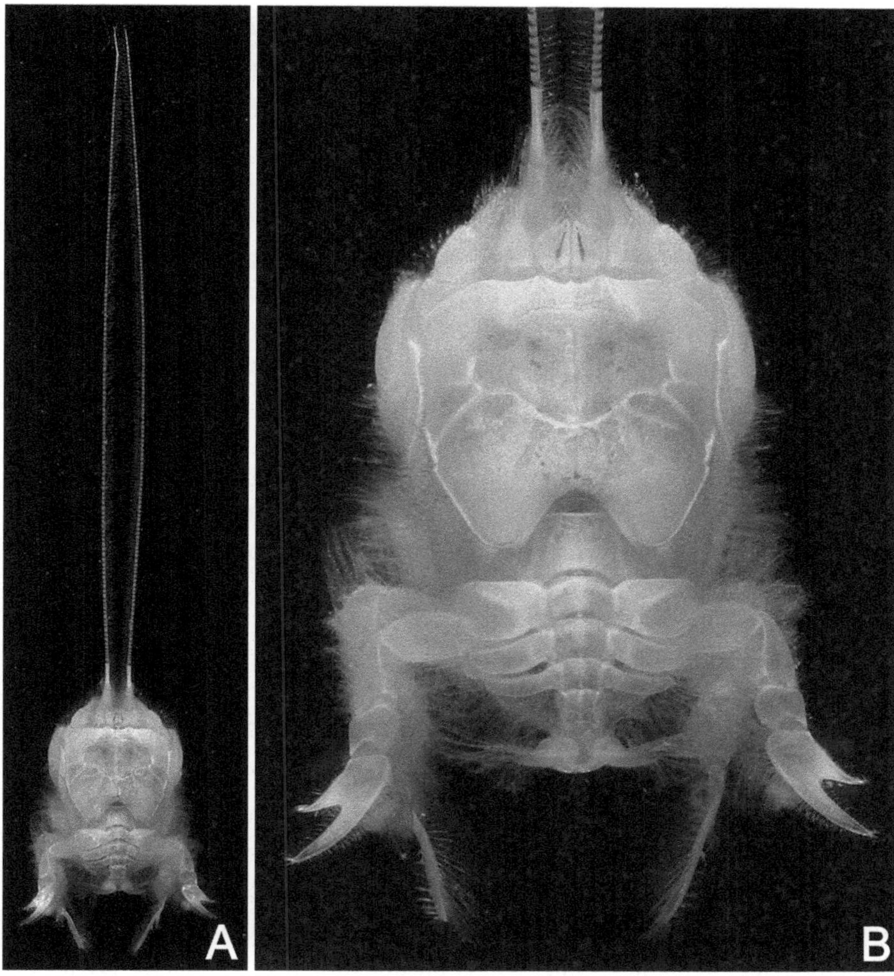

Fig. 6. *Paralbunea takedai* sp. nov., holotype, male (cl 4.7 mm), RUMF-ZC-1526. A, entire animal, dorsal view; B, carapace, abdomen, and cephalic and thoracic appendages, dorsal view.

Colouration in life. — Body and appendages entirely bright white. Carapace and abdomen with iridescent tinge. Distal segment of ocular peduncle with narrow, slightly oblique, longitudinal black patch near dorsomesial margin.

Etymology. — The new species is named in honour of Dr. Masatsune Takeda, for his outstanding contribution to the systematics of decapod crustaceans, especially pilumnid crabs. It is a pleasure to dedicate this anomuran sand crab to him.

Distribution. — So far known only from Okinawa Island, southwestern Japan, and possibly from the Philippines (see Remarks); 5 m depth.

Remarks. — The male holotype is small in size (cl 4.7 mm) but has distinct gonopores on the coxae of the fifth pereopods. Thus, it is considered mature.

The genus *Paralbunea* Serène, 1977 includes four species: *P. dayriti*; *P. intermedia* (Balss, 1916); *P. manihinei* Serène, 1977; and *P. paradoxa* (Gordon, 1938). Among them, *P. dayriti*, *P. manihinei*, and *P. paradoxa* have been reported from the Indo-West Pacific, and *P. dayriti* was the sole species known from Japanese waters (Boyko, 2002). *Paralbunea takedai* sp. nov. is the fifth member of the genus and the second Japanese species.

The new species is closely allied to *P. paradoxa* in having the following characters: carapace without spines on each side of the anterior margins, absence of CG8, subtriangular ocular plate, maxilliped III carpus with short dorsodistal projection, pereopods III and IV dactyli with acutely produced heels, and pereopod IV dactylus with distinctly concave indent. However, comparison with a redescription of the female holotype of *P. paradoxa* by Boyko (2002) reveals that *P. takedai* differs from *P. paradoxa* in the configuration of the distal segment of the ocular peduncle which is considered as a key character in albuneid taxonomy (see Boyko, 2002; Osawa et al., 2010). Sexual dimorphism in the ocular segment has not been reported for any albuneid species. The distomesial margin of the distal segment is distinctly produced anteriorly and terminates in a small blunt tip in *P. takedai*, whereas it is obtusely angled but never produced anteriorly in *P. paradoxa*. The new species is further distinguished from *P. paradoxa* in comparatively longer and more slender meri of the pereopods II–IV. For example, the merus of the pereopod IV is 2.2 times as long as high in *P. takedai*, instead of 1.9 times longer in *P. paradoxa*. The mesogastric region of the carapace has very short setose punctae on the lateral parts in *P. paradoxa*, but such punctae are absent in the new species.

Boyko (2002) and Boyko & Harvey (2009) concluded that *P. dayriti* and *P. paradoxa* are sister species based on their phylogenetic analyses. Because of the close similarity between *P. paradoxa* and *P. takedai*, these two species can become a sister species-pair to *P. dayriti*. However, *P. paradoxa* and *P. takedai* are immediately distinguished from *P. dayriti* by the armature on the anterior margins of the carapace and the shape of the heels of the dactyli of the pereopods III and IV (see Boyko, 2002; Osawa et al., 2010).

Paralbunea paradoxa was originally described by Gordon (1938) on the basis of a female holotype from Singapore, and was subsequently reported by Serène & Umali (1965) from Bataan and Zambales Provinces in Luzon Island, the Philippines. Serène & Umali (1965) did not report the sex of their specimens but observed morphological variations of the ocular peduncles. The distal segments of the ocular peduncles in some specimens are subquadrate in shape, agreeing with those of the holotype of *P. paradoxa*; whereas in the other specimens, the segments are stated to be pointed at the distomesial ends, better fitting the morphology in the present new species. Serène & Umali's (1965) material may partially contain *P. takedai*, although a re-examination of their specimens is needed.

As shown in the above description of *P. takedai*, the corneas are not delineated but the distal ocular segment has a narrow, slightly oblique, longitudinal black patch near the dorsomesial margin. The shape of the pigmented part was identical in both fresh and preserved conditions of the sole holotype, and thus it is not due to an artificial factor introduced through the preservative process. Serène & Umali (1965, fig. 10A, B) reported similar pigmentation for *P. dayriti* and *P. paradoxa* and considered it as a vestigial cornea, although the shape of the pigmentation somewhat varied in their material. The presence of the black-pigmented part suggests that the ocular peduncles still function as a visual organ in *P. dayriti*, *P. paradoxa*, and *P. takedai*. Further detailed studies on more material are needed for clarifying the structure and evolution of eyes in albuneids.

ACKNOWLEDGEMENTS

Field research in Oura Bay, Okinawa Island, was supported by a fund for "Nansei (Ryukyu) Islands Biodiversity Evaluation Project" from WWF (World Wide Fund for Nature) Japan. We thank T. Komai (Natural History Museum and Institute, Chiba) and Y. Yamada (Uruma City) for providing us with the interesting specimens for study. The authors are also grateful to C. B. Boyko (American Museum of Natural History), J. Okuno (Coastal Branch of Natural History Museum and Institute, Chiba), and an anonymous reviewer for their valuable comments and suggestions on this paper.

REFERENCES

BALSS, H., 1916. Crustacea II: Decapoda Macrura und Anomura (außer Fam. Paguridae). Beiträge zur Kenntnis der Meeresfauna Westafrikas, **2**: 11-46.

BENEDICT, J. E., 1904. A new genus and two new species of crustaceans of the family Albuneidae from the Pacific Ocean; with remarks on the probable use of the antennule in *Albunea* and *Lepidopa*. Proceedings of the United States National Museum, **27**(1367): 621-625.

BOYKO, C. B., 2000. The Hippoidea (Decapoda, Anomura) of the Marquises Islands, with description of a new species of *Albunea*. Zoosystema, **22**: 107-116.

— —, 2002. A worldwide revision of the recent and fossil sand crabs of the Albuneidae Stimpson and Blepharipodidae, new family (Crustacea: Decapoda: Anomura: Hippoidea). Bulletin of the American Museum of Natural History, **272**: 1-396.

BOYKO, C. B. & A. W. HARVEY, 1999. Crustacea Decapoda: Albuneidae and Hippidae of the tropical Indo-West Pacific region. In: A. CROSNIER (ed.), Résultats des Campagnes MUSORSTOM, **20**. Mémoires du Museum National d'Histoire Naturelle, Paris, **180**: 379-406.

— — & — —, 2009. Phylogenetic systematics and biogeography of the sand crab families Albuneidae and Blepharipodidae (Crustacea: Anomura: Hippoidea). Invertebrate Systematics, **23**: 1-18.

BOYKO, C. B. & P. A. MCLAUGHLIN, 2010. Annotated checklist of anomuran decapod crustaceans of the world (exclusive of the Kiwaoidea and families Chirostylidae and Galatheidae of the Galatheoidea). Part IV. Hippoidea. The Raffles Bulletin of Zoology, (Supplement) **23**: 139-151.

DANA, J. D., 1852. Crustacea. United States Exploring Expedition during the years 1838, 1839, 1840, 1841, 1842, under the command of Charles Wilkes, U. S. N., **13**: i–viii, 1-685 [1852]. (C. Sherman, Philadelphia).

GORDON, I., 1938. A comparison of the two genera *Albunea* and *Lepidopa* (Crustacea, Anomura), with description of a new species from Singapore. Bulletin of the Raffles Museum, **14**: 186-197.

LINNAEUS, C., 1758. Systema naturae per regna tria naturae, secundum classes, ordines, genera, species, cum characteribus, differentis, synonymis, locis. 1 (Edition 10): 1-824. (Laur. Salvii, Holmiae).

MIERS, E. J., 1878. Revision of the Hippidea. Journal of the Linnean Society, (Zoology) **14**: 312-336, pl. 5.

OSAWA, M., C. B. BOYKO & T.-Y. CHAN, 2010. Part I. Hippoidea (Mole crabs). In: T.-Y. CHAN (ed.), Crustacean Fauna of Taiwan: Crab-like anomurans (Hippoidea, Lithodoidea and Porcellanidae): 1-41. (National Taiwan Ocean University, Keelung).

OSAWA, M. & Y. FUJITA, 2007. Sand crabs of the genus *Albunea* (Crustacea: Decapoda: Anomura: Albuneidae) from the Ryukyu Islands, southwestern Japan, with the description of a new species. Species Diversity, **12**: 127-140.

SERÈNE, R., 1977. Crustacés hippidés et brachyoures des îles Séchelles. Revue de Zoologie Africaine, **91**: 45-68.

SERÈNE, R. & A. F. UMALI, 1965. A review of Philippine Albuneidae, with descriptions of two new species. Philippine Journal of Science, **94**: 87-116, 6 pls.

STIMPSON, W., 1858. Prodromus descriptionis animalium evertebratorum, quae in expeditione ad oceanum pacificum septentrionalem, a republica federata missa, Cadwaladaro Ringgold et Johanne Rodgers ducibus, observavit et descripsit. Pars VII. Crustacea Anomura. Proceedings of the Academy of Natural Sciences of Philadelphia, **10**: 225-252.

WEBER, F., 1795. Nomenclator entomologicus secundum entomologiam systematicam ill. Fabricii adjectis speciebus recens detectis et varietatibus: i–viii, 1-172. (Chilonii et Hamburgi).

First received 25 August 2011.
Final version accepted 10 November 2011.

A NEW SPECIES OF THE HERMIT CRAB GENUS *DIOGENES* DANA, 1851 (DECAPODA, ANOMURA, DIOGENIDAE) FROM LOMBOK, INDONESIA

BY

DWI LISTYO RAHAYU[1])
Marine Bio-Industry Technical Implementation Unit, Research Center for Oceanography, Indonesian Institute of Sciences (LIPI), Teluk Kodek, Kec. Pemenang, Lombok Utara 83352, NTB, Indonesia

ABSTRACT

A new hermit crab species of the family Diogenidae, *Diogenes takedai* sp. nov., is described from sea grass beds of Lombok Island, Indonesia. The new species appears closest to *D. tirmiziae* Siddiqui & McLaughlin, 2003, in having banded ocular peduncles and pereopods and similar armatures of the antennal peduncle and left chela, but is distinguished from that species by the tapered ocular peduncles and moderately narrow corneas, and the propodi of the second and third pereopods without spines on the dorsal margin. The new species is compared to other similar congenerics from Indo-West Pacific. The species of *Diogenes* known from Indonesian waters are briefly reviewed.

INTRODUCTION

Hermit crab species of the genus *Diogenes* Dana, 1851, are defined primarily by the presence of intercalary rostral process between the ocular acicles. This genus has been relatively well studied in last two decades with several new species described (Morgan & Forest, 1991; Rahayu & Forest, 1995; Rahayu, 1996; McLaughlin & Clark, 1997; McLaughlin & Holthuis, 2001; Rahayu & Hortle, 2002; Siddique & McLaughlin, 2003; Asakura, 2006; Asakura & Godwin, 2006; Asakura & Tachikawa, 2010). Currently this genus contains 59 species distributed in the Indo-West Pacific and eastern Atlantic (McLaughlin et al., 2010).

[1]) e-mail: dwilistyo@yahoo.com

Recently an undescribed species was discovered from Lombok Island, Indonesia. The new species described herein as *Diogenes takedai* sp. nov. has very short antennal acicles, a character shared only with *D. dorotheae* Morgan & Forest, 1991, *D. guttatus* Henderson, 1888, *D. holthuisi* Asakura & Tachikawa, 2010, and *D. tirmiziae* Siddiqui & McLaughlin, 2003.

The material examined in this study was collected from sea grass beds in intertidal areas of Lombok Island. The holotype and paratypes are deposited in the Museum Zoologi Bogor, Indonesian Institute of Sciences, Indonesia (MZB), and selected paratypes are also deposited in the Zoological Reference Collection of the Raffles Museum of Biodiversity Research, National University of Singapore (ZRC). Specimen size is indicated by length of the shield, measured from the tip of the rostrum to the mid-point of the posterior margin, measured to the nearest 0.1 mm. Terminology used in the descriptions generally follows that of McLaughlin & Clark (1997). The abbreviation, ovig, refers to ovigerous female. An ovigerous female of *D. holthuisi* (shield length 1.3 mm) from Kume Island, Japan, deposited in the Ryukyu University Museum, Fujukan, Japan (RUMF-ZC-1644) is used as a comparative material.

TAXONOMY

Family DIOGENIDAE Ortmann, 1892
Genus *Diogenes* Dana, 1851
Diogenes takedai sp. nov.
(figs. 1, 2)

Material examined. — Holotype: female, 2.5 mm, Kecinan, Lombok, 08°24.368′S 116°03.561′E, 3 May 2011 (MZB Cru 3261). Paratypes: 1 male, 2.1 mm, same locality as holotype (MZB Cru 3262); 1 ovig. 2.7 mm, 1 male, 1.6 mm, Sira, Lombok, 08°21.641′S 116°06.272′E, 21 July 2009 (ZRC 2011.0995); 4 males, 1.8-2.3 mm, Pandanan, Lombok, 25 October 2009 (MZB Cru 3263).

Description. — Shield (fig. 1A) slightly longer than broad; rostrum broadly rounded, anterior margin between rostrum and lateral projection somewhat concave; anterolateral angle with small spine, lateral margins with few transverse spinulose or tuberculate ridges, extending slightly onto lateral surfaces of shield; posterior margin truncate or slightly concave; dorsal surface with few small spines and tufts of sparse setae. Lateral projections bluntly triangular, each with small marginal spine. Branchiostegites each with row of 2-7 small spines on dorsal margin.

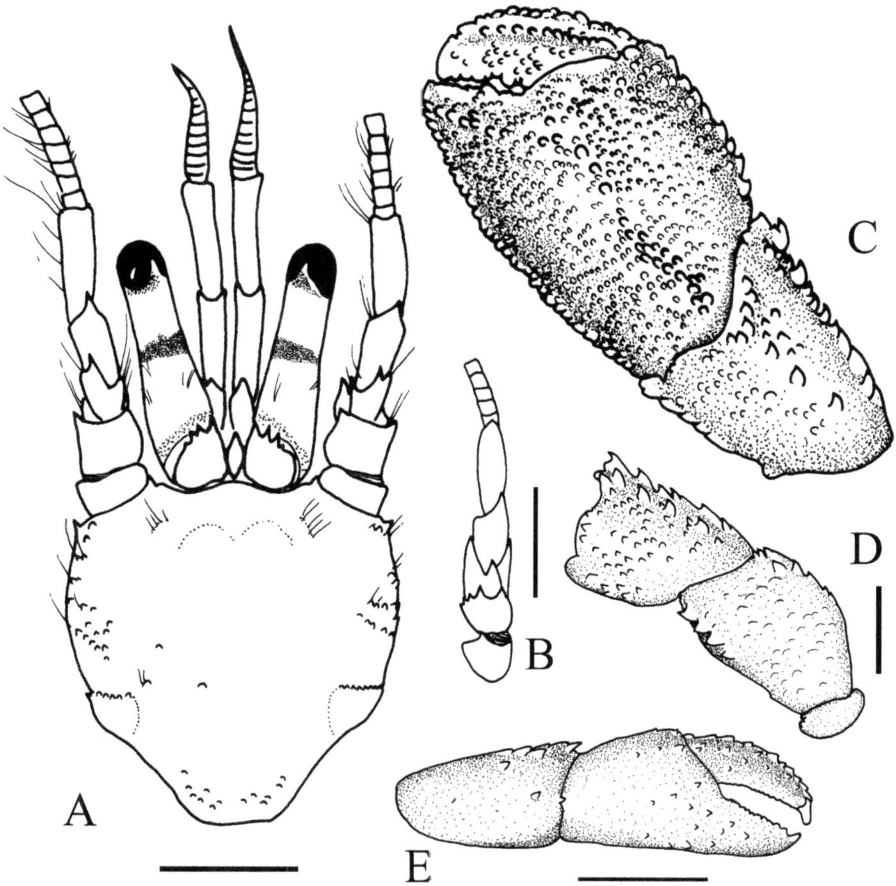

Fig. 1. *Diogenes takedai* sp. nov., holotype, female, 2.5 mm. A, shield and cephalic appendages, dorsal view; B, left antennal peduncle, ventral view; C, chela and carpus of left cheliped, outer view; D, carpus and merus of left cheliped, lateral view; E, chela and carpus of right cheliped, outer view. Scales = 1 mm. A, setae and color pattern partially omitted; B–E setae and color pattern omitted.

Ocular peduncles (fig. 1A) slender and cylindrical, about 0.7-0.8 length of shield, inflated basally, tapering to moderately narrow corneas, each with tufts of very sparse setae dorsomesially and dorsolaterally; corneas not dilated, its diameter about 0.2 of peduncular length. Ocular acicles broad, subtriangular, basally separated by approximately 0.3 or less width of 1 acicle, with 3 or 4 spinules on mesial half of terminal margin, mesial spine largest. Intercalary rostral process simple, stout proximally, tapering to pointed tip, not reaching tips of mesial spines of ocular acicles, with few long setae; no ventral spine.

Antennular peduncles (fig. 1A) moderately slender, when fully extended overreaching corneas by 0.3-0.5 length of ultimate segments. Ultimate and penultimate segments unarmed, but with few short setae. Basal segment with few long setae.

Antennal peduncles (fig. 1A, B) overreaching corneas by 0.3 length of fifth segment. Fifth segment with tufts of long setae ventrodistally, and row of long setae ventrally. Fourth segment with pointed dorsodistal margin and few short setae. Third segment with tuft of long setae on ventrodistal margin. Second segment with dorsolateral margin produced into small but prominent spine, 3 or 4 additional spines on ventrolateral distal margin (not visible in dorsal view); dorsomesial distal angle with or without small spine, mesial and lateral margins with tufts of setae. First segment unarmed. Antennal acicle short, reaching only to proximal third of fourth segment, with simple terminal spine; mesial margins each with 1-3 small spines and tufts of setae. Antennal flagellum long, slightly overreaching tip of left cheliped; each article with 3 or 4 long setae on ventral surface.

Third maxilliped denticulate on mesial margin of coxal plate; 3 small spines on mesial margin of basis; ischium with crista dentata composed of 3 or 4 small corneous teeth. Merus, carpus and propodus approximately same length, long and dense setae on mesial margin, lateral margin with less dense setae; dactyl slightly shorter than propodus, dense setae ventrally.

Left cheliped (fig. 1C, D) much larger than right. Dactyl slightly shorter than upper margin of palm, arched, cutting edge with row of large calcareous teeth and few tufts of short setae, terminating in small calcareous claw; outer surface with longitudinal sulcus medially flanked by row of blunt tubercles parallel to upper margin, and irregular row of tubercles near cutting edge; upper margin with double rows of prominent tuberculate spines separated by shallow narrow sulcus, tuberculate spines decreasing in size distally; inner surface with longitudinal row of tubercles accompanied by sparse short setae. Fixed finger with outer surface somewhat flattened on lower half, covered with small blunt tubercles, longitudinal short ridge of larger blunt tubercles adjacent to cutting edge; lower margin delimited by double rows of acute calcareous tubercles, sparsely interspersed with short setae, forming straight line with similarly armed lower margin of palm; inner surface smooth but with row of tufts of short setae on midline and near cutting margin; cutting edge with row of small calcareous teeth, 1 large tooth proximally; terminating in large calcareous claw. Palm with convex outer surface covered with small, drop-like tubercles, larger tubercles near upper margin; longitudinal rows of moderately large tubercles on midline, beginning proximally and continuing distalward, but not reaching to articulation with fixed finger; upper margin with irregular double rows of moderately large, acute and subacute tubercles; lower margin delimited by row of acute tubercles; inner surface tuberculate and with sparse tufts of setae. Carpus (fig. 1C, D) slightly longer than palm, subequal to merus in length; upper margin with row of prominent spines and sparse long setae, distal spine strongest; outer face convex, covered with well developed tubercles interspersed with sparse short setae, irregular longitudinal row of large spines medially, row of prominent spines also on distomedian margin; lower margin with row of tuberculate spines interspersed with setae; inner surface with flattened tubercles and sparse setae. Merus (fig. 1D) subtriangular in dorsal view; dorsodistal margin with few small spines, dorsal surface with rows of tuberculate spines and tufts of moderately long setae; lateral face tuberculate, ventrolateral margin with row of small spines distally and with low protuberances proximally, accompanied by long setae; mesial face sparsely tuberculate, with moderately long setae, ventromesial margin with row of tubercles and long setae; ventral surface tuberculate. Ischium with row of small acute tubercles and few long setae on ventromesial margin.

Right cheliped (fig. 1E) reaching midlength of palm of left cheliped; no hiatus between dactyl and fixed finger. Dactyl slightly longer than palm, arched; upper margin with row of small, calcareous spines, with numerous long setae; outer surface with irregular rows of small tuberculate spines, obscured by moderately dense long setae; cutting edge with row of small calcareous teeth, terminating in moderately large calcareous claw overlapped by fixed finger. Palm with row of tufts of long setae on upper surface, unarmed; outer surface with sparse small tubercles, obscured by long, usually simple but sometimes plumose setae; outer surface of fixed finger with irregular row of small tuberculate spines obscured by long simple setae; lower margin with row of small subacute spines and long setae; cutting edge with row of small calcareous teeth, terminating in moderately large calcareous claw. Carpus with row of large spines on upper margin obscured by long setae; outer face with 2 large spines medially on distal half, accompanied by sparse long setae; lower margin not delimited; inner and lower surfaces with sparse long setae. Merus with numerous long, usually simple but sometimes plumose setae arising from low protuberances or tubercles on dorsal margin; ventromesial and ventrolateral margins each with row of small spines and long, simple and plumose setae. Ischium unarmed, but with few long setae.

Second and third pereopods (fig. 2A, B) slender. Dactyls subequal to or 1.1-1.2 times longer than propodi in length, somewhat curved, but not twisted,

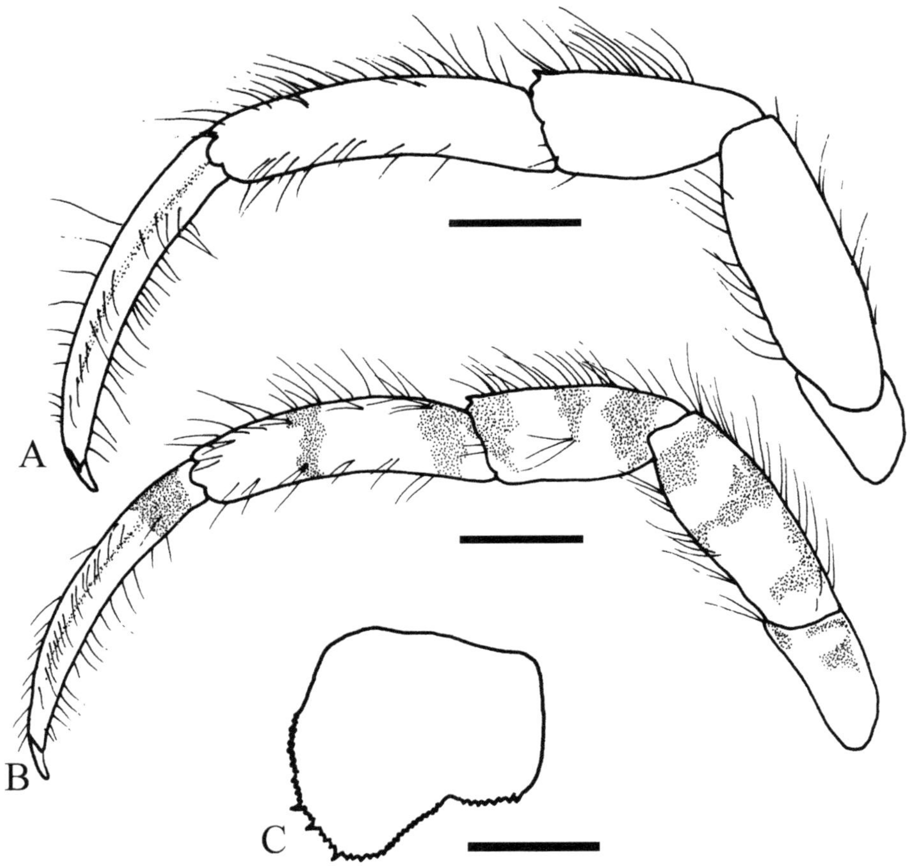

Fig. 2. *Diogenes takedai* sp. nov., holotype, female, 2.5 mm. A, left second pereopod, lateral view; B, left third pereopod, lateral view; C, telson, dorsal view. Scales: A and B = 1 mm; C = 0.5 mm. A, setae partially omitted, color pattern omitted; B, setae partially omitted; C, setae omitted.

terminating in moderately small corneous claw; dorsal and ventral surfaces with numerous long, simple setae; lateral and mesial surfaces each with shallow sulcus and row of long setae along midline. Propodi unarmed; dorsal margins each with long setae; ventral margins with short setae; lateral and mesial faces each with row of sparse setae. Carpi each with dorsodistal spine and long simple setae on dorsal margin; lateral and mesial surfaces with sparse setae. Meri unarmed, but with tufts of long plumose setae on dorsal and ventral margins. Ischia with long plumose setae dorsally and ventrally.

Fourth pereopod semichelate. Anterior lobe of sternite of third pereopods divided by shallow, longitudinal median groove, each half with 1 small spine accompanied by long setae.

Telson (fig. 2C) with shallow median cleft; terminal margin rounded, left longer and slightly wider than right, left terminal margin with row of tiny spines, continued onto lateral margin, 3 somewhat larger spines at or near lateral angle; right terminal margin with row of tiny spines, not continued onto lateral margin; long setae at outer angle continued onto lateral margin.

Colouration in life. — In general whitish cream or pinkish white. Shield with 2 pairs of dark brown spots on anterior half. Ocular peduncles white each with dark brown band medially and light brown band proximally, dark brown spot at base of cornea; ocular acicle each with dark brown spot at base. Antennular peduncles whitish cream; penultimate segment with dark brown transverse band distally and proximally; ultimate segment with greenish brown transverse band distally, and dark brown transverse band medially and proximally; flagella pinkish white, with greenish brown bands. Antennal peduncles each with distal and proximal dark brown spots on lateral margin of fifth segment; fourth segment with dark brown transverse band distally; third segment with dark brown transverse band ventrodistally; first segment dark brown; antennal acicle whitish cream, with dark brown spot proximally; antennal flagella alternately banded with dark brown and whitish cream on every 3 or 4 articles. Left cheliped white, palm with greenish brown spot on upper margin proximally; carpus with dark brown spot on median area of upper margin, outer surface and lower margin, dark brown transverse band proximally; merus with distal and subdistal dark brown transverse bands on each dorsal, lateral and mesial surface, proximal dark brown spot on dorsal surface. Right cheliped white, dactyl and palm each with dark brown transverse band proximally; carpus with transverse band medially and proximally; merus as in left cheliped. Second and third pereopods white to pinkish white; dactyls each with dark brown transverse band proximally; propodi, carpi and meri each with greenish brown transverse band distally, and dark brown with green tinge medially and proximally; median transverse bands on carpi and meri only on dorsal, mesial and lateral surfaces, not continued onto ventral surfaces.

Etymology. — This species is named in honor of an eminent carcinologist, Dr. Masatsune Takeda, in recognition of his enormous contribution to the study of decapods crustaceans.

Habitat. — This species was collected in intertidal sea grass bed area with sand, dead coral and shell fragments substrates.

Distribution. — Known only from Lombok Island, Indonesia.

Remarks. — Among the species of *Diogenes* that possess short antennal acicles, the new species most resembles *D. tirmiziae* Siddiqui & McLaughlin,

2003, known only from Pakistan in having transverse bands on the ocular peduncles and pereopods, three or four additional spines on the ventrolateral distal margin (not visible in dorsal view) of the second segment of the antennal peduncle, and drop-like tubercles on the outer surface of the left chela. However, the ocular peduncles of *D. takedai* are inflated basally and taper to moderately narrow cornea, while in *D. tirmiziae*, the ocular peduncles are the same width proximally and distally. A more significant difference is found on the armature of the propodi of the second and third pereopods which are entirely devoid of spines and tubercles in *D. takedai* sp. nov., whereas in *D. tirmiziae*, a row of small spines is present on the dorsal margin of the propodi of the second and third pereopods, and tiny tubercles or protuberances are also present on the lateral and mesial surfaces of the propodi of the second and third pereopods. Furthermore in *D. takedai*, the outer surface of the fixed finger of the left cheliped is flattened dorsolaterally, covered with small blunt tubercles, and a short ridge of larger tubercles is present near the cutting edge, while in *D. tirmiziae*, the outer surface of the fixed finger is somewhat convex, with covering of small drop-like tubercles.

Differences from *D. guttatus* distributed on northern Australia and Andaman Sea are obvious on the armature of the left cheliped, ambulatory pereopods, and telson. The palm of the left cheliped of *D. guttatus* is covered with circular, mushroom-shaped, flattened tubercles (Siddiqui & McLaughlin, 2003, fig. 4B; McLaughlin, 2004, fig. 2A), the carpi of the second and third pereopods each has a row of small spines on the dorsal margin, and the telson is armed with large spines on the terminal margin of the posterior lobes. In *D. takedai*, the palm of the left cheliped is covered with drop-like tubercles, the carpi of the second and third pereopods possess only dorsodistal small spine, and the telson is armed with tiny spines on the terminal margin of each posterior lobe.

Direct comparison with *D. holthuisi* known only from Japanese waters shows differences in the shapes of the antennal acicles and the telson, and the armature of the branchiostegites. The antennal acicles of *D. holthuisi* are armed only with a single spine on each dorsodistal and dorsomesial margin, and the telson has a few large spines on the terminal margin of each lobe, while in *D. takedai*, the antennal acicles are armed with one to three mesial spines and one terminal spine, and the telson bears a row of tiny spinules on the terminal margin of each lobe. Moreover, the branchiostegites of *D. holthuisi* are armed with a row of very large spines on the dorsal margin and additional two or three strong spines on the posterior carapace; in *D. takedai*, the branchiostegites are armed with a row of two to seven small spines, and the posterior carapace is devoid of spines.

Diogenes dorotheae known from Australian and Indonesian waters may be distinguished from *D. takedai* by the presences of four large spines which are arranged in a short longitudinal row extending from the proximoventral angle of the palm of the left cheliped and a row of spines on the dorsal margin of the carpus of the second and third pereopods. Furthermore, the morphologies of the ocular peduncles and telson are also different between *D. takedai* and *D. dorotheae*. The ocular peduncles taper to somewhat narrow corneas in *D. takedai* instead of inflated distally and proximally in *D. dorotheae*. The telson has a row of minute spinules on the terminal margin of each rounded lobe in *D. takedai* whereas it is armed with strong spines on the terminal margin of each subtriangular lobe in *D. dorotheae*.

Including this new species, the number of species in the genus *Diogenes* is up to 60, 21 of which are found in Indonesian waters. Five species, *D. foresti* Rahayu & Hortle, 2002, *D. moosai* Rahayu & Forest, 1995, *D. spinicarpus* Rahayu & Forest, 1995, *D. takedai* sp. nov. and *D. viridis* Haig & Ball, 1988, are at this moment found exclusively in Indonesian waters (Haig & Ball, 1988; Rahayu & Forest, 1995; Rahayu & Hortle, 2002). *Diogenes biramus* Morgan, 1987, and *D. dorotheae*, which have been originally described from the northern and northwestern Australia are found only in the eastern part of Indonesia (Maluku and Papua) (Rahayu & Forest, 1995; Rahayu & Hortle, 2002). Other species previously reported from Indonesian waters, viz., *D. alias* McLaughlin & Holthuis, 2001 (as *D. diogenes* (Herbst, 1791)), *D. avarus* Heller, 1865, *D. custos* (Fabricius, 1798), *D. dubius* (Herbst, 1804), *D. fasciatus* Rahayu & Forest, 1995, *D. goniochirus* Forest, 1956, *D. jousseaumei* (Bouvier, 1897) (as *D. stenops* Morgan & Forest, 1991), *D. jubatus* (Nobili, 1903) (as *D. platyops* Rahayu & Forest, 1995), *D. klaasi* Rahayu & Forest, 1995, *D. leptocerus* Forest, 1956, *D. mixtus* Lanchester, 1902 (as *D. intermedius* De Man, 1892), *D. pallescens* Whitelegge, 1897, *D. rectimanus* Miers, 1884, and *D. tumidus* Rahayu & Forest, 1995, are also distributed in the Indian Ocean, northern Australia, Malaysia Archipelago, South China Sea and Vietnam (Forest, 1956; Rahayu & Forest, 1995; Rahayu, 2000; McLaughlin, 2002; Rahayu & Hortle, 2002; Siddiqui et al., 2004).

ACKNOWLEDGEMENTS

I thank Dr. Peter K. L. Ng from National University of Singapote for the invitation to participate in the volume dedicated to Dr. Matsasune Takeda.

The careful reviews of Dr. Rafael Lemaitre from Smithsonian Institution and an anonymous reviewer are much appreciated. I also thank the personnel of Marine Bio-Industry Technical Implementation Unit, Research Center for Oceanography-Indonesian Institute of Sciences for assisting in the collection of material.

REFERENCES

ASAKURA, A., 2006. Shallow water hermit crabs of the families Pylochelidae, Diogenidae, Paguridae (Crustacea: Decapoda: Anomura) from the Sea of Japan, with a description of a new species of *Diogenes*. Bulletin of the Toyama Science Museum, **29**: 23-103.

ASAKURA, A. & S. GODWIN, 2006. *Diogenes patae* n. sp. a new species of hermit crab (Crustacea, Decapoda, Anomura, Diogenidae) from American Samoa. Zoosystema, **28**: 457-463.

ASAKURA, A. & H. TACHIKAWA, 2010. *Diogenes holthuisi*, a new species of hermit crab (Decapoda: Anomura: Diogenidae) from shallow water of the Ogasawara (Bonin) Islands, Japan. In: C. H. J. M. FRANSEN, S. DE GRAVE & P. K. L. NG (eds.), Studies on Malacostraca: Lipke Bijdeley Holthuis Memorial Volume. Crustaceana Monographs, **14**: 133-144. (Brill, Leiden).

BOUVIER, E.-L., 1897. Sur deux Paguriens nouveaux trouvés par M. Coutière dans le récif madréporiques à Djibouti. Bulletin du Muséum national d'Histoire naturelle, Paris, **3**: 228-233.

DANA, J. D., 1851. Conspectus Crustaceorum quae in orbis terrarum circumnavigatione, Carolo Wilkes e classe Reipublicae Foederatae duce, lexit et descripcit. Paguridea. Proceedings of the Academy of Natural Sciences, Philadelphia, **5**: 267-272.

DE MAN, J. G., 1892. Decapoden des Indischen Archipels. In: M. WEBER (ed.), Zoologisch Ergebnisse einer Reise in Niederländisch Ost-Indien, **2**: 265-527, pls. 15-29.

FABRICIUS, J. C., 1798. Supplementum Entomologiae systematicae: i–iv, 1-572. (Hafniae).

FOREST, J., 1956. Les Pagures du Viet-Nam. I. Le genre *Diogenes* Dana. Bulletin du Muséum national d'Histoire naturelle, Paris, (2) **28**: 524-532.

HAIG, J. & E. E. BALL, 1988. Hermit crabs from northern Australian and eastern Indonesian waters (Crustacea Decapoda: Anomura: Paguroidea) collected during the 1975 Alpha Helix Expedition. Records of the Australian Museum, **40**: 151-196.

HELLER, C., 1865. Die Crustaceen. Reise der öesterreichisechen Fregatte Novara um die Erde in den Jahren 1857-1859 untr den Befehlen des Commodore B. von Wüllertort-Urbair. Zoologischer, **2**(3): 1-280, pls. 1-25. (Wien).

HENDERSON, J. R., 1888. Report on the Anomura collected by H.M.S. Challenger during the years 1873-76. Scientific Results of the Exploratory Voyage of HMS Challenger, (Zoology) **27**: 1-221.

HERBST, J. F. W., 1791-96. Versuch einer Naturgeschichte der Krabben und Krebse nebst einer systematischen Beschreibung ihrer verschiedenen Arten. **2**: i–viii, 1-226. (Gottlieb August Lange, Berlin und Stralsund).

— —, 1804. Versuch einer Naturgeschichte der Krabben und Krebse etc. etc., **3**(4): 1-49. (Gottlieb August Lange, Berlin und Stralsund).

LANCHESTER, W. F., 1902. On the Crustacea collected during the "Skeat Expedition" to the Malay Peninsula. Proceedings of the Zoological Society of London, **1902**: 363-381.

McLaughlin, P. A., 2002. A review of the hermit crab (Decapoda: Anomura: Paguridea) fauna of southern Thailand, with particular emphasis on the Andaman Sea, and descriptions of three new species. In: N. Bruce, M. Berggren & S. Bussawarit (eds.), Proceedings of the international workshop on the Crustacea in the Andaman Sea, Phuket Marine Biological Center 29 November–20 December, 1998. Phuket Marine Biological Center Special Publication, **23**(2): 385-460.

— —, 2004. A description of the first complete specimen of *Diogenes guttatus* Henderson, 1888 (Decapoda: Anomura: Paguroidea; Diogenidae). Zootaxa, **466**: 1-8.

McLaughlin, P. A. & P. F. Clark, 1997. A review of the *Diogenes* (Crustacea, Paguroidea) hermit crabs collected by Bedford and Lanchester from Singapore, and from the 'Skeat' Expedition to the Malay Peninsula, with a description of a new species and notes on *Diogenes intermedius* De Man, 1892. Bulletin of the Natural History Museum, London, (Zoology) **63**: 33-49.

McLaughlin, P. A. & L. B. Holthuis, 2001. In pursuit of J. F. W. Herbst's species of *Diogenes* (Anomura: Paguroidea: Diogenidae). Journal of Crustacean Biology, **21**: 257-273.

McLaughlin, P. A., T. Komai, R. Lemaitre & D. L. Rahayu, 2010. Annotated checklist of Anomuran Decapod Crustaceans of the world (Exclusive of the Kiwaoidea and families Chirostylidae and Galatheidae of the Galatheoidea). Part I — Lithodoidea, Lomisoidea and Paguroidea. In: M. E. Y. Low & S. H. Tan (eds.), Checklists of Anomuran Decapod Crustaceans of the world (Exclusive of the Kiwaoidea and families Chirostylidae and Galatheidae of the Galatheoidea) and Marine Lobsters of the World. Raffles Bulletin of Zoology, (Supplement) **23**: 5-107.

Miers, E. J., 1884. Crustacea. Part I. The collections from Melanesia. In: Report on the zoological collections made in the Indo-Pacific Ocean during the voyage of H.M.S. 'Alert' 1881-1882: 178-322, pls. 18-32. (British Museum (Natural History), London).

Morgan, G. J., 1987. Hermit crabs (Decapoda, Anomura: Coenobitidae, Diogenidae, Paguridae) of Darwin and Port Essington, Northern Australia. The Beagle, Records of the Northern Territory Museum of Art and Sciences, **4**(1): 165-186.

Morgan, G. J. & J. Forest, 1991. Seven new species of hermit crabs from Northern and Western Australia (Decapoda, Anomura, Diogenidae). Bulletin du Muséum national d'Histoire naturelle, Paris, (4) (A) **12**: 649-689.

Nobili, G., 1903. Crostacei di Singapore. Bollettino dei Musei di Zoologia ed Anatomia comparata della R. Università di Torino, **18**(455): 1-39.

Ortmann, A., 1892. Die Decapodes-Krebse des Strassburger Museum, mit besonderer Berücksichtigung der von Herrn Dr. Döderlein bei Japan und bei Liu-Kiu Inseln gesammelten und zur Zeit im Strassburger Museum aufbewahrten Formen. IV. Die Abtheilungen Galatheidea und Paguridea. Zoologische Jahrbücher, Abtheilung für Systematik, Geographie und Biologie der Thiere, **6**: 241-326.

Rahayu, D. L., 1996. Notes on littoral hermit crabs (excluding Coenobitidae) (Crustacea: Decapoda: Anomura) mainly from Singapore and Peninsular Malaysia. Raffles Bulletin of Zoology, **44**(2): 335-355.

— —, 2000. Hermit crabs from the South China Sea (Crustacea: Decapoda: Anomura: Diogenidae: Paguridae: Parapaguridae). Raffles Bulletin of Zoology, (Supplement) **8**: 377-404.

Rahayu, D. L. & J. Forest, 1995. Le genre *Diogenes* (Crustacea, Decapoda, Diogenidae) en Indonésie, avec la description de six espèces nouvelles. Bulletin du Muséum national d'Histoire naturelle, Paris, **16**(2-4): 383-415.

RAHAYU, D. L. & K. G. HORTLE, 2002. The genus *Diogenes* (Decapoda, Anomura, Diogenidae) from Irian Jaya, Indonesia with description of a new species. Crustaceana, **75**(3-4): 609-619.

SIDDIQUI, F. A. & P. A. MCLAUGHLIN, 2003. A new species of the hermit crab genus *Diogenes* (Decapoda: Anomura: Paguroidea: Diogenidae) from Pakistan, with a comparative diagnosis of *D. guttatus* Henderson, 1888. Proceedings of the Biological Society of Washington, **116**(4): 956-966.

SIDDIQUI, F. A., O. B. KAZMI & P. A. MCLAUGHLIN, 2004. Review of the Pakistani species of *Diogenes* Dana, 1951 (Decapoda Anomura Paguroidea Diogenidae). Tropical Zoology, **17**: 155-200.

WHITELEGGE, T., 1897. The fauna of Funafuti [6]. The Crustacea. Australian Museum Memoirs, **3**: 7-151.

First received 23 August 2011.
Final version accepted 21 December 2011.

GRIFFINIA TAKEDAI, A NEW SPECIES OF DEEP SEA MAJOID CRAB (DECAPODA, BRACHYURA, EPIALTIDAE) FROM THE PHILIPPINES

BY

BERTRAND RICHER DE FORGES[1,3]) and PETER K. L. NG[2,4])
[1]) 5, rue Félix Franchette, 98800 Nouméa New Caledonia
[2]) Tropical Marine Science Institute and Department of Biological Sciences, National University of Singapore, Kent Ridge, Singapore 119260, Republic of Singapore

ABSTRACT

A new species of deep-sea majoid is described from the eastern Philippines. *Griffinia takedai* n. sp. (Epialtidae) is the fourth species in this genus to be described, and it differs from congeners in its setose carapace, elongate rostral spines, as well as the well-developed supraorbital and hepatic spines. The new species is diagnosed, and a key to the genus is presented.

INTRODUCTION

It is well known that the Philippine archipelago is a hot spot of marine biodiversity. Forming the core of the "Coral Triangle", this area has been intensively sampled since the 19th century but continues to surprise biologists studying this region. The deep-sea fauna of the east coast of Luzon, however, has remained relatively neglected, even after the major surveys by the American vessel "Albatross" in 1908-1910 and three French MUSORSTOM cruises in 1976, 1981 and 1985. Using a research vessel from the Philippine Fisheries Institute, this shortcoming was recently further addressed when Aurora Bay on the Pacific coast of Luzon was sampled by the AURORA cruise in 2007. From these collections, several deep-sea brachyuran crabs have already been reported (e.g., Richer de Forges & Ng, 2008, 2009a, b; Galil & Ng, 2009; Ahyong & Ng, 2009, 2011). With regard to majoid crabs, Richer de Forges & Ng (2009a, b)

[3]) e-mail: b.richerdeforges@gmail.com
[4]) Corresponding author; e-mail: peterng@nus.edu.sg

have already reported on a number of epialtids, namely *Oxypleurodon stimpsoni* Miers, 1886, *O. bipartitum* (Guinot & Richer de Forges, 1986), *O. auritum* (Rathbun, 1916) and *O. annulatum* Richer de Forges & Ng, 2009.

In this paper, we describe a new species of *Griffinia* Richer de Forges, 1994, from the area. Specimens examined are deposited in the Crustacean Collection of the National Museum of the Philippines (NMCR) and the Zoological Reference Collection of the Raffles Museum of Biodiversity Research, National University of Singapore (ZRC). Measurements provided are in millimeters. The following abbreviations are used: cl = carapace length; cb = carapace width; G1 = male first gonopod; G2 = male second gonopod; P2–P5 = pereiopods 2-5, respectively; prl = postrostral carapace length.

TAXONOMY

Superfamily MAJOIDEA Samouelle, 1819
Family EPIALTIDAE MacLeay, 1838
Subfamily EPIALTINAE MacLeay, 1838
Griffinia Richer de Forges, 1994

Antilibinia — Rathbun, 1918: 12 (part); Hale, 1927: 133 (part); Griffin, 1966: 267 (part); Griffin & Tranter, 1986: 70 (part) (not *Antilibinia* MacLeay, 1838: 56).
Pisidarum Serène & Vadon, 1981: 119, 124, 128, 140, pl. 4F (nomen nudum).
Griffinia Richer de Forges, 1994: 65; Ng et al., 2008: 100.

Remarks. — Serène & Vadon (1981) first used the genus name *Pisidarum* for an unidentified species from the Philippines. Although they did not state anywhere in the paper that they regarded *Pisidarum* as a new genus, this name had never been used before. The manner in which Serène & Vadon (1981) used "*Pisidarum*", first in a list of species found in each station (p. 119), then in a list of taxa found (p. 124), a short discussion of each taxon (p. 128) and finally as a caption for his figures (p. 140, pl. 4F), makes it clear that it was used as a generic and not a vernacular name. Despite providing characters for the genus (Serène & Vadon, 1981: 128), their failure to designate a named species makes *Pisidarum* Serène & Vadon, 1981, a nomen nudum. Richer de Forges (1994) described *Griffinia* in his study of deep-sea material from Australia and Indonesia, designating *Antilibinia lappacea* Rathbun, 1918, as the designated type species. Richer de Forges (1994: 65), however, had noted that *Griffinia* is the same taxon as *Pisidarum*, but the latter name was not available. Three species of *Griffinia* are currently known, all from the Western Pacific: *G. gilloloensis* (Rathbun, 1916), *G. polita* (Griffin & Tranter, 1986) and *G. lappacea* (Rathbun, 1918). *Griffinia takedai* n. sp. is the fourth species, and interestingly, all species are currently only known from the Philippines, with the type locality of three of them being in this country. A key to the genus is provided below.

Griffinia takedai n. sp.
(figs. 1-4)

Material examined. — Holotype — male (cl 9.6 mm, prl 8.0 mm, cb 5.8 mm) (NMCR), eastern Luzon, Philippines, station CP 2658, 15°59.45′N 121°50.68′E, 422-431 m, coll. AURORA cruise, 20 May 2007. Paratype — 1 male (cl 14.3 mm, prl 11.4 mm, cb 8.5 mm) (ZRC), eastern Luzon, Philippines, station CP 2695, 14°45.28′N 123°38.36′E, 377 m, coll. AURORA cruise, 26 May 2007.

Diagnosis. — Relatively small species (adult size cl 9.6 mm) with distinctly pyriform carapace shape (figs. 1, 2A). Carapace dorsal surface covered with numerous long dispersed simple, straight to gently curved setae which do not obscure margins, with relatively dense tomentum of short setae between them; surfaces almost smooth, mostly unarmed (figs. 1, 2A, 3A). Rostrum about 1/3 postrostral carapace length; rostral spines relatively short, diverging, gently curved outwards in distal half, forming distinct V-shape (figs. 1, 2A, B). Supraocular spine strong, sharp, relatively short, pointing anteriorly; preorbital, postorbital and hepatic spines relatively narrow, laterally compressed; postorbital spine short, blunt; hepatic spine well developed but relatively short, straight, spiniform, not reaching orbit, directed laterally (figs. 1, 2A, B). Basal antennal article rectangular, stout, fused to carapace, with large external distal spine; second and third antennal articles straight, elongate, cylindrical, slender, not reaching tip of rostrum (figs. 1, 2A, B, 3A). Third maxilliped relatively short; ischium with no visible sulcus; merus with distinct auriculiform anteroexternal angle (fig. 2B). Border of carapace with row of 9 tubercules with flattened or rounded tips, 4 visible ventrally on border of pterygostomian area (figs. 2A, B, 3A). Thoracic sternites 1-4 fused, median suture between sternites 1 and 2 indicated by shallow depressions; sternites 2 and 3 indicated by lateral clefts (fig. 2C, D). Adult male cheliped with inflated chelae; merus with sharp distal tooth, 1 proximal tubercle, 1 subproximal tubercle on proximal one-third; carpus with 2 low carinae on upper surface, outer margin with broad lobiform crest; propodus carinated on superior border, with relatively short fingers, cutting borders with series of low teeth (figs. 1, 3B). Ambulatory legs relatively short, surfaces smooth; P2 (and merus) longest, slightly longer than cheliped; merus with distal tooth which is subovate with tip tapering suddenly; dactylus curved, long, tip sharp; inferior border of propodi

Fig. 1. *Griffinia takedai* n. sp. A, holotype male (cl 9.6 mm) (NMCR); B, paratype male (cl 14.3 mm) (ZRC).

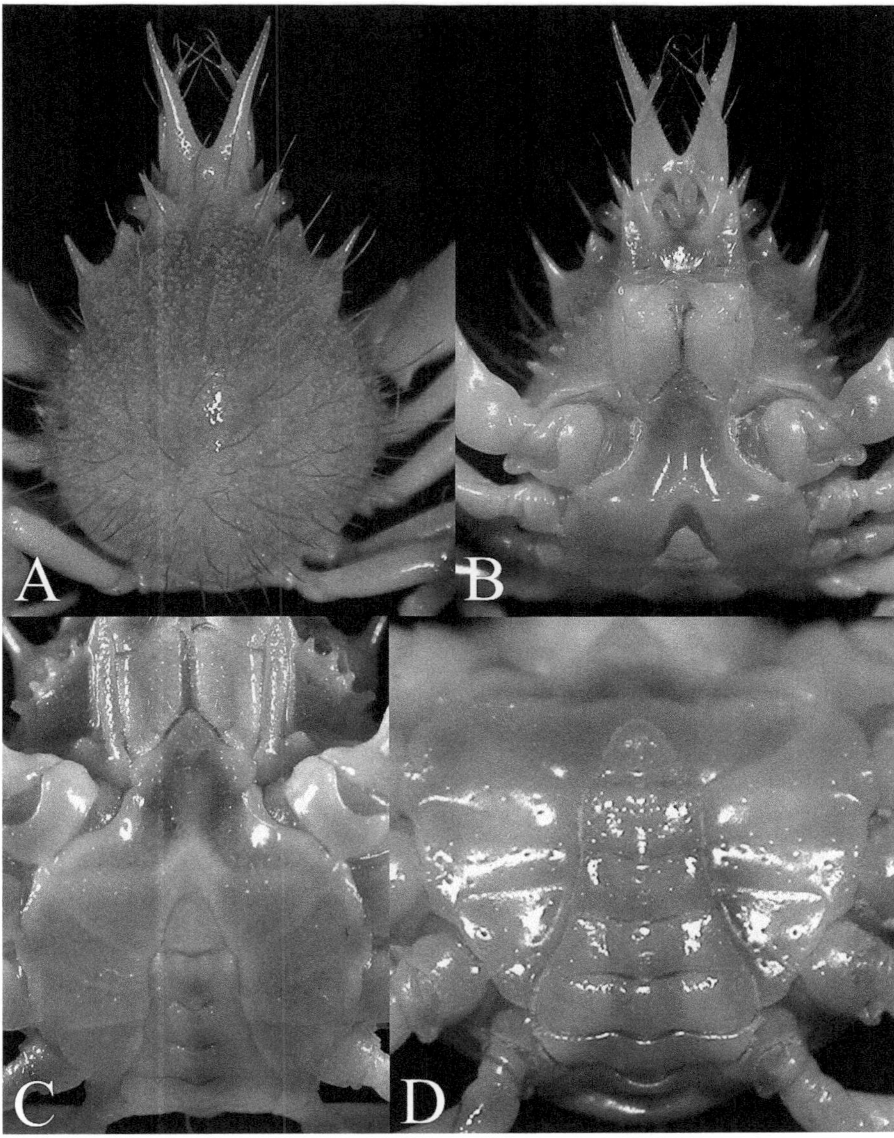

Fig. 2. *Griffinia takedai* n. sp. A, B, D, paratype male (cl 14.3 mm) (ZRC); C, holotype male (cl 9.6 mm) (NMCR). A, dorsal carapace view; B, frontal view showing third maxillipeds and anterior thoracic sternum; C, anterior thoracic sternum and anterior male abdomen; D, male abdomen.

with row of short median setae, may be medially interrupted in those on P4 and P5; dactylo-propodal lock formed by expanded distal part of propodus almost completely covering base of dactylus in dorsal view (figs. 1, 3C, 4D, E).

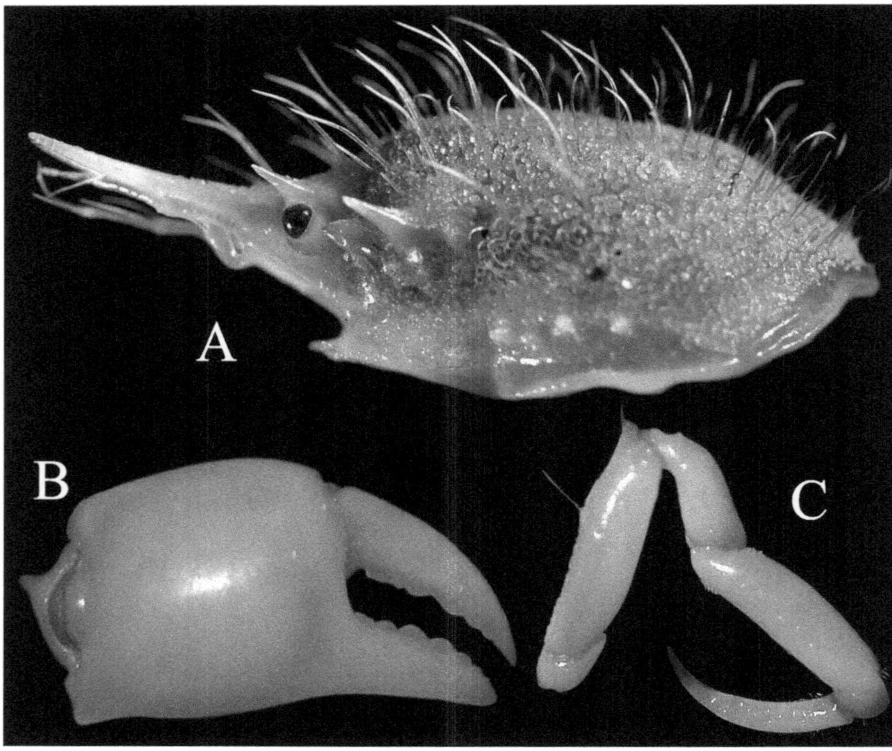

Fig. 3. *Griffinia takedai* n. sp. A, holotype male (cl 9.6 mm) (NMCR); B, C, paratype male (cl 14.3 mm) (ZRC). A, lateral view of carapace; B, outer view of right chela; C, right P5.

Male abdomen with 6 free somites and telson (fig. 2C, D). Press-button locking mechanism of male abdomen peg-like, on anterior part of thoracic sternite 5, adjacent to sternite 4. G1 straight, laterally flattened subdistally, forming flap-like structure, subdistal part entire, slightly constricted near apex, tip sharp (fig. 4A, B); G2 short, broadly C-shaped, with short flagellum (fig. 4C).

Colour. — In life, the carapace is orange and the legs are a clear greenish-white.

Etymology. — The species is established to honour an esteemed colleague and friend for many years, Masatsune Takeda, who devoted his professional life to the study of Crustacea and has described numerous deep-sea species over a long and illustrious career.

Remarks. — *Griffinia takedai* n. sp. is closest to *G. lappacea* (Rathbun, 1918) in general appearance. However, *G. lappacea* has proportionately longer rostral spines, longer supraocular spines and a much longer and curved blade-like hepatic spine (relatively shorter, straight and spiniform in *G. takedai*) (cf. Rathbun, 1918: 12, 13, pl. 7 fig. 3; Richer de Forges, 1994: 67-69, figs. 3, 4B,

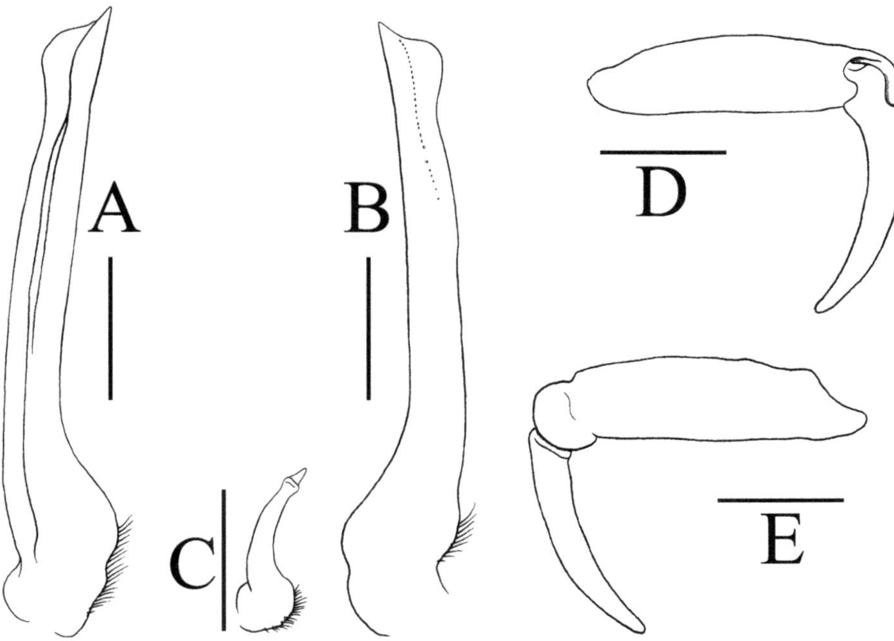

Fig. 4. *Griffinia takedai* n. sp. Holotype male (cl 9.6 mm) (NMCR). A, ventral view of G1; B, dorsal view of G1; C, ventral view of G2; D, ventral view of dactylus and propodus of P5 (setae not drawn); dorsal view of dactylus and propodus of P5 (setae not drawn). Scales A–C = 0.5 mm; D, E, 1.0 mm.

C). In addition, the G1 of *G. lappacea* has a small subdistal cleft (Richer de Forges, 1994: fig. 3D) whilst that of *G. takedai* is entire (fig. 4A, B). *Griffinia lappacea* has been reported from South Australia (type locality), northwestern Australia, Kei Islands in Indonesia and Philippines (Richer de Forges, 1994).

The Philippine record of *G. lappacea* was based on a single female specimen reported by Serène & Vadon (1981) as "*Pisidarum* sp." and although Richer de Forges (1994) referred it to *Griffinia*, he was not able to examine the actual specimen. However, the Philippine specimen (Serène & Vadon, 1981: pl. 4F) also differs from *G. lappacea* s. str. (cf. Richer de Forges, 1994: fig. 3) in having the hepatic spines long and straight, not blade-like and curved. The Australian and Indonesian material of *G. lappacea* is thus unlikely to be conspecific with the "*Pisidarum* sp." of Serène & Vadon (1981) from the Philippines. In fact, Serène & Vadon's (1981: 124) "*Pisidarum* sp." more closely resembles *G. takedai*. The Philippines specimen has relatively longer hepatic spines which reach past the orbits (shorter, not reaching orbits in *G. takedai*, figs. 1, 2A); the supraorbital spine is relatively longer and curved outwards (relatively shorter, straight in *G. takedai*, figs. 1, 2A, 3A); and the rostrum is proportionately longer, being more than half of the postrostral carapace length (relatively shorter in *G. takedai*, being less than half the postrostral carapace length, figs. 1, 2A, 3A) (cf. Serène & Vadon, 1981: pl. IV fig. F). These differences suggest we are dealing with two species. In addition, Serène & Vadon's (1981) specimen was collected from the western-central Philippines in the South China Sea (13°49.2′N 120°01.8′E), whilst the types of *G. takedai* are from the eastern Philippines in the Pacific. If Serène & Vadon's (1981) specimen can be found and re-examined, it is likely to represent an undescribed species. Serène & Vadon's (1981) material should be in MNHN but their specimen could not be found when the authors searched for it on separate occasions. Unfortunately, it is known that some of this material is missing (see Ng & Chen, 2005).

KEY TO *GRIFFINIA*

1. Rostrum very short, about 1/6 postrostal carapace length; postorbital and hepatic lobes very short, subequal .... *G. gilloloensis* (Gillolo Islands, Philippines)
– Rostrum about 1/3 postrostral carapace length or greater; hepatic lobe much longer than postorbital lobe.. 2
2. Carapace with scattered setae, surface appears almost glabrous; rostral spines short, broad, dorsoventrally flattened; preorbital, postorbital and hepatic lobes broad, dorsoventrally flattened.............................. *G. polita* (Mindanao, Philippines)
– Carapace with numerous long, straight setae; rostral spines slender; preorbital, postorbital and hepatic spines narrow, laterally compressed 3
3. Rostral spines shorter than half postrostral carapace length; hepatic spines long, not reaching orbits, spiniform; G1 with entire subdistal part .... *G. takedai* (eastern Luzon, Philippines)
– Rostral spines distinctly longer than half postrostral carapace length; hepatic spines long, reaching orbits, G1 not as above 4
4. Rostral spines straight; hepatic spines curving inwards; G1 with small subdistal cleft ... . *G. lappacea* (Australia and Kei Islands, Indonesia)
– Rostral spines gently curving outwards; hepatic spines straight; G1 not known .... *G.* sp. (western-central Philippines)

ACKNOWLEDGEMENTS

The material was obtained by the AURORA 2007 Expedition to Luzon organized by the Muséum national d'Histoire naturelle (MNHN, Paris), University of San Carlos (USC, Cebu); National Museum of the Philippines (NMCR, Manila), Philippines Bureau of Fisheries Research (BFAR, Manila), Raffles Museum of Biodiversity Research, National University of Singapore (RMBR) and the National Taiwan Ocean University (NTOU, Keelung). We are grateful to Philippe Bouchet (MNHN) (expedition leader), Danilo Largo (USC), Ludivina Labe (BFAR), Marivene Manuel-Santos (NMCR) for help in organizing this expedition and to Chan Tin-Yam (NTOU) for photographing the fresh material. The expedition was supported by grants from the Total Foundation, French Ministry of Foreign Affairs, MNHN, RMBR and NTOU. The first author was supported by a research fellowship from the RMBR. We are also grateful to Marcos Tavares and Colin McLay for their many helpful comments with this manuscript.

REFERENCES

AHYONG, S. T. & P. K. L. NG, 2009. The Cymonomidae of the Philippines (Crustacea: Decapoda: Brachyura), with descriptions of four new species. Raffles Bulletin of Zoology, (Supplement) **20**: 233-246.

— — & — —, 2011. Cyclodorippoid crabs from the Philippines collected by the PANGLAO 2004-2005 and AURORA 2006 expeditions. Zoologische Anzeiger, **250**(4): 479-487.

GALIL, B. S. & P. K. L. NG, 2009. Calappoidea and Leucosioidea (Crustacea: Decapoda: Brachyura) from Luzon, Philippines, with descriptions of two new species of *Mursia*. Zootaxa, **2085**: 45-60.

GRIFFIN, D. J. G., 1966. A review of the Australian Majid Spider Crabs (Crustacea, Brachyura). Australian Zoologist, **13**(3): 259-298, 6 text-figs., pls. XV–XVII.

GRIFFIN, D. J. G. & H. A. TRANTER, 1986. The Decapoda Brachyura of the Siboga Expedition. Part VIII. Majidae. Siboga-Expeditie Monographie, **39**c4: 1-335, pls. 1-22.

GUINOT, D. & B. RICHER DE FORGES, 1986. Crustacés Décapodes: Majidae (genres *Platymaia*, *Cyrtomaia*, *Pleistacantha*, *Sphenocarcinus* et *Naxioides*). In: Résultats des Campagnes MUSORSTOM 1 et 2. Tome 2. Mémoires du Muséum national d'Histoire naturelle, 1985 (1986) Zoology, (A) **133**: 83-178.

HALE, H. M., 1927. The crustaceans of South Australia. Part I: 1-201, figs. 1-202. (Harrison Weir, Government Printer, Adelaide).

MACLEAY, W. S., 1838. On the brachyurous decapod Crustacea. Brought from the Cape by Dr. Smith. In: A. SMITH (ed.), Illustrations of the Annulosa of South Africa; consisting chiefly of figures and descriptions of the objects of natural history collected during an expedition into the interior of South Africa, in the years 1834, 1835, and 1836; fitted out by 'The Cape of Good Hope Association for Exploring Central Africa': together with a summary of African Zoology, and an inquiry into the geographical ranges of species in that quarter of the globe, published under the Authority of the Lords Commissioners of Her Majesty's Treasury, Invertebrata. IV [1849]: 53-71, pls. 2, 3. (Smith, Elder & Co., London).

MIERS, E. J., 1886. Part II. Report on the Brachyura collected by H.M.S. *Challenger* during the years 1873-76. In: Report on the Scientific Results of the Voyage of H.M.S. Challenger during the years 1873-1876 under the command of Captain George S. Nares, N.R., F.R.S. and the late Captain Frank Tourle Thomson, R.N. prepared under the Superintendence of the late Sir C. Wyville Thomson, Knt., F.R.S. &c. Regius Professor of Natural history in the University of Edinburgh of the civilian scientific staff on board and now of John

Murray one of the naturalists of the Expedition, (Zoology) **17**: i–l, 1-362, pls. 1-29. (Her Majesty's Stationary Office, London, Edinburgh and Dublin).

NG, P. K. L. & H.-L. CHEN, 2005. On two species of euxanthine crabs from the South China Sea, including a description of a new species of *Crosnierius* (Crustacea: Decapoda: Brachyura: Xanthidae). Proceedings of the Biological Society of Washington, **118**(2): 319-325.

NG, P. K. L., D. GUINOT & P. J. F. DAVIE, 2008. Systema Brachyurorum: Part I. An annotated checklist of extant brachyuran crabs of the world. Raffles Bulletin of Zoology, (Supplement) **17**: 1-286.

RATHBUN, M. J., 1916. New species of crabs of the families Inachidae and Parthenopidae. Scientific results of the Philippine cruise of the fisheries steamer "Albatross", 1907-1910. (No.34). Proceedings of the U.S. National Museum, **50**: 527-559.

— —, 1918. Report on the spider crabs. Report on the crabs obtained by the F.I.S. "Endeavour" on the coasts of Queensland, New South Wales, Victoria, South Australia and Tasmania. In: Biological results of the fishing experiments carried on by the F.I.S. "Endeavour," 1909-14, **5**(1): 1-29, pls. 1-15.

RICHER DE FORGES, B., 1994. A new genus of deep-sea majid crab: *Griffinia* gen. nov. (Crustacea, Decapoda, Brachyura). The Beagle, Records of the Museums and Art Galleries of the Northern Territory, **11**: 65-72, figs. 1-4.

RICHER DE FORGES, B. & P. K. L. NG, 2008. New western Pacific records of Homolidae De Haan, 1839, with descriptions of new species of *Homolochunia* Doflein, 1904, and *Latreillopsis* Henderson, 1888 (Crustacea: Decapoda: Brachyura). Zootaxa, **1967**: 1-35.

— — & — —, 2009a. On the majoid genera *Oxypleurodon* Miers, 1886, and *Sphenocarcinus* A. Milne-Edwards, 1875 (Crustacea: Brachyura: Epialtidae), with descriptions of two new genera and five new species. Raffles Bulletin of Zoology, (Supplement) **20**: 247-266.

— — & — —, 2009b. New genera, new species and new records of Indo-West Pacific spider crabs (Crustacea: Brachyura: Epialtidae: Majoidea). Zootaxa, **2025**: 1-20.

SAMOUELLE, G., 1819. The entomologist's useful compendium; or an introduction to the knowledge of British insects, etc: 1-496. (London).

SERÈNE, R. & C. VADON, 1981. Crustacés Décapodes: Brachyoures. Liste préliminaire, description de formes nouvelles et remarques taxonomiques. In: Résultats des Campagnes MUSORSTOM, 1. Philippines (18-29 mars 1976). Mémoires ORSTOM, Paris, **91**: 117-140.

First received 8 August 2011.
Final version accepted 9 December 2011.

A NEW SPECIES OF THE GENUS *MICROPROSTHEMA* STIMPSON, 1860 (STENOPODIDEA, SPONGICOLIDAE) FROM THE SOCIETY ISLANDS, FRENCH POLYNESIA

BY

TOMOMI SAITO[1,3]) and ARTHUR ANKER[2,4])

[1]) Usa Marine Biological Institute, Kochi University, 194 Inoshiri, Usa, Tosa 781-1164, Japan
[2]) Instituto de Ciências do Mar — Labomar, Universidade Federal do Ceará (UFC), Avenida da Abolição 3207, Meireles, CEP 60.165-081, Fortaleza, Ceará, Brazil

ABSTRACT

A new species of the spongicolid genus *Microprosthema* Stimpson, 1860 is described and illustrated based on material collected in Moorea, Society Islands, French Polynesia, increasing the total number of known species of *Microprosthema* to 13, 6 of which are found in the Indo-West Pacific. *Microprosthema takedai* sp. nov. is distinguishable from its congeners by the less numerous teeth on the carapace; the third pereopod with several teeth and a crista on the dorsal margin of the palm, and with a broad groove on the carpus; the absence of dorsal teeth on the carpus of the first and second pereopods; and the absence of lateral teeth on the sixth pleonal somite.

RÉSUMÉ

Une nouvelle espèce du genre *Microprosthema* Stimpson, 1860 (Spongicolidae) est décrite à la base de spécimens recoltés à Moorea, Îles de la Société de la Polynésie Française, augmentant le nombre total des espèces connues dans ce genre à 13, 6 d'entre elles étant présentes dans l'Indo-Ouest Pacifique. *Microprosthema takedai* sp. nov. peut être séparée des autres espèces du genre par les dents de la carapace étant peu nombreuses; le troisième péréiopode portant plusieurs dents et une crista (crête) sur la paume, et avec un sillon assez large sur le carpe; l'absence des dents dorsales sur le carpe du premier et second péréiopodes; et enfin, l'absence des dents latérales sur le sixième somite abdominal.

[3]) Corresponding author; e-mail: t-saito@kochi-u.ac.jp
[4]) e-mail: arthuranker7@yahoo.com

 Studies on Eumalacostraca: 285-301

INTRODUCTION

The stenopodidean family Spongicolidae Schram, 1986 presently contains six genera, *Globospongicola* Komai & Saito, 2006, *Microprosthema* Stimpson, 1860, *Paraspongicola* de Saint Laurent & Cleva, 1981, *Spongicola* De Haan, 1844, *Spongicoloides* Hansen, 1908 and *Spongiocaris* Bruce & Baba, 1973. Most members of this family are obligate symbionts of deep-water hexactinellid sponges, typically occupying the atrium of their hosts. However, the species of the genus *Microprosthema* constitute an exception among the spongicolids, with all 12 hitherto known species (table I) occurring in shallow-water habitats, mainly on coral and rocky reefs and adjacent seagrass and rubble flats. Five species of *Microprosthema* have been previously described from the Indo-West Pacific: *M. fujitai* Saito & Okuno, 2011; *M. lubricum* Saito & Okuno, 2011; *M. plumicorne* (Richters, 1880); *M. scabricaudatum* (Richters, 1880) and *M. validum* Stimpson, 1860 (Holthuis, 1946; Baba et al., 1968; Devaney & Bruce, 1987; Saito & Okuno, 2011; see also table I). In the present study, a sixth Indo-West Pacific species is described based on material collected in Moorea, Society Islands, French Polynesia, mainly during October–November 2008 (Biocode Moorea 2008).

MATERIAL AND METHODS

The postorbital carapace length is abbreviated as cl in the text. The term tooth (or sharp tooth) is here used for fixed, usually sharp, spine-like projections on the carapace, pleon, antenna, third maxilliped, and third pereopod. Counts of the teeth on the lateral margins of the antennal scale and uropods include the terminal tooth. The material examined in this study is deposited in the following institutions: Natural History Museum and Institute, Chiba, Japan (CBM); Coastal Branch of the Natural History Museum and Institute, Chiba, Japan (CMNH); Florida Museum of Natural History, University of Florida, Gainesville, USA (FLMNH UF); National Museum of Nature and Science, Tokyo, Japan (NSMT); National Museum of Natural History, Smithsonian Institution, Washington DC, USA (USNM) and Kitakyushu Museum of Natural History and Human History, Kitakyushu, Japan (KMNH), which now includes specimens formerly deposited in the Zoological Laboratory, Faculty of Agriculture, Kyushu University, Japan (ZLKU). The abbreviation "fcn" stands for field collection number (also corresponding to a photo voucher of the specimen).

TABLE I

Geographical ranges of the known species of *Microprosthema* Stimpson, 1860

Name	Geographical range	Reference
Microprosthema emmiltum Goy, 1987	Eastern Pacific: Galapagos; Panama; Mexico	Goy, 1987
Microprosthema fujitai Saito & Okuno, 2011	Indo-West Pacific: Japan (Ryukyu Is.)	Saito & Okuno, 2011
Microprosthema granatense Criales, 1997	Western Atlantic: Colombia	Criales, 1997
Microprosthema inornatum Manning & Chace, 1990	Central Atlantic: Ascension I.	Manning & Chace, 1990
Microprosthema jareckii Martin, 2002	Western Atlantic: British Virgin Is.	Martin, 2002
Microprosthema looensis Goy & Felder, 1988	Western Atlantic: Florida Keys	Goy & Felder, 1988
Microprosthema lubricum Saito & Okuno, 2011	Indo-West Pacific: Japan (Ryukyu Is.); Japan; French Polynesia (Society Is.)	Saito & Okuno, 2011
Microprosthema manningi Goy & Felder, 1988	Western Atlantic: Belize; Mexico; Florida	Goy & Felder, 1988
Microprosthema plumicorne (Richters, 1880)	Indo-West Pacific: Mauritius	Richters, 1880
Microprosthema scabricaudatum (Richters, 1880)	Indo-West Pacific: Japan; Mauritius; Marshall Is.; Papua New Guinea	Richters, 1880; Holthuis, 1946; Baba et al., 1968; Devaney & Bruce, 1987
Microprosthema semilaeve (von Martens, 1872)	Western Atlantic: SE USA; Bahamas; Mexico; Colombia; Honduras; Panama; NE Brazil	von Martens, 1872; Alves et al., 2008; A. Anker, pers. obs.
Microprosthema takedai sp. nov.	Indo-West Pacific: French Polynesia (Society Is.)	present study
Microprosthema validum Stimpson, 1860	Indo-West Pacific: Japan; Australia; Pakistan	Stimpson, 1860; Holthuis, 1946

Comparative material. — *Microprosthema emmiltum* Goy, 1987: holotype, female, cl 5.2 mm (USNM 231363), paratype, female, cl 3.8 mm (USNM 231364), Ecuador: Galapagos Islands, Isla Santa Maria.

Microprosthema fujitai Saito & Okuno, 2011: holotype, female, cl 3.0 mm (NSMT-Cr 21265), allotype, male, cl 2.0 mm (NSMT-Cr 21266), Japan: Ryukyu Islands, Okinawa.

Microprosthema granatense Criales, 1997: holotype, male, cl 3.5 mm (USNM 275993), Colombia: Granate Bay.

Microprosthema inornatum Manning & Chace, 1990: holotype, male, cl 1.8 mm (USNM 221894), South-Central Atlantic: Ascension Island.

Microprosthema looensis Goy & Felder, 1988: holotype, ovigerous female, cl 4.3 mm (USNM 233999), USA: Florida, Looe Key.

Microprosthema lubricum Saito & Okuno, 2011: holotype, ovigerous female, cl 3.3 mm (CMNH-ZC 01067), Japan: Ryukyu Islands, Kerama Islands; ovigerous female, cl 2.4 mm (FLMNH UF Arthropoda 16363), French Polynesia: Society Islands, Moorea.

Microprosthema manningi Goy & Felder, 1988: holotype, male, cl 4.3 mm (USNM 233997), allotype, ovigerous female, cl 5.9 mm (USNM 233998), Belize: Carrie Bow Cay.

Microprosthema scabricaudatum (Richters, 1880): female, cl 2.7 mm (CBM-ZC 3241), Japan: Ryukyu Islands, Kumejima; 2 males, cl 2.8-3.0 mm, ovigerous female, cl 3.0 mm, juv., cl 1.5 mm (ZLKU 11094), Japan: Ryukyu Islands, Amami-oshima.

Microprosthema semilaeve (von Martens, 1872): male, cl 4.5 mm (USNM 184533), USA: Florida, Dry Tortugas.

Microprosthema validum Stimpson, 1860: male, cl 2.0 mm (ZLKU 11098), Japan: Ryukyu Islands, Ishigaki-jima; male, cl 3.1 mm, 2 females, cl 2.9-3.1 mm (NSMT-Cr 8176), Hong Kong; female, cl 4.2 mm (CMNH-ZC 00972), male, cl 2.9 mm (CMNH-ZC 00973), Japan, Ryukyu Islands, Kumejima.

TAXONOMIC ACCOUNT

Genus *Microprosthema* Stimpson, 1860

Microprosthema takedai sp. nov.

(figs. 1-7)

Type material. — French Polynesia, Society Islands, Moorea. Holotype: ovigerous female, cl 1.7 mm (FLMNH UF Arthropoda 16271), Moorea Lagoon, west of Cook's Bay, approximately between Cook's Bay entrance and Sheraton Hotel, 17°47′67″S 149°83′01″W, outer reef slope, in crevices of large piece of coral rubble, depth 4-7 m, 4 Nov. 2008, leg. A. Anker, S. McKeon, J. Moore [fcn BMOO-5118]. Allotype: male, cl 2.3 mm (FLMNH UF Arthropoda 16504), Motu Tiahura, 17°29′17″S 149°54′58″W, shallow sand flat with rubble and massive corals, under rubble and coral bommies, depth 0.5-1 m, 18 Nov. 2008, leg. A. Anker [fcn BMOO-5571]. Paratypes: 1 ovigerous female, cl 2.1 mm (FLMNH UF Arthropoda 10053), Pt. Faupo, 17°49′64″S 149°75′28″W, narrow oceanic reef flat, under rock in 10-20 cm sand in horizontal burrow, depth 0.5-1 m, 20 Jun.–1 Jul. 2006, leg. G. Paulay, C. McKeon [fcn BMOO-377]; 1 female, cl 1.6 mm (FLMNH UF Arthropoda 15432), Moorea lagoon off Cook's Bay, 17°55′67″S 149°87′44″W, mid-barrier reef, on sand under rubble, depth 2-3 m, 11 Oct. 2008, leg. A. Anker [fcn BMOO-2246]; 1 female (missing both P3), cl 1.5 mm (FLMNH UF Arthropoda 15448), north coast of Moorea, off Sheraton Hotel, 17°55′67″S 149°87′44″W, outer reef slope, crevices in coral rubble, depth 10-17 m, 12 Oct. 2008, leg. C. McKeon, G. Paulay, J. Moore, A. Anker [fcn BMPP-2264]; 1 female, cl 2.5 mm (FLMNH UF Arthropoda 15506), Moorea lagoon at Haapiti, 17°55′67″S 149°87′44″W, shallow sand flat off mangroves, depth 0.5-1 m, 13 Oct. 2008, leg. A. Anker [fcn BMOO-2371]; 1 male, cl 1.5 mm (FLMNH UF Arthropoda 15651), north coast of Moorea, off Sheraton Hotel, 17°48′04″S 149°85′29″W, outer reef slope, under rocks and rubble, depth 10-12 m depth, 16 Oct. 2008, leg. G. Paulay, J. Moore, C. McKeon [fcn BMOO-2918]; 1 male, cl 1.5 mm (FLMNH UF Arthropoda 16269), same collection data as for the holotype [fcn BMOO-5116].

Description. — Females. Rostrum (figs. 1, 2A, B) slender distally, proximally broad, forming supraorbital eave, directed slightly downwards, falling slightly short of distal margin of antennal scale, 0.54-0.65 of carapace length; dorsal carina armed with 6 large acute teeth, their size decreasing proximally, proximal 2 teeth situated posterior to orbital margin; ventral carina with 1 small acute subdistal tooth, directed ventrally; lateral carinae usually with small tooth at midlength.

Carapace (figs. 1, 2A, B) with postrostral median ridge extending to epigastric region; postorbital region armed with 2-3 pairs of stout, acute, anteriorly directed teeth; orbital margin concave, inferior orbital angle obsolete; antennal tooth large, acuminate; pterygostomial margin rounded, with several sharp teeth of various sizes; gastric region with 3-5 pairs of postrostral submedian teeth; cervical groove distinct, posterior margin armed with cincture of several, acute, anteriorly directed teeth of various sizes; hepatic tooth present; anterolateral teeth relatively small, scattered; branchial and post-cervical regions sometimes sparsely with small teeth.

Sixth thoracic sternite (fig. 2C) with medially jointed pair of slender, contiguous lobes; lateral and mesial margins unarmed; ventral surface concave, unarmed. Seventh sternite with paired, broad subrectangular plates; each distolateral angle rounded; lateral and anteromesial margins unarmed; ventral surface concave, unarmed. Eighth sternite with paired smaller trapezoid plates; each distolateral angle feebly angular; lateral and anteromesial margins unarmed; ventral surface concave, unarmed.

Pleonal somites (figs. 1, 2D) not sculptured. First pleonal somite short, divided in two sections by distinct transverse carina; anterior section with pleuron unarmed laterally, posteroventrally ending in process supported by short transverse carina; posterior section with pleuron unarmed laterally; posteroventral margin armed with a few small teeth. Second to fifth pleura unarmed laterally; ventral margins produced, terminating in large tooth, and armed with a few anterior and posterior teeth. Second pleonal somite with rather distinct transverse carina, ending at the middle of pleuron. Third pleonal

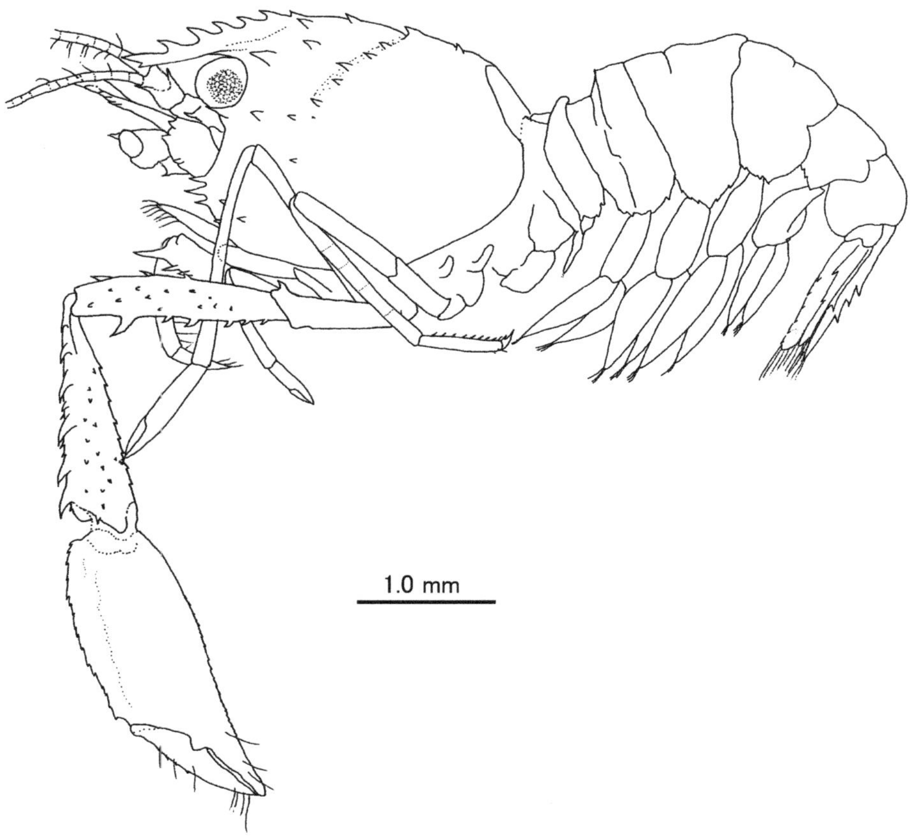

Fig. 1. *Microprosthema takedai* sp. nov. Holotype, female (cl 1.7 mm) from Moorea, French Polynesia (FLMNH UF Arthropoda 16271): habitus, lateral view.

somite longest, with short transverse groove anteriorly. Fourth pleonal somite with short transverse groove. Sixth pleonal somite unarmed.

Telson (fig. 2D) oblong, broad, slightly constricted near base, feebly tapering distally, about 1.7 times as long as maximum width; dorsal surface with shallow median groove flanked by dorsolateral carinae each bearing a row of 3 strong, acute, symmetrically situated teeth, and 2 pairs of proximal and subproximal small teeth; lateral margins each armed proximally with 1 small submarginal tooth and with 1 strong tooth at mid-length; posterior margin subquadrate, armed with 3 small teeth.

Eyes (fig. 2A, B): well developed, cornea hemispherical, pigmented; eyestalk armed with several denticles on dorsal and frontal regions. Antennular peduncle (figs. 2A, B, 3A) reaching mid-length of antennal scale; proximal article subequal to distal 2 articles combined, armed with a few mesial row

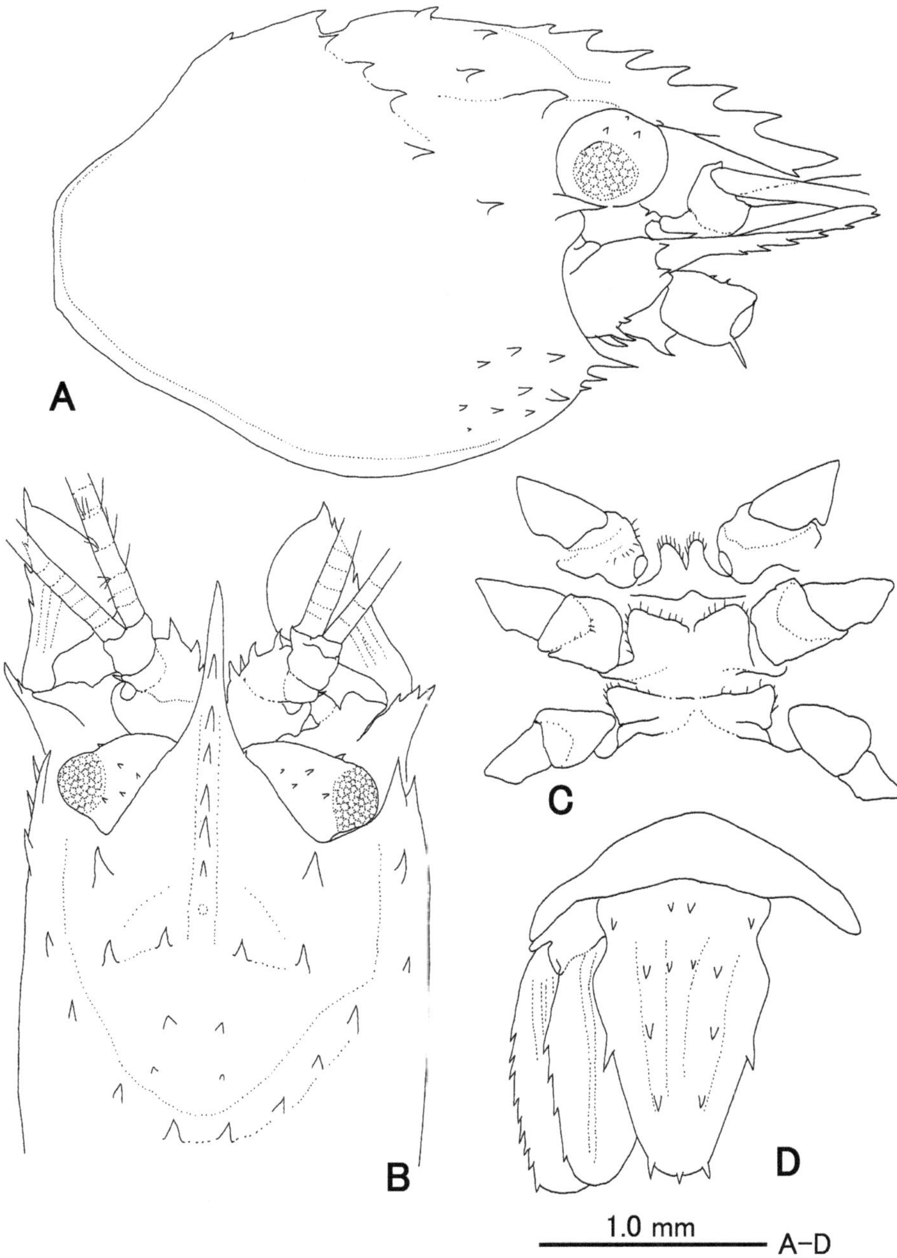

Fig. 2. *Microprosthema takedai* sp. nov. Holotype, female (cl 1.7 mm) from Moorea, French Polynesia (FLMNH UF Arthropoda 16271): A, carapace and cephalic appendages, lateral view; B, same, dorsal view; C, sixth to eighth thoracic sternites, ventral view; D, sixth pleonal somite, telson and left uropod, dorsal view. Setae omitted.

of strong sharp teeth; stylocerite acute, curved mesially; intermediate article armed with a few strong mesial teeth and a distolateral tooth; distal article unarmed; flagella slender, long, 3.6 times longer than carapace. Antenna (figs. 2A, B, 3B, C) stout, basicerite bearing 1 or 2 moderately large, sharp teeth at distolateral angle and with a few additional teeth of various sizes on ventrolateral margin; mesial margin with laminate process; antennal scale 2.3 times longer than broad, semicircular, lateral margin nearly straight, armed with 6 or more teeth, dorsal surface with 2 longitudinal carinae; carpocerite short, reaching level of distal margin of antennular peduncle, armed with a few strong teeth on mesial and lateral margins; flagella slender, long, 6.2 times as long as than carapace.

Mandible (fig. 3D, E) robust, with palp consisting of 3 articles; distal article oval, furnished with dense setae; intermediate article subequal to distal article in length, distally with tuft of setae; molar and incisor processes clearly separated. Maxillule (fig. 3F) robust, with simple, slender, distally tapering endopod; coxal endite suboval, with submarginal row of stiff setae on outer surface; basial endite moderately broad, truncate distally, with several slender spines and sparse, long spiniform setae. Maxilla (fig. 3G) with curved, slender endopod; both coxal and basial endites bilobed; scaphognathite well developed, anterior lobe subquadrate distally, posterior lobe short, widening posteromesially. First maxilliped (fig. 3H) with broad endopod, tapering distally; coxal endite bilobed; basial endite large, subtriangular, with concave mesial margin; exopod with well-developed flagellum; epipod large, distinctly bilobed. Second maxilliped (fig. 3I) with normal endopod; dactylus tapering distally; propodus anteromesially truncated, ~1.2 times as long as dactylus; carpus cup-shaped; merus about twice as long as carpus, oblong, mesial margin pectinate; ischium and basis incompletely fused; coxa demarcated; epipod elongate, with well-developed, lamellate podobranch; exopod with well-developed flagellum. Third maxilliped (fig. 3J) overreaching tip of antennal scale by lengths of dactylus and propodus; dactylus tapering distally, lateral margins of dactylus and propodus furnished with long setae, distomesial surface with shallow depression furnished with dense grooming setae; carpus subequal to propodus in length, unarmed, ventrodistal margin produced; merus armed with dorsolateral and lateral rows each with 3 strong sharp teeth, those of lateral row decreasing in size proximally; ischium compressed, armed with ventromesial row of several small teeth and dorsolateral row of several strong teeth, decreasing in size proximally; exopod with well-developed, simple flagellum.

First pereopod (fig. 3K) short, slender, unarmed, overreaching tip of antennal scale by length of dactylus, with well-developed grooming apparatus;

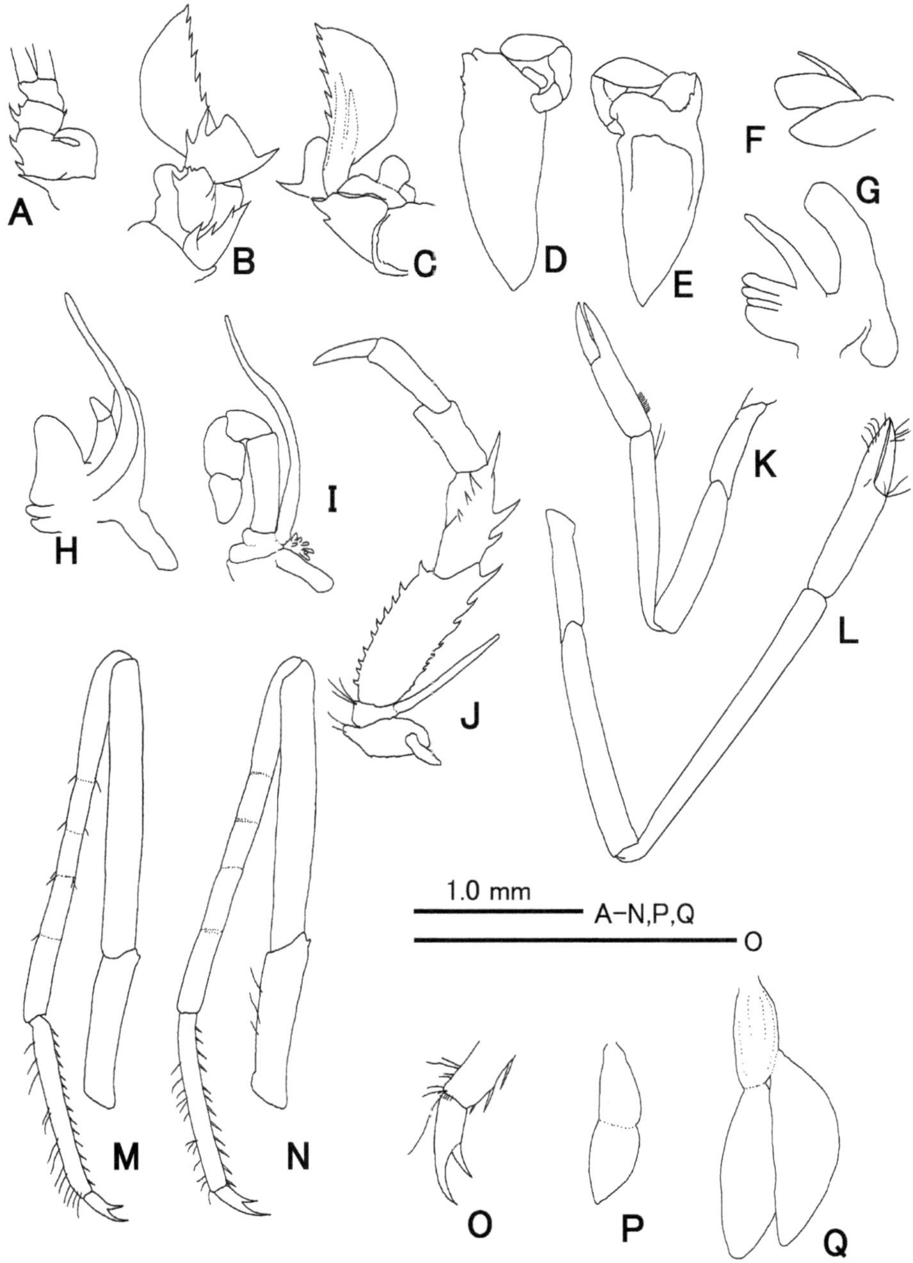

Fig. 3. *Microprosthema takedai* sp. nov. Holotype, female (cl 1.7 mm) from Moorea, French Polynesia (FLMNH UF Arthropoda 16271): A, left antennule, ventral view; B, left antenna, dorsal view; C, left antenna, ventral view; D, left mandible, ventral view; E, same, dorsal view; F, left maxillule, mesial view, spines omitted; G, left maxilla, mesial view; H, left first maxilliped, mesial view; I, left second maxilliped, ventral view; J, left third maxilliped, ventral view; K, right first pereopod, lateral view; L, left second pereopod, lateral view; M, left fourth pereopod, lateral view; N, left fifth pereopod, lateral view; O, same, dactylus; P, left first pleopod, dorsal view; Q, left second pleopod, lateral view. Setae omitted.

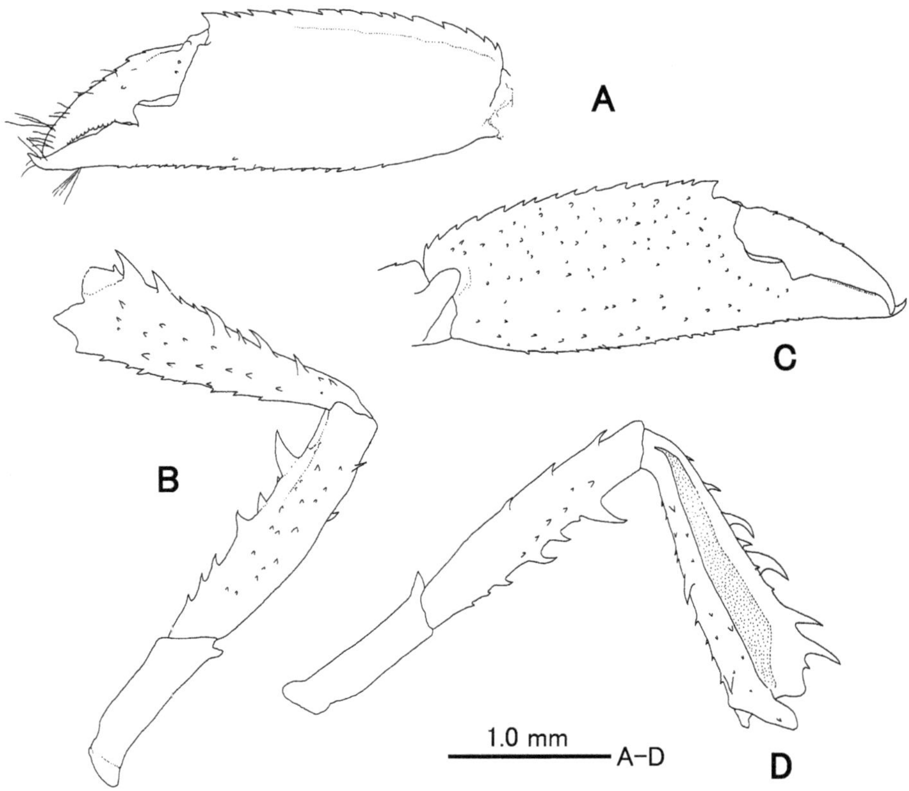

Fig. 4. *Microprosthema takedai* sp. nov. Holotype, female (cl 1.7 mm) from Moorea, French Polynesia (FLMNH UF Arthropoda 16271): A, left third pereopod, chela, lateral view; B, same, carpus and merus, lateral view; C, same, chela, mesial view; D, same, carpus, merus and ischium, mesial view.

dactylus ~0.4 chela length; palm subcylindrical; carpus slender, ~1.4 chela length; merus ~0.8 carpal length; ischium ~0.6 meral length.

Second pereopod (fig. 3L) longer than first pereopod, unarmed, overreaching tip of antennal scale by lengths of chela and distal half of carpus; dactylus ~0.4 chela length; palm subcylindrical; carpus ~1.8 times as long as chela; merus ~0.7 carpal length; ischium ~0.6 meral length.

Third pereopod (fig. 4A–D) strongest, overreaching tip of antennal scale by lengths of chela and carpus; chela about 1.8 times as long as carapace; dactylus ~0.4 chela length, with hook-like tip, armed with 6 small acute teeth on dorsal margin, cutting edge proximally with large triangular tooth, remaining edge unarmed; fixed finger generally similar to dactylus, cutting edge proximally with large blunt teeth opposed into hiatus on dactylus; palm subcylindrical, ~1.9 times as long as wide, dorsal margin armed with several strong teeth forming narrow crista, ventral margin armed with row of several fairly small teeth, lateral surface unarmed, mesial surface with scattered small blunt teeth and granules; carpus slightly widened distally, about 3.8 times as long as wide, approximately 0.7 chela length, dorsal margin armed with several, widely spaced, strong teeth of various sizes, distal-most tooth strongest, dorsomesial surface with longitudinal groove, ventrolateral and ventromesial margins each armed with irregular row of several small teeth; merus about 0.9 carpal length, dorsal margin armed with row of 3 small teeth, ventral margin with row of 7 strong sharp teeth, distal-most tooth strongest, lateral surface with scattered small teeth and granules, mesial surface with 1 row of several small teeth; ischium ~0.6 meral length, dorsodistal angle armed with 1 strong tooth.

Fourth and fifth pereopods (fig. 3M–O) similar in shape and length, long, slender; dactylus compressed laterally, about 4 times as long as wide, biunguiculate; propodus about 3 times as long as dactylus, not subdivided, ventral surfaces armed with several movable spines; carpus about twice as long as propodus, subdivided into 5 articles; merus ~0.8 carpal length, entire; ischium 0.6 meral length, with small blunt distodorsal tooth.

First pleopod (fig. 3P) uniramous, shorter than other pleopods. Second to fifth pleopods biramous. Second pleopod (fig. 3Q) with basipodite shorter than exopodite, mesial and lateral margins unarmed. Third to fifth pleopods generally similar, decreasing in size posteriorly.

Uropod (fig. 2D) with stout protopod, its lateral margin terminating in blunt process; exopod relatively broad, overreaching posterior margins of telson and endopod, lateral margin nearly straight, armed with row of 8 or 9 acute teeth, dorsal surface with 2 longitudinal carinae, unarmed; endopod tapering distally, lateral margin armed with 3 or 4 acute teeth, distal part unarmed, dorsal surface with a longitudinal carina, unarmed.

Gill formula as shown in table II. Egg diameter ~0.5 mm.

TABLE II

Gill formula of *Microprosthema takedai* sp. nov.

	Maxillipeds			Pereopods				
	1	2	3	1	2	3	4	5
Pleurobranchs	–	–	1	1	1	1	1	1
Arthrobranchs	1	1	2	2	2	2	2	–
Podobranch	–	1	–	–	–	–	–	–
Epipods	1	1	1	1	1	1	1	–
Exopods	1	1	1	–	–	–	–	–

Males. Rostrum (fig. 5A) 0.58-0.83 carapace length. Carapace armed with sharp teeth on postorbital and anterolateral regions, more numerous than in females. Sixth thoracic sternite (fig. 5B) with a pair of slender, anteriorly directed lobes, their lateral and mesial margins armed with several small acute teeth, distolateral angles produced; seventh sternite with a pair of broad, triangular lobes, distolateral angles produced, lateral and anteromesial margins armed with several small acute teeth; eighth sternite with pair of trapezoid lobes, distolateral angles produced, lateral and anteromesial margins armed with several small acute teeth. Chela of third pereopod (fig. 5C–F) about 2.4 times as long as carapace; dactylus ~0.4 chela length; palm broad, ~1.5 times as long as wide, mesial surface armed with scattered small teeth and granules; carpus somewhat broadened distally, about 2.9 times as long as wide, about half-length of chela; merus subequal to carpus in length.

Colouration (figs. 6, 7). — Body and appendages with whitish semi-transparent background, and with pale golden-orange tinge, more intense in older females; carapace and abdomen with small red chromatophores forming diffuse shorter or longer, semi-circular bands on epigastric region of the carapace, and transverse bands or larger dorsal patches on abdominal somites; telson and uropods mostly transparent whitish in males and young females or pale-yellow to golden-yellowish in older females; eyestalks colourless; antennal scale with pale yellow tinge; antennal flagella conspicuously orange-reddish; third maxillipeds mostly transparent; first and second pereopods semitransparent with reddish chromatophores; third pereopod chelae varying from hyaline-white with pale orange tinge (males, young females) to pale golden-orange (older females), fingers always white distally; carpus and merus varying from white with pale orange areas to more uniform pale golden-orange; fourth and fifth pereopods transparent with reddish chromatophores typically forming longitudinal lines; eggs or maturing ovaries in females dull green.

Etymology. — Named after Dr. Masatsune Takeda (recently retired from the National Museum of Nature and Science, Tokyo) in recognition of his numerous important contributions to the taxonomy of Crustacea Decapoda.

Type locality. — Moorea, Society Islands, French Polynesia.

Distribution. — Presently known only from the Society Islands of French Polynesia: Moorea and Tahiti.

Ecology. — Most specimens were collected on shallow mixed sand-rubble reef flats within Moorea Lagoon, deep under large coral rocks or massive corals, at depths of 0.5-4 m; some specimens were extracted from crevices of coral rubble brought up from depths of 10-12 m.

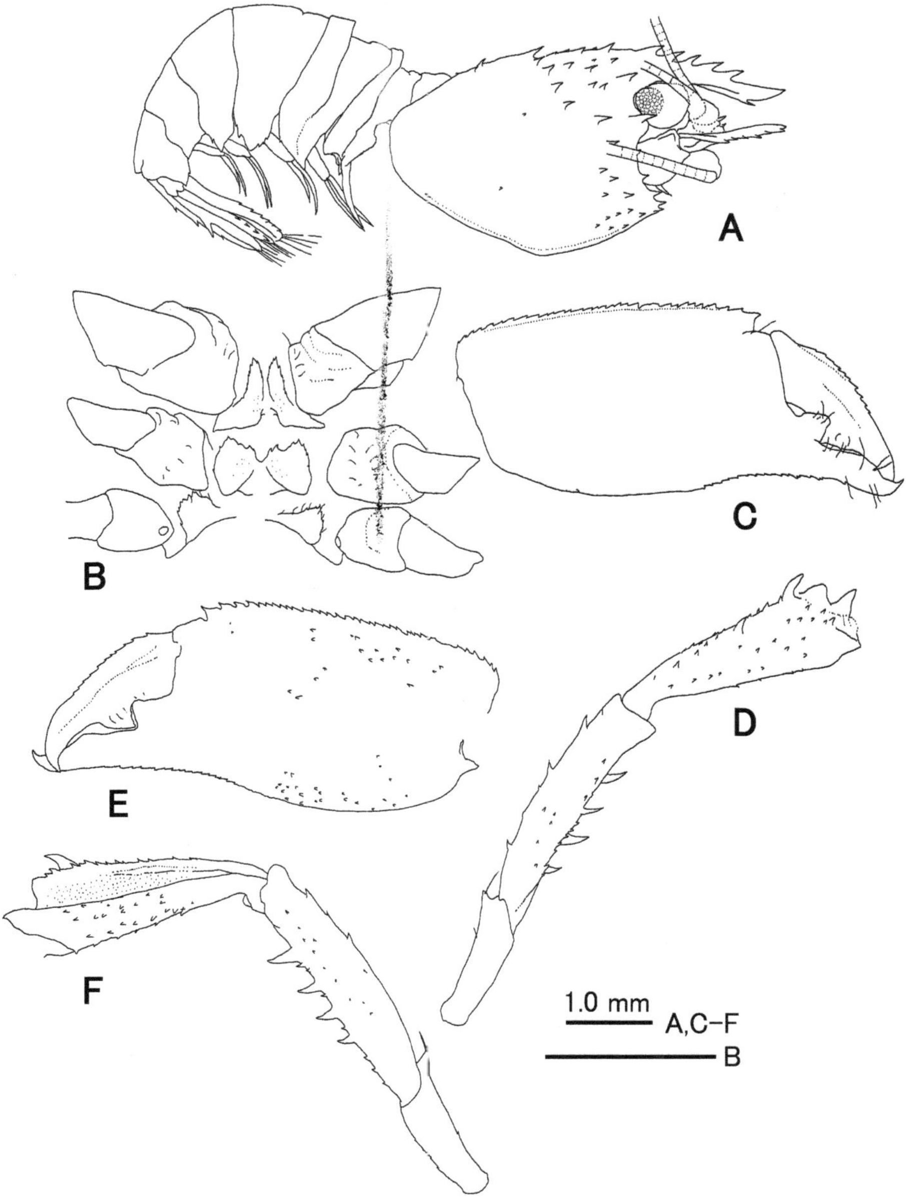

Fig. 5. *Microprosthema takedai* sp. nov. Allotype, male (cl 2.3 mm) from Moorea, French Polynesia (FLMNH UF Arthropoda 16504): A, carapace and pleon, lateral view; B, sixth to eighth thoracic sternites, ventral view; C, right third pereopod, chela, lateral view; D, same, carpus, merus and ischium, lateral view; E, same, chela, mesial view; F, same, carpus, merus and ischium, mesial view.

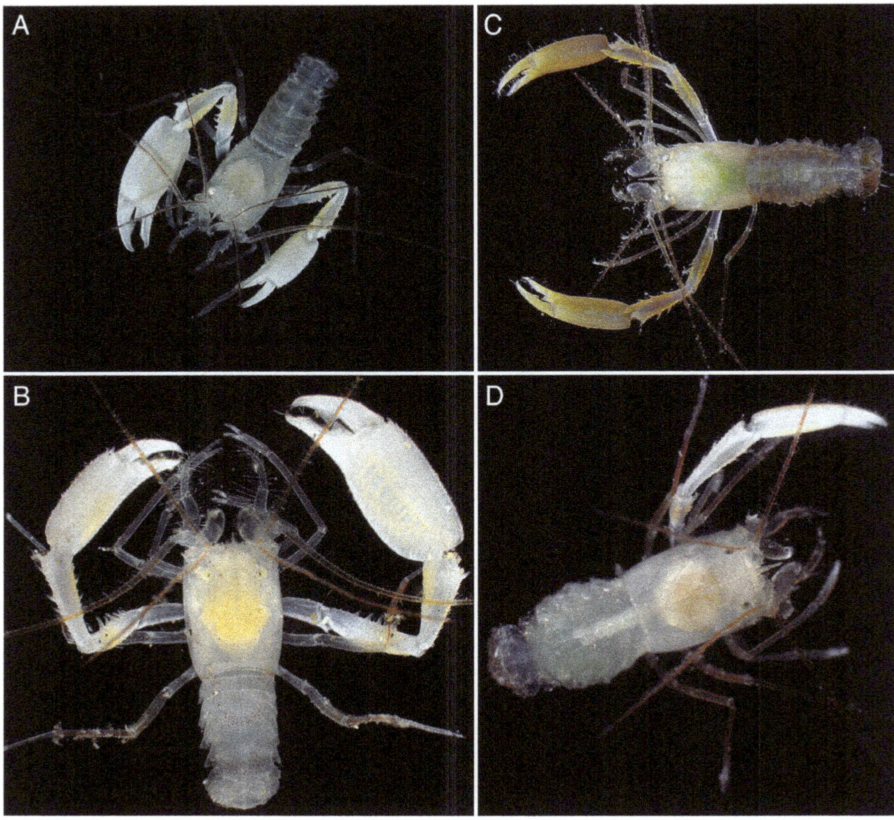

Fig. 6. *Microprosthema takedai* sp. nov., colour pattern of living individuals from Moorea, French Polynesia: A, B, allotype, male (cl 2.3 mm) from Motu Tiahura (FLMNH UF Arthropoda 16504); C, paratype, female (cl 2.5 mm) from Haapiti (FLMNH UF Arthropoda 15506); D, holotype, female (cl 1.7 mm) from west of Cook's Bay (FLMNH UF Arthropoda 16271).

Remarks. — *Microprosthema takedai* sp. nov. is morphologically most similar to the Indo-West Pacific *M. fujitai* and *M. validum*, sharing with them the following features: (1) pleonal somites without teeth; (2) antennal scale semicircular in shape; (3) mandibular palp unarmed and divided into three articles; (4) first and second pereopods lacking dorsal teeth on the carpus and (5) third pereopodal chela armed with dorsal and ventral rows of teeth. However, *M. takedai* sp. nov. differs from both *M. fujitai* and *M. validum* in the armature of the dorsal surface of the carapace; more precisely, in the new species, the teeth on the carapace are stronger and less numerous than in *M. fujitai* and *M. validum*. The new species differs specifically from *M. validum* by the absence of a median carina on the third pleonal somite, which is very

Fig. 7. *Microprosthema takedai* sp. nov., a specimen photographed in situ in Tahiti by Philippe Bacchet.

distinct in *M. validum*. The Polynesian species also differs from *M. validum* by the generally much paler colour of the carapace and abdomen, especially in males (see colour description of *M. validum* in Baba et al., 1968 and colour photograph in Kuiter & Debelius, 2009). A direct comparison between *M. takedai* sp. nov. and the type specimens of *M. fujitai* revealed several additional differences between these two species. For instance, *M. takedai* sp. nov. is lacking a short transverse row of teeth on the lateral surface of the sixth pleonal somite, which is present in *M. fujitai*. Furthermore, in *M. takedai* sp. nov., the palm of the third pereopod is somewhat broader (~1.5-1.9 as long as wide) than *M. fujitai* (~2.3-2.4 as long as wide). Finally, the groove on the dorsomesial surface of the carpus of the third pereopod is relatively broad in *M. takedai* sp. nov., being much narrower in *M. fujitai*. The two species also differ in the colour of the antennal and antennular flagella, which are reddish in *M. takedai* sp. nov. (fig. 6) vs. white in *M. fujitai* (see colour photograph in Saito & Okuno, 2011: fig. 7).

ACKNOWLEDGEMENTS

The first author wishes to express his gratitude to Dr. Masatsune Takeda (NSMT) for his continuous guidance. The second author's fieldwork in Moorea was sponsored by the Biocode Moorea project through the Gordon and Betty Moore Foundation and supported by the Richard B. Gump South Pacific Research Station of the University of California, Berkeley. Dr. Gustav Paulay, Seabird McKeon and Jenna Moore (FLMNH) helped collecting additional specimens of the new species. Research permits in Moorea were issued by the Délégation à la Recherche of the Government of French Polynesia. Philippe Bacchet (Papéete, Tahiti) provided a beautiful underwater photograph of the new species. Our cordial thanks are also extended to Dr. Tomoyuki Komai (CBM) and Dr. Junji Okuno (CMNH) for loan of specimens used as comparative material, technical advice and encouragement. Dr. Yoshihisa Fujita (University Education Center, University of the Ryukyus and Marine Learning Center), Dr. Hironori Komatsu (NSMT), Dr. Michitaka Shimomura (KMNH), and Janice C. Walker (USNM) provided or arranged the loan of additional comparative material. The second author also thanks CAPES of the Brazilian Government for financial support in the form of a postdoctoral fellowship.

REFERENCES

ALVES, M. DE. L., M. RAMOS-PORTO & G. F. S. VIANA, 2008. Checklist of the Decapods (Crustacea) from the Fernando de Noronha Archipelago, Brazil. Zootaxa, **1881**: 43-68.

BABA, K., Y. NAKASONE & M. TAKEDA, 1968. Two species of *Microprosthema* found on the coral reefs of the Ryukyu Islands (Stenopodidae, Crustacea). Ohmu, Occasional Papers of Zoological Laboratory, Faculty of Agriculture, Kyushu University, **1**: 173-181.

BRUCE, A. J. & K. BABA, 1973. *Spongiocaris*, a new genus of stenopodidean shrimp from New Zealand and South African waters, with a description of two new species (Decapoda Natantia, Stenopodidea). Crustaceana, **25**: 153-170.

CRIALES, M. M., 1997. *Microprosthema granatense*, new species, from the southern Caribbean, with a key to shrimps of the genus *Microprosthema* from the western Atlantic and a new record of *Odontozona libertae* (Decapoda: Stenopodidea). Journal of Crustacean Biology, **17**: 538-545.

DEVANEY, D. S. & A. J. BRUCE, 1987. Crustacea Decapoda (Penaeidea, Stenopodidea, Caridea, and Palinura) of Enewetak Atoll. In: D. M. DEVANEY, E. S. REESE, B. L. BURCH & P. HELFRICH (eds.), The natural history of Enewetak Atoll, Volume II. Biogeography and systematics: 221-233. (U. S. Department of Energy, Oak Ridge).

GOY, J. W., 1987. *Microprosthema emmiltum*, new species, and other records of stenopodidean shrimps from the Eastern Pacific (Crustacea: Decapoda). Proceedings of the Biological Society of Washington, **100**: 717-725.

GOY, J. W. & D. L. FELDER, 1988. Two new species of *Microprosthema* from the Western Atlantic (Crustacea: Decapoda: Stenopodidea). Journal of Natural History, **22**: 1277-1292.

HAAN, W., DE, 1844. Crustacea, Fascicle 6-7. In: P. F. VON SIEBOLD (ed.), Fauna Japonica sive Descriptio Animalium, quae in Itinere per Japoniam, Jussu et Auspiciis Superiorum, qui Summum in India Batava Imperium Tenent, Suscepto, Annis 1823-1830 Collegit, Notis, Observationibus et Adumbrationibus Illustravit: pls. 38, 43-46, 48, 51-55, I–N. (Lugduni-Batavorum, Leiden).

HANSEN, H. J., 1908. Crustacea Malacostraca. I. The Danish Ingolf-Expedition, vol. **3**(2): 1-120, pls. 1-5.

HOLTHUIS, L. B., 1946. Biological results of the Snellius Expedition. XIV. The Decapoda Macrura of the Snellius Expedition. I. The Stenopodidae, Nephropsidae, Scyllaridae and Palinuridae. Temminckia, **7**: 1-178.

KOMAI, T. & T. SAITO, 2006. A new genus and two new species of Spongicolidae from the southwestern Pacific. Tropical Deep-sea Benthos, **24**: 265-284.

KUITER, R. & H. DEBELIUS, 2009. World atlas of marine fauna: 1-725. (IKAN, Frankfurt).

MANNING, R. B. & F. A. CHACE, 1990. Decapod and stomatopod Crustacea from Ascension Island, South Atlantic Ocean. Smithsonian Contributions to Zoology, **503**: i–v, 1-91.

MARTENS, E., VON, 1872. Über cubanische Crustaceen nach den Sammlungen Dr. J. Gundlach's. Archiv für Naturgeshichte, **38**: 77-147.

MARTIN, J. W., 2002. *Microprosthema jareckii*, a new species of stenopodidean shrimp (Crustacea: Decapoda: Stenopodidea: Spongicolidae) from Guana Island, British Virgin Islands. Proceedings of the Biological Society of Washington, **115**: 108-117.

RICHTERS, F., 1880. Decapoda. In: K. MÖBIUS (ed.), Beiträge zur Meeresfauna der Insel Mauritius und der Seychellen: 139-178.

SAINT LAURENT, M. DE & R. CLÉVA, 1981. Crustacés décapodes: Stenopodidea. In: Résultats des Campagnes MUSORSTOM. 1. Philippines. Mémoirs du Muséum national d'Histoire naturelle, **91**: 151-188.

SAITO, T. & J. OKUNO, 2011. Two new species of the genus *Microprosthema* Stimpson, 1860 (Stenopodidea: Spongicolidae), from the Ryukyu Islands. Bulletin of National Museum of Nature and Science, (A) (Supplement) **5**: 83-99.

SCHRAM, F. R., 1986. Euzygida. In: F. R. SCHRAM (ed.), Crustacea: 276-285. (Oxford University Press, New York).

STIMPSON, W., 1860. Prodromus descriptionis animalium evertebratorum, quae in Expeditione ad Oceanum Pacificum Septentrionalem, a Republica Federata missam Cadwaladaro Ringgold et Johanne Rodgers ducibus, observavit et descripsit. Pars VIII. Crustacea Macrura. Proceedings of the Academy of Natural Sciences of Philadelphia, **10**: 22-47.

First received 18 August 2011.
Final version accepted 10 November 2011.

A NEW SPECIES OF *STEGIDOTEA* POORE, 1985 (ISOPODA, CHAETILIIDAE) FROM JAPAN

BY

MICHITAKA SHIMOMURA[1,3]) and KEIICHI KAKUI[2])

[1]) Kitakyushu Museum of Natural History and Human History, 2-4-1 Higashida, Yahatahigashi-ku, Kitakyushu, Fukuoka 805-0071, Japan

[2]) Department of Natural History Sciences, Faculty of Science, Hokkaido University, Sapporo, Hokkaido 060-0810, Japan

ABSTRACT

Stegidotea takedai n. sp. is described from off Koshikijima Islands, Kyushu, western Japan, from a depth of 495-500 m. The new species differs from congeners in having a body about 2.2 times as long as wide; frontal margin of the head concave; sublateral projections of the pereon well-developed; articles 2 and 3 of the maxillipedal palp completely distinct; and the exopod of the uropod moderately short, being 0.7 times as long as endopod.

INTRODUCTION

The isopod family Chaetiliidae Dana, 1849, has 44 species in 12 genera, ranging in depth from the intertidal to 2500 m worldwide (Schotte et al., 2011). *Stegidotea* Poore, 1985, is a small genus of six species, which has previously been known only from shallow to deep waters of the southern Pacific (Poore, 1985, 1990, 1991). The new species described here is from off Koshikijima Islands, Kyushu, western Japan, and is the first record of the genus from the northern Pacific.

Isopods were collected during the Survey of Benthos in Western Japan conducted by Nagasaki University in 2009. The gear used for the collection was a beam trawl (mouth opening 3 m; mesh size 4 mm). Sediment samples were sorted on board in sea water through a sieve with a mesh size of 0.5 mm. The processed samples were then fixed and preserved in 70% ethanol.

[3]) Corresponding author; e-mail: shimomura@kmnh.jp

Appendages were dissected, mounted on glass microslides and observed using a compound microscope (Nikon E600). The total length was measured from the tip of the head to the end of the pleotelson. Terminology follows Poore (1985). The type specimen is deposited in the National Museum of Nature and Science, Tokyo (NSMT).

TAXONOMY

Family CHAETILIIDAE Dana, 1849
Genus **Stegidotea** Poore, 1985

Stegidotea Poore, 1985: 161; 1990: 101; 1991: 140.

Type species. — *Stegidotea pinnata* Poore, 1985; by original designation.

Included species. — *Stegidotea scabra* Poore, 1985, *S. latipoda* Poore, 1990, *S. longipes* Poore, 1991, *S. carinata* Poore, 1991, *S. forcipes* Poore, 1991, *S. takedai* n. sp.

Remarks. — Poore (1985) established the genus for two new species, *S. pinnata* from Bass Strait, from depths of 42-140 m, and *S. scabra* from off Victoria, from depths of 58-95 m, southeastern Australia. Poore (1990) later redefined the genus with the description of *S. latipoda* from western Australia, from a depth of 78 m. Three more species, *S. longipes*, *S. carinata* and *S. forcipes*, were described by Poore (1991) from New Caledonia, from depths of 535-1410 m.

Stegidotea takedai n. sp.
(figs. 1-5)

Material examined. — Holotype. Male, 4.0 mm (NSMT-Cr 21870), dissected and mounted on glass slides, off Koshikijima Islands, Kyushu, 32°08.941′N 129°32.148′E–32°08.951′N 129°30.514′E, 495-500 m, sand, beam trawl, TS Nagasaki-Maru, N295 cruise, Station A-1, 22 November 2009.

Description. — Body (fig. 1A, B) about 2.2 times as long as wide, dorsal-ventrally convex. Integument rough, scattered denticles, especially visible on pleon. Head (fig. 1A) 0.38 times as long as wide, with concave frontal margin, with well-defined ocular lobes lacking eye pigments and ommatidia. Pereonites each with triangular mid-dorsal projection; pereonites 1 and 2 with 2 pairs of triangular sublateral projections; pereonites 3-6 with 1 pair of triangular sublateral projections; pereonite 7 without triangular sublateral projections. Pleon less than half total length: pleonite 1 with 1 mid-dorsal projection; pleonites 2 and 3 fused mid-dorsally; lateral part of pleonites 2 and

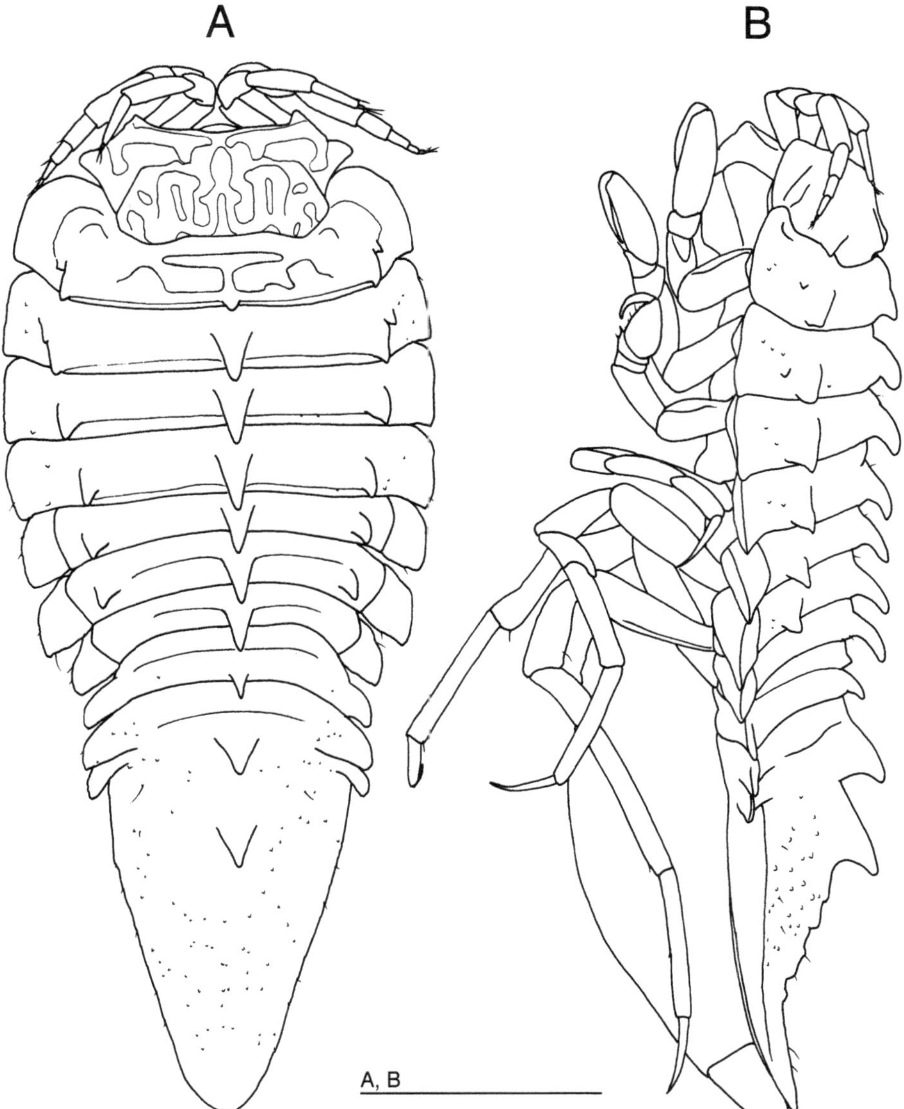

Fig. 1. *Stegidotea takedai* n. sp., holotype male: A, habitus, dorsal; B, habitus, lateral. Scale bar: 1 mm.

3 curved posteriorly. Pleotelson 1.5 times as long as wide, with 2 mid-dorsal projections, tapering to sharply rounded apex. Coxal plates of pereonites 5-7 curved posteriorly, with blunt apex.

Antenna 1 (fig. 2A) with many scales: peduncular article 1 with 1 plumose seta laterally, and with many fine and 1 simple setae medially; article 2 0.58

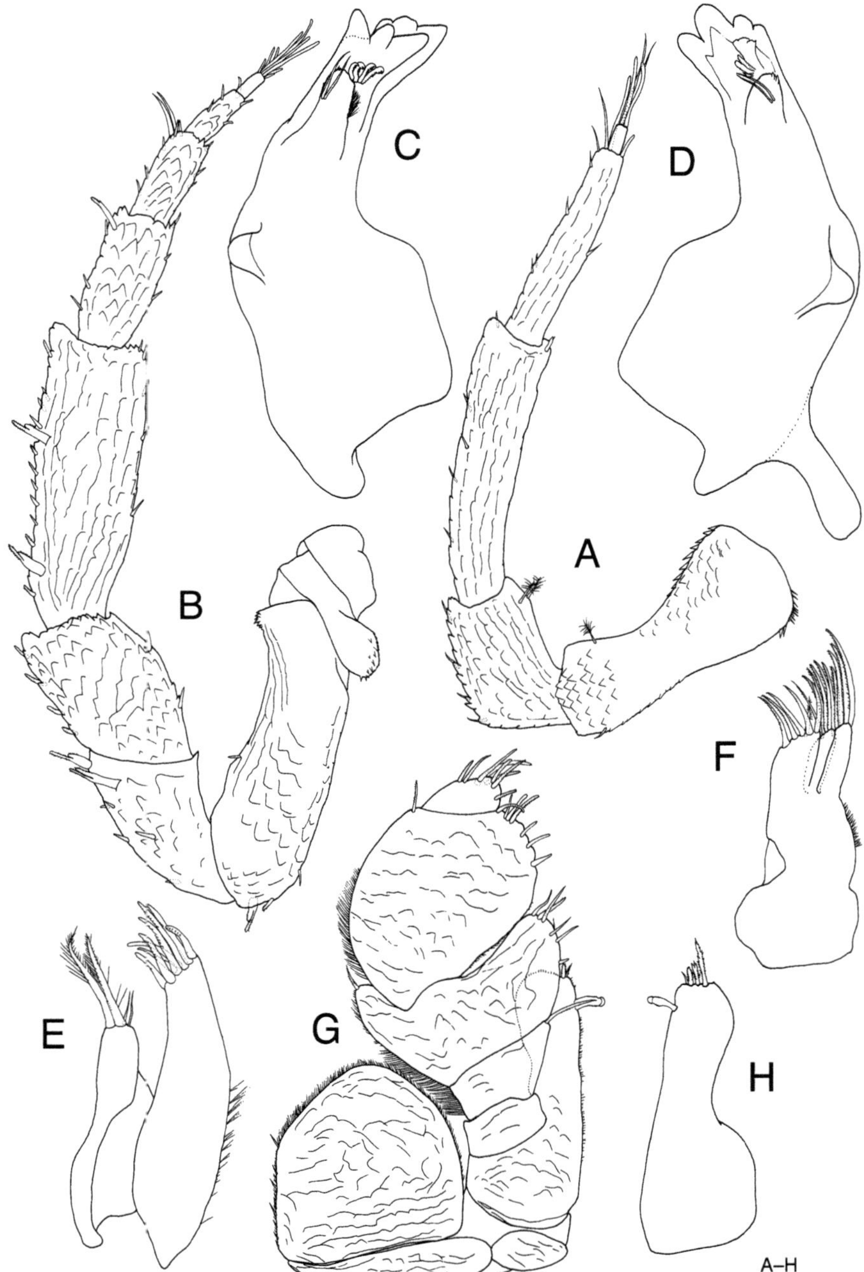

Fig. 2. *Stegidotea takedai* n. sp., holotype male: A, right antenna 1, dorsal; B, right antenna 2, dorsal; C, left mandible, medial; D, right mandible, medial; E, right maxilla 1, dorsal; F, right maxilla 2, dorsal; G, right maxilliped, ventral; H, endite of right maxilliped, dorsal. Scale bar: 0.1 mm.

times as long as article 1, with 2 filoplume-like setae laterally and 4 simple setae medially; article 3 1.8 times as long as article 2, with 1 stout seta distal-laterally and 5 simple setae medially. Flagellum consisting of 1 large and 1 minute articles; article 1 0.78 times as long as peduncular article 3, with 1 simple seta laterally and 6 simple setae medially, and 2 aesthetascs distally; article 2 with 2 long setae and 2 aesthetascs apically.

Antenna 2 (fig. 2B) with many simple scales: peduncular article 1 without setae; article 2 with medial projection bearing many small scales; article 3 longest of all articles, with 1 simple seta laterally and 1 simple, 1 plumose and 1 bifid plumose setae medially; peduncular article 4 triangulate, with 5 simple and 2 bifid setae medially; peduncular article 5 as long as article 4, with 2 simple setae laterally and 4 plumose setae medially. Flagellum 0.83 times as long as peduncular articles combined: article 1 with 5 simple setae laterally and 6 simple, 5 plumose and 2 bifid plumose setae medially; article 2 half length of article 1, with 3 simple setae laterally and 3 simple and 1 bifid plumose setae medially; article 3 0.68 times as long as article 2, with 3 simple setae laterally and 7 simple setae medially; article 4 with 2 simple setae laterally, 2 simple setae medially and 1 simple seta distally; article 5 minute, with 9 simple setae apically.

Left mandible (fig. 2C): lacinia mobilis larger than incisor, with 3 cusps; incisor with 3 cusps; spine row of 4 pectinate setae; molar process rounded, with 2 long setae and many fine setae. Right mandible (fig. 2D): lacinia mobilis minute, with 3 cusps; incisor with 4 cusps; spine row of 3 pectinate setae; molar process rounded, with 2 long setae and 4 short setae.

Maxilla 1 (fig. 2E): medial lobe with 2 plumose setae and 2 simple setae apically; lateral lobe with 2 simple and 6 denticulate setae apically, with many fine setae laterally.

Maxilla 2 (fig. 2F): medial lobe with 1 plumose and 5 pectinate setae apically; middle lobe with 5 pectinate setae; lateral lobe with 5 pectinate setae.

Maxilliped (fig. 2G, H) with many scales ventrally and numerous fine setae marginally: endite reaching halfway along palp article 3, with 1 coupling hook medially, and with 1 simple and 3 plumose setae distally; palp article 1 without setae; article 2 slightly longer than article 1, with 1 medial seta; article 3 about twice as long as article 2, with 6 medial and 1 lateral setae; article 4 longest article, with 7 medial and 1 lateral setae; article 5 with 8 marginal setae; epipod ovate, about 0.73 times length of endite.

Pereopods (figs. 3A–5A) with dense cover of simple scales. Pereopod 1 (fig. 3A): basis with 6 simple setae ventrally and 7 simple setae dorsally;

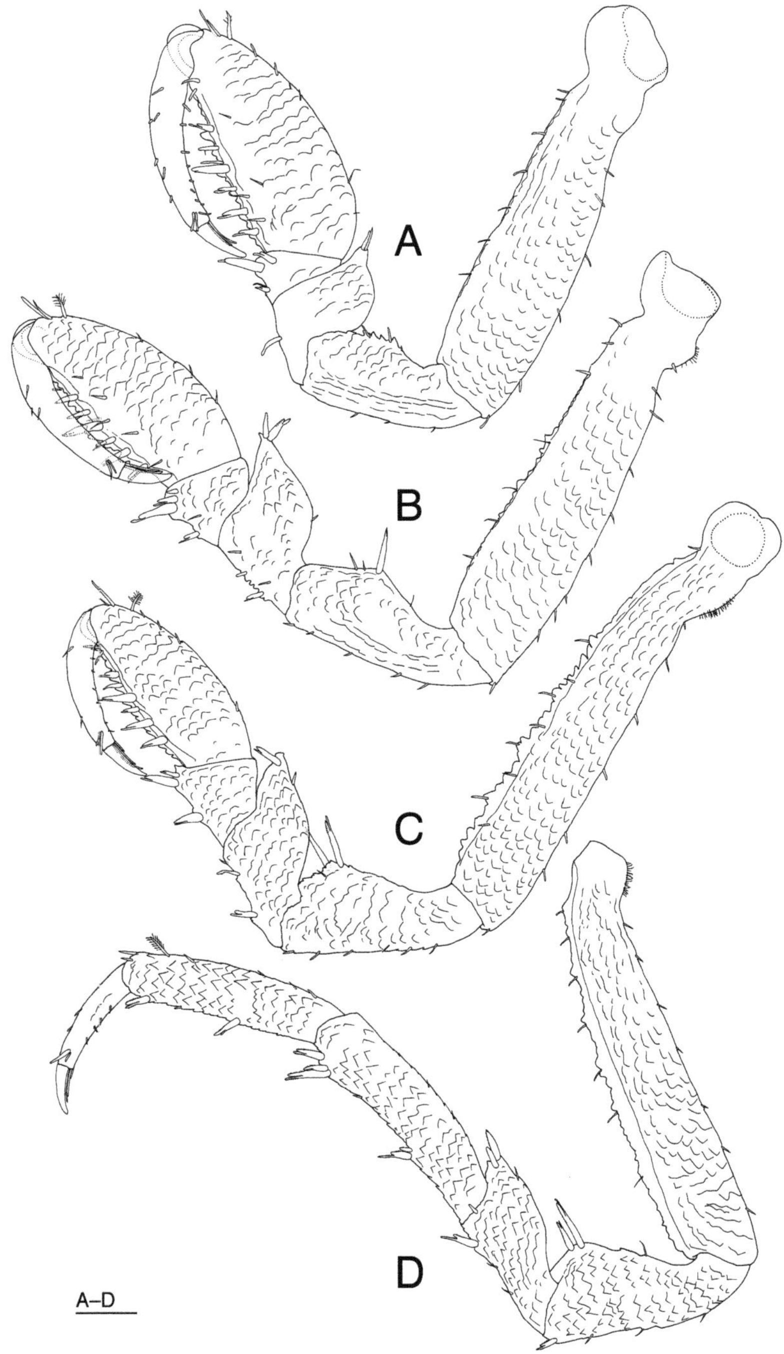

Fig. 3. *Stegidotea takedai* n. sp., holotype male: A, right pereopod 1, medial; B, right pereopod 2, medial; C, right pereopod 3, medial; D, right pereopod 4, medial. Scale bar: 0.1 mm.

ischium 0.38 times length of basis, denticulate dorsally, with 3 simple setae ventrally and 2 simple setae dorsally; merus trapezoidal, with 1 simple seta ventrally and 2 simple setae dorsally; carpus triangulate, with 3 bifid setae and 1 simple seta ventrally; propodus 1.7 times as long as ischium, denticulate ventrally, with 6 bifid and 5 simple setae ventrally, 2 simple setae medially and 8 simple and 1 filoplume-like setae dorsally; dactylus with 8 ventral, 6 dorsal, 3 subdistal, 2 distal setae and 1 unguis. Pereopod 2 (fig. 3B) slightly longer than pereopod 1: basis denticulate dorsally, with 10 simple setae ventrally, some fine setae distal-proximally ventral-proximally and 7 simple setae dorsally; ischium 0.48 times length of basis, with 5 simple setae ventrally, 3 simple and 1 bifid setae dorsally; merus trapezoidal, denticulate ventrally, with 3 simple and 1 bifid setae ventrally, and 1 simple and 2 bifid setae dorsally; carpus triangular, with 2 simple and 3 bifid setae ventrally; propodus as long as ischium, denticulate ventrally, with 2 simple and 6 bifid setae ventrally, 4 simple setae medially and 5 simple and 1 filoplume-like setae dorsally; dactylus with 1 ventral, 3 dorsal, 2 subdistal, 2 distal setae and 1 unguis. Pereopod 3 (fig. 3C) as long as pereopod 2, narrower than pereopod 2: basis denticulate dorsally, with 7 simple setae ventrally, some fine setae distal-proximally and 9 simple setae dorsally; ischium 0.42 times length of basis, with 6 simple setae ventrally, 1 simple and 2 bifid setae dorsally; merus trapezoidal, with 1 simple and 2 bifid setae ventrally, and 1 simple and 1 bifid setae dorsally; carpus trapezoidal, with 3 bifid setae ventrally; propodus as long as ischium, denticulate ventrally, with 3 simple and 5 bifid setae ventrally, and 6 simple and 1 filoplume-like setae dorsally; dactylus with 7 ventral, 2 dorsal, 2 subdistal, 3 distal setae and 1 unguis. Pereopod 4 (fig. 3D) slightly longer than pereopod 3: basis denticulate dorsally, with 7 simple setae ventrally, some fine setae distal-proximally and 7 simple setae dorsally; ischium about half length of basis, with 3 simple and 1 bifid setae ventrally, 1 simple and 2 bifid setae dorsally; merus 0.66 times as long as ischium, with 2 simple and 1 bifid setae ventrally, 1 simple and 2 bifid setae dorsally, 1 simple seta distal-medially; carpus 1.84 times as long as merus, with 2 simple and 4 bifid setae ventrally; propodus 0.87 times as long as carpus, with 1 simple and 2 bifid setae ventrally, 6 simple and 1 filoplume-like setae dorsally; dactylus with 4 ventral, 2 dorsal, 2 subdistal, 3 distal setae and 1 unguis. Pereopod 5 (fig. 4A) longer than pereopod 4: basis denticulate dorsally, with 10 simple setae ventrally, some fine setae distal-proximally and 9 simple setae dorsally; ischium 0.62 times length of basis, with 2 stout, simple and 6 narrow, simple setae ventrally, 2 simple and 1 bifid setae dorsally; merus 0.67 times as long as ischium, with 2 simple and 1 bifid setae ventrally, 2 simple

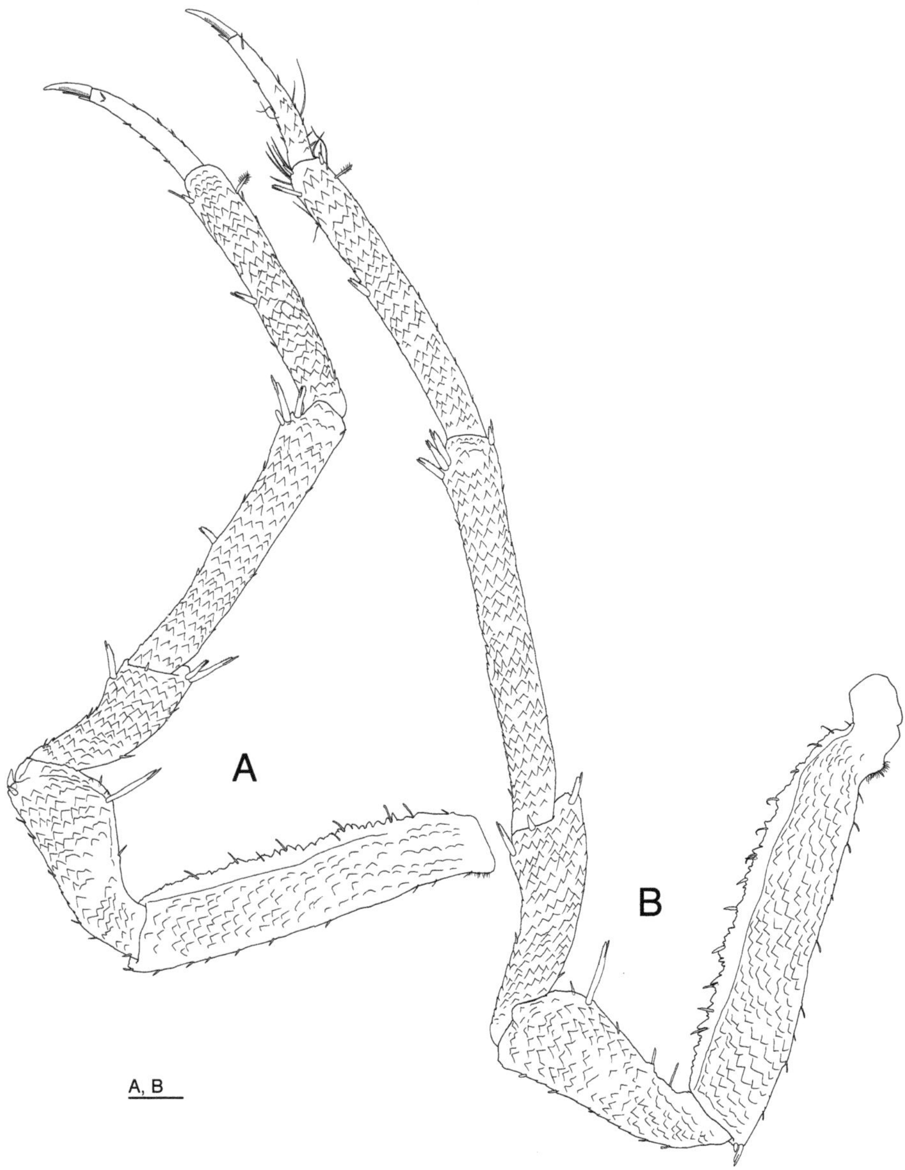

Fig. 4. *Stegidotea takedai* n. sp., holotype male: A, right pereopod 5, medial; B, right pereopod 6, medial. Scale bar: 0.1 mm.

and 2 bifid setae dorsally, 1 simple seta distal-medially; carpus 2.2 times as long as merus, with 1 stout, simple and 3 narrow, simple setae, and 3 bifid setae ventrally, and 4 simple setae dorsally; propodus 0.86 times as long as carpus, with 1 simple and 2 bifid setae ventrally, 5 simple and 1 filoplume-like setae dorsally; dactylus with 4 ventral, 4 dorsal, 2 subdistal, 3 distal setae and 1 unguis. Pereopod 6 (fig. 4B) longest: basis denticulate dorsally, with 10 simple and 1 bifid setae ventrally, some fine setae distal-proximally and 14 simple setae dorsally; ischium about half length of basis, with 5 simple setae ventrally, 4 simple and 1 bifid setae dorsally; merus as long as ischium, with 1 simple and 1 bifid setae ventrally, 1 simple and 1 bifid setae dorsally, 1 simple seta distal-medially; carpus 1.6 times as long as merus, with 2 simple and 3 bifid setae ventrally, and 1 simple and 1 bifid setae dorsally; propodus 0.78 times as long as carpus, with 9 simple and 2 bifid setae ventrally, 6 simple, 1 bifid and 1 filoplume-like setae dorsally; dactylus with 6 ventral, 9 dorsal, 1 subdistal, 3 distal setae and 1 unguis. Pereopod 7 (fig. 5A) slightly shorter than pereopod 6: basis denticulate dorsally, with 6 simple and 1 bifid setae ventrally, some fine setae distal-proximally and 12 short, simple and 1 long, simple setae dorsally; ischium about half length of basis, denticulate dorsally, with 4 simple and 1 bifid setae ventrally, 3 simple and 1 bifid setae dorsally; merus as long as ischium, with 2 simple and 2 bifid setae ventrally, 2 simple and 2 bifid setae dorsally, 1 simple seta distal-medially; carpus 1.6 times as long as merus, with 3 simple and 3 bifid setae ventrally, and 1 simple and 2 bifid setae dorsally; propodus 0.83 times as long as carpus, with 1 simple and 2 bifid setae ventrally, 4 simple and 1 filoplume-like setae dorsally; dactylus with 4 ventral, 3 dorsal, 3 distal setae and 1 unguis.

Pleopod 1 (fig. 5B): peduncle with 1 long and many fine setae, 4 coupling hooks medially, and 4 simple and many fine setae laterally; endopod ovate, slightly shorter than exopod, with 10 long, plumose setae and many fine setae marginally; exopod with 15 long, plumose and many fine setae marginally. Pleopod 2 (fig. 5C): peduncle with many fine setae and 3 hooks medially, and 2 plumose and many fine setae laterally; endopod with 8 plumose and many fine setae marginally; stylet long, fused with endopod at its base; exopod with 18 long, plumose and many fine setae marginally. Pleopod 3 (fig. 5D): peduncle with many fine setae and 4 hooks medially, 1 plumose and many fine setae laterally, and 2 simple setae dorsally; endopod quadrate, slightly shorter than exopod, with 14 long, plumose setae and numerous fine setae marginally; exopod remaining suture line laterally, with 20 long, plumose and many fine setae marginally, and 4 simple setae dorsally. Pleopod 4 (fig. 5E): peduncle with 1 plumose and many fine setae medially, 1 plumose and many fine setae laterally; endopod as long as exopod, without setae; exopod article 1 with many denticles medially and 3 plumose setae laterally; exopod article 2 with many denticles medially, 7 plumose and 1 simple setae laterally. Pleopod 5

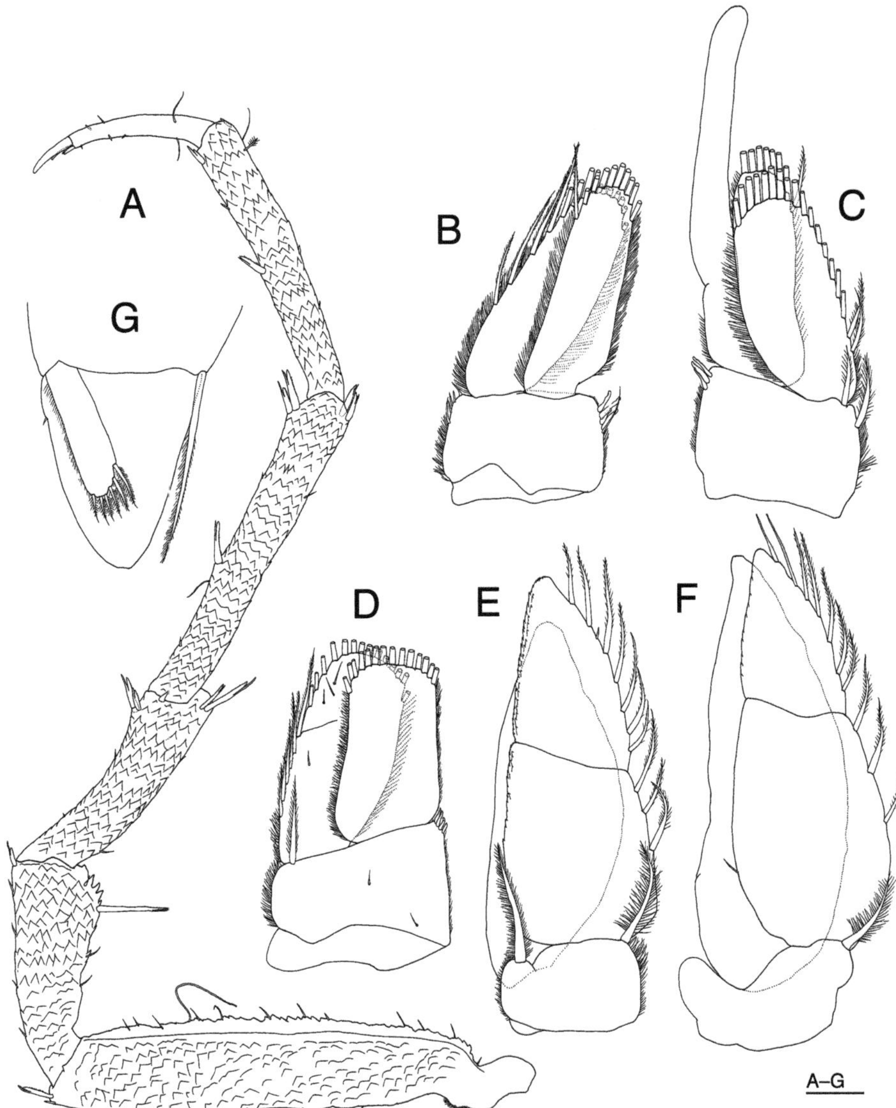

Fig. 5. *Stegidotea takedai* n. sp., holotype male: A, right pereopod 7, medial; B, right pleopod 1, dorsal; C, right pleopod 2, ventral; D, right pleopod 3, dorsal; E, right pleopod 4, ventral; F, right pleopod 5, ventral; G, left uropod, ventral. Scale bar: 0.1 mm.

(fig. 5F): peduncle with 1 plumose seta laterally; endopod slightly longer than exopod, without setae; exopod article 1 with 1 simple seta medially and 3 plumose setae laterally; exopod article 2 with many denticles medially, and 5 plumose and 2 simple setae laterally.

Uropod (fig. 5G): peduncle with 1 plumose seta laterodistally; endopod triangular, apically rounded, with 1 short, simple seta medially; exopod moderately short, 0.7 times as long as endopod, parallel-sided, with 6 plumose setae distally.

Colour. — Colour in life yellowish white.

Etymology. — The new species is named after the eminent carcinologist Dr. Masatsune Takeda.

Remarks. — *Stegidotea takedai* n. sp. is readily identified from all congeners by the combination of following characters: concave anterior margin to the head, long pereopods 5-7, pereonites 1-7 each with mid-dorsal projection, pleon with 2 mid-dorsal projections and uropodal exopod 0.7 times as long as endopod with 6 plumose setae distally.

Stegidotea takedai is most similar to *S. longipes* (from depths of 1395-1410 m off New Caledonia) in having long pereopods 5-7. *Stegidotea takedai*, however, differs from *S. longipes* by the following characters: body about 2.2 times as long as wide (about twice as long as wide in *S. longipes*); frontal margin of head concave (nearly straight in *S. longipes*); articles 2 and 3 of maxillipedal palp distinct (partly fused in *S. longipes*); and exopod of uropod moderately short, 0.7 times as long as endopod (moderately long, 0.9 times as long as endopod in *S. longipes*).

ACKNOWLEDGEMENTS

We express sincere thanks to the captain and crew of TS Nagasaki-Maru of Nagasaki University, for their cooperation. We also thank two anonymous reviewers for improving this manuscript with their comments. Many thanks are due to Prof. J. Hashimoto of Nagasaki University for his kind help in the sampling of the specimens. This research was supported in part by the Ministry of Education, Science, Sports and Culture, Grant-in-Aid for Young Scientists (B), No. 21770100.

REFERENCES

DANA, J. D., 1849. Conspectus crustaceorum quae in orbis terrarum circumnavigatione, Carolo Wilkes e classe Reipublicae, Foederate Duce, lexit et descripsit. American Journal of Sciences and Arts, **8**: 424-428.

POORE, G. C. B., 1985. Australian chaetiliids (Crustacea: Isopoda: Valvifera): a new genus, new species and remarks on the family. Memoirs of the Museum Victoria, **46**: 153-172.

— —, 1990. Two new species of isopod crustaceans belonging to Australian endemic genera (Serolidae and Chaetiliidae). Memoirs of the Museum Victoria, **51**: 99-107.

— —, 1991. Crustacea Isopoda: Deep-sea Chaetiliidae (Valvifera) from New Caledonia and the Philippines. In: A. CROSNIER (ed.), Résultats des Campagnes Musorstom, Volume 9. Mémoires du Museum national d'Histoire naturelle, Paris, (A) **152**: 139-153.

SCHOTTE, M., B. F. KENSLEY & S. SHILLING, 2011. World list of Marine, Freshwater and Terrestrial Crustacea Isopoda. (National Museum of Natural History Smithsonian Institution, Washington, D.C., U.S.A.). Available at http://invertebrates.si.edu/isopod/ (accessed 30 August 2011).

First received 8 September 2011.
Final version accepted 22 December 2011.

A NEW SPECIES OF THE GENUS *DULICHIELLA* (AMPHIPODA, MELITIDAE) FROM THE OGASAWARA ISLANDS, JAPAN

BY

KO TOMIKAWA[1,3]) and HIRONORI KOMATSU[2])

[1]) Graduate School of Education, Hiroshima University, 1-1-1 Kagamiyama, Higashi-Hiroshima, Hiroshima 739-8524, Japan

[2]) Department of Zoology, National Museum of Nature and Science, 4-1-1 Amakubo, Tsukuba, Ibaraki 305-0005, Japan

ABSTRACT

A new amphipod *Dulichiella takedai* sp. nov. is described from the Ogasawara Islands, Japan, only the second species of the genus from Japan. *Dulichiella takedai* is morphologically similar to *D. appendiculata* (Say, 1818) from the USA and *D. cuvettensis* Appadoo & Myers, 2005 from Mauritius, but *D. takedai* differs from *D. appendiculata* in having slender setae on urosomite 1 shorter than the spines, a developed fourth spine of the distolateral crown of larger male gnathopod 2 and the bases and meri of male pereopods 5-7 lacking bunches of long, slender setae. *Dulichiella takedai* differs from *D. cuvettensis* in having a minutely serrate posteroventral margin of epimeral plate 3, a short mandibular palp article 3 with more than eight slender setae, and equally developed third and fourth spines on the distolateral crown of larger male gnathopod 2.

INTRODUCTION

The genus *Dulichiella* Stout, 1912 sensu Lowry & Springthorpe (2007) is distributed in tropical and temperate marine waters. To date, 14 species of the genus are known from the world's oceans (Lowry & Springthorpe, 2007). In Japanese waters, only one species, *D. tomioka* Lowry & Springthorpe, 2007, has been recorded, from Tomioka Bay, Kumamoto Prefecture (Hirayama & Kikuchi, 1979 as *D. appendiculata* (Say, 1818); Hirayama, 1986 as *D. appendiculata*; Lowry & Springthorpe, 2007).

During a field survey of littoral marine invertebrates off the Ogasawara Islands, Tokyo, Japan in 2009, one of the authors (HK) collected specimens of

[3]) Corresponding author; e-mail: tomikawa@hiroshima-u.ac.jp

an undescribed species of *Dulichiella*. In this paper, this undescribed species is described based on these specimens.

MATERIAL AND METHODS

The amphipods were collected using an ORI dredge (50 cm wide). Specimens were fixed and preserved in 80% ethanol on site. Dissected appendages were mounted on glass slides in gum-chloral medium. All drawings were made with the aid of a camera lucida attached to a differential interference microscope (Olympus BH-2). Type specimens are deposited at the National Museum of Nature and Science, Tokyo (NSMT). Terminology follows Bradbury & Williams (1999).

SYSTEMATICS

Dulichiella takedai sp. nov.

[New Japanese name: Ogasawara-uchidenokozuchi]

(figs. 1-5)

Material examined. — Holotype: ♂, 4.1 mm (NSMT-Cr 21841), off Minami-shima Island, Ogasawara Islands, Tokyo, Japan, 27°2′N 142°7′E, depth 136 m, 10 July 2009, collected by H. Komatsu. Paratypes: two ♀♀, 5.6 mm, 4.6 mm (NSMT-Cr 21842, 21843), data same as holotype.

Etymology. — The new species is named in honour of Dr. Masatsune Takeda who has contributed much to the taxonomy of crustaceans in Japan. The name is a noun in the genitive singular.

Description. — Male (holotype, NSMT-Cr 21841). Head (fig. 1A): lateral cephalic lobe broad, truncated, anteroventral corner with 2 slender setae. Pleonite/urosomite dorsal spine formula 7-6-7-5-4-2, with slender setae (fig. 4F–K). Epimeral plates 1 and 2 heavily damaged. Epimeral plate 3 (fig. 4E): posteroventral corner strongly produced, posteroventral margin minutely serrate distally, ventral and posterior margins with 4 robust and 2 slender setae, respectively.

Antenna 1 (fig. 1B): peduncular articles 1 : 2 : 3 in ratio 1.0 : 1.3 : 0.3, posterior margin of peduncular article 1 with 3 robust setae; accessory flagellum 4-articulate; primary flagellum 28-articulate, each article with aesthetasc, calceoli absent. Antenna 2 (fig. 1C): 0.6 times as long as antenna 1; gland cone

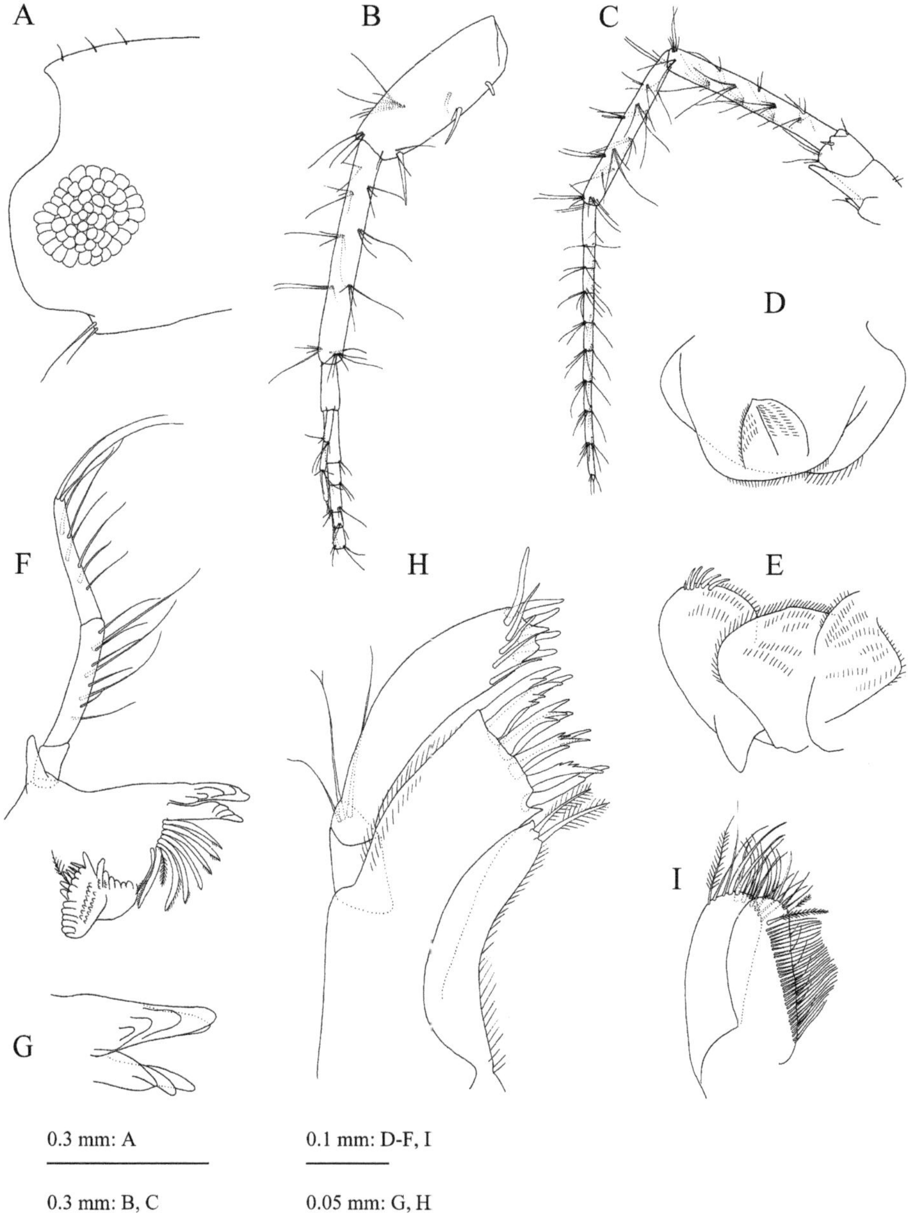

Fig. 1. *Dulichiella takedai* sp. nov., holotype male, 4.1 mm, NSMT-Cr 21841. A, head, lateral view; B, peduncular articles 1-3, primary flagellar articles 1-5, and accessory flagellum of antenna 1, medial view; C, antenna 2, medial view; D, upper lip, anterior view; E, lower lip (partly damaged), ventral view; F, left mandible, medial view; G, incisor and lacinia mobilis of right mandible, lateral view; H, maxilla 1, dorsal view; I, maxilla 2, dorsal view.

of peduncular article 2 reaching to end of peduncular article 3; peduncular article 4 almost as long as peduncular article 5; flagellum 10-articulate, calceoli absent.

Upper lip (fig. 1D): ventral margin with fine setae. Mandible (fig. 1F, G): left and right incisors 5-dentate, respectively; left and right lacinia mobilis 4 and 3-dentate, respectively; accessory robust setae with row of 7 slender blades; molar triturative, with plumose seta; palp 3-articulate, articles 1 : 2 : 3 in ratio 1.0 : 3.2 : 3.0; article 1 bare; article 2 with 9 setae; article 3 with 8 setae. Lower lip (fig. 1E): outer lobes broad, shoulder of outer lobes rounded, setulose; inner lobes present. Maxilla 1 (fig. 1H): inner plate slender, with 2 apical plumose setae; outer plate rectangular, with 9 serrate robust setae; palp 2-articulate, article 1 with 3 slender setae on outer distal corner, apical part of palp article 2 with 4 robust and 5 slender setae. Maxilla 2 (fig. 1I): medial margin of inner plate lined with setae, oblique setal row of inner plate with 32 setae. Maxilliped (fig. 2A): inner plate exceeding distal margin of palp article 1, with 3 robust setae on innerdistal corner; outer plate reaching half of palp article 2; palp 4-articulate, medial margins of palp article 2 and 3 lined with setae, palp articles 3 and 4 facially setulose.

Gnathopods 1 and 2 greatly asymmetric. Gnathopod 1 (fig. 2B, C): coxa anteroventral corner not produced, anterior margin weakly convex, posteroventral corner with small notch; merus and carpus facially setulose, carpus 1.3 times as long as propodus; palm of propodus convex, palm defined by posterodistal corner, with posterodistal robust setae, palmar margin with bifid robust setae; dactylus not reaching beyond palmar margin. Left (larger) gnathopod 2 (fig. 2D, E): coxa anteroventral corner not produced, anterior margin weakly concave, posteroventral corner with small notch; propodus distolateral crown with 4 rounded spines, posterodistal corner produced, upturned, posterodistal corner of dactylus fitting into corner; dactylus apically blunt. Right (smaller) gnathopod 2 (fig. 2F, G): coxa anterior margin weakly convex; posterodistal corner of merus acutely produced; palm straight, posterodistal corner with robust setae; dactylus not reaching beyond palmar margin, posterior margin with slender setae.

Pereopod 3 (fig. 3A, B): coxa subrectangular, ventral margin straight, posteroventral corner with small notch; merus 0.8 times as long as propodus; dactylar unguis anterior margin with 2 accessory spines. Pereopod 4 (fig. 3C, D): posterior margin of coxa widely excavate, ventral margin straight, posteroventral corner with small notch; merus 0.7 times as long as propodus; dactylar unguis anterior margin with 2 accessory spines. Pereopod 5 (fig. 3E):

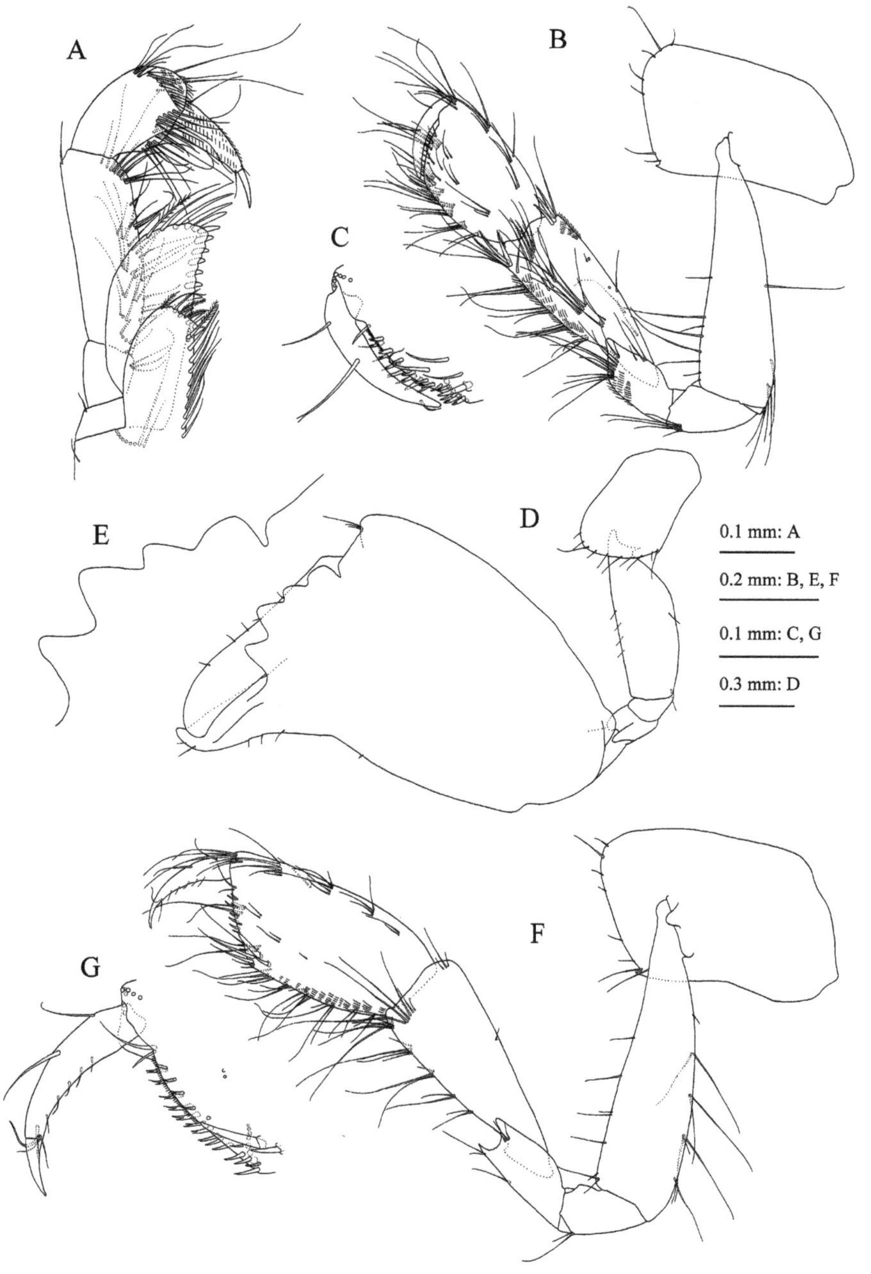

Fig. 2. *Dulichiella takedai* sp. nov., holotype male, 4.1 mm, NSMT-Cr 21841. A, maxilliped, dorsal view; B, gnathopod 1, medial view; C, palmar margin and dactylus of gnathopod 1, medial view, some setae omitted; D, left (larger) gnathopod 2, lateral view; E, spines on distolateral crown of gnathopod 2, lateral view; F, right (smaller) gnathopod 2, medial view; G, palmar margin and dactylus of gnathopod 2, medial view, some setae omitted.

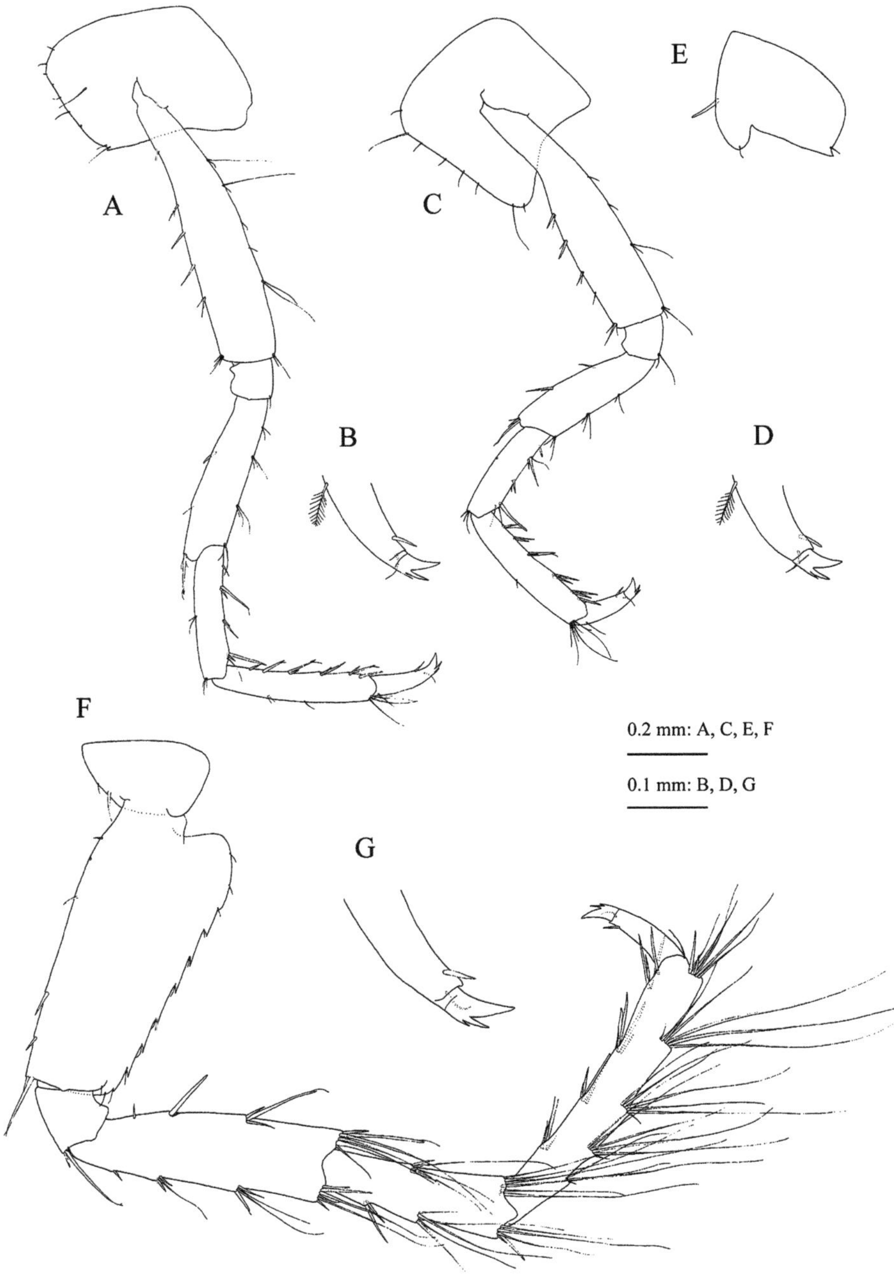

Fig. 3. *Dulichiella takedai* sp. nov., holotype male, 4.1 mm, NSMT-Cr 21841. A, pereopod 3, medial view; B, dactylus of pereopod 3, medial view; C, pereopod 4, medial view; D, dactylus of pereopod 4, medial view; E, coxa of pereopod 5, lateral view; F, pereopod 7, medial view; G, dactylus of pereopod 7, medial view.

coxa anterior margin with robust seta; basis to dactylus broken off. Pereopod 6: broken off. Pereopod 7 (fig. 3F, G): anterior margin of coxa with 1 robust and 1 small seta; posterior margin of basis serrate, basis and merus without bunches of long slender setae; carpus and propodus with bunches of long slender setae; dactylar unguis anterior margin with 2 accessory spines. Coxal gills: on gnathopod 2 and pereopods 3-6. Sternal gills: absent.

Pleopods (fig. 4A): peduncle without setae, inner distal corner with 2 retinacula (fig. 4B); rami long, proxiomesial part of inner ramus with bifid plumose setae.

Uropod 1 (fig. 4L): peduncle almost as long as outer ramus, with robust, basofacial and dorsal setae; outer ramus slightly shorter than inner one, with 5 robust setae on outer and inner margins, respectively; inner ramus with robust innerventral seta, minute setae along with inner margin, outer and inner margins with 3 and 4 robust setae, respectively. Uropod 2 (fig. 4M): peduncle short, length 0.8 times as long as outer ramus, with robust dorsal setae; outer ramus 0.9 times as long as inner one, with 3 outer and 4 inner robust setae, respectively, inner margin along with minute setae; inner ramus with 3 robust outer and 4 robust outer marginal setae, inner margin along with minute setae. Uropod 3: broken off. Telson (fig. 4N): deeply cleft, longer than wide, with dorsal robust setae.

Female (paratype, NSMT-Cr 21842). Pleonite/urosomite dorsal spine formula 7-7-7-5-4-2. Epimeral plate 1 (fig. 4C): posteroventral corner weakly pointed, ventral and posterior margins with 1 robust and 1 slender seta, respectively. Epimeral plate 2 (fig. 4D): posteroventral corner pointed, posteroventral margin smooth, ventral margin with 4 robust setae and posterior margin with slender seta. Mandible: palp article 3 with 9-10 setae. Gnathopod 1 (fig. 5A, B) and gnathopod 2 (fig. 5C, D) similar, not asymmetric: setae on coxa, basis, carpus and propodus denser and longer than these of male. Pereopod 5 (fig. 5E): anterior margin of coxa without robust seta; posterior margin of basis serrate and weakly convex; bunches of long slender setae without on basis, merus and carpus. Pereopod 6 (fig. 5F, G): anterior margin of coxa with 2 robust and 1 slender setae; posterior margin of basis serrate; bunches of long slender setae without basis and merus. Pereopod 7 (fig. 5H): posterior margin of basis serrate, expanded proximally; merus to dactylus broken off.

Distribution. — The new species is known only from the type locality.

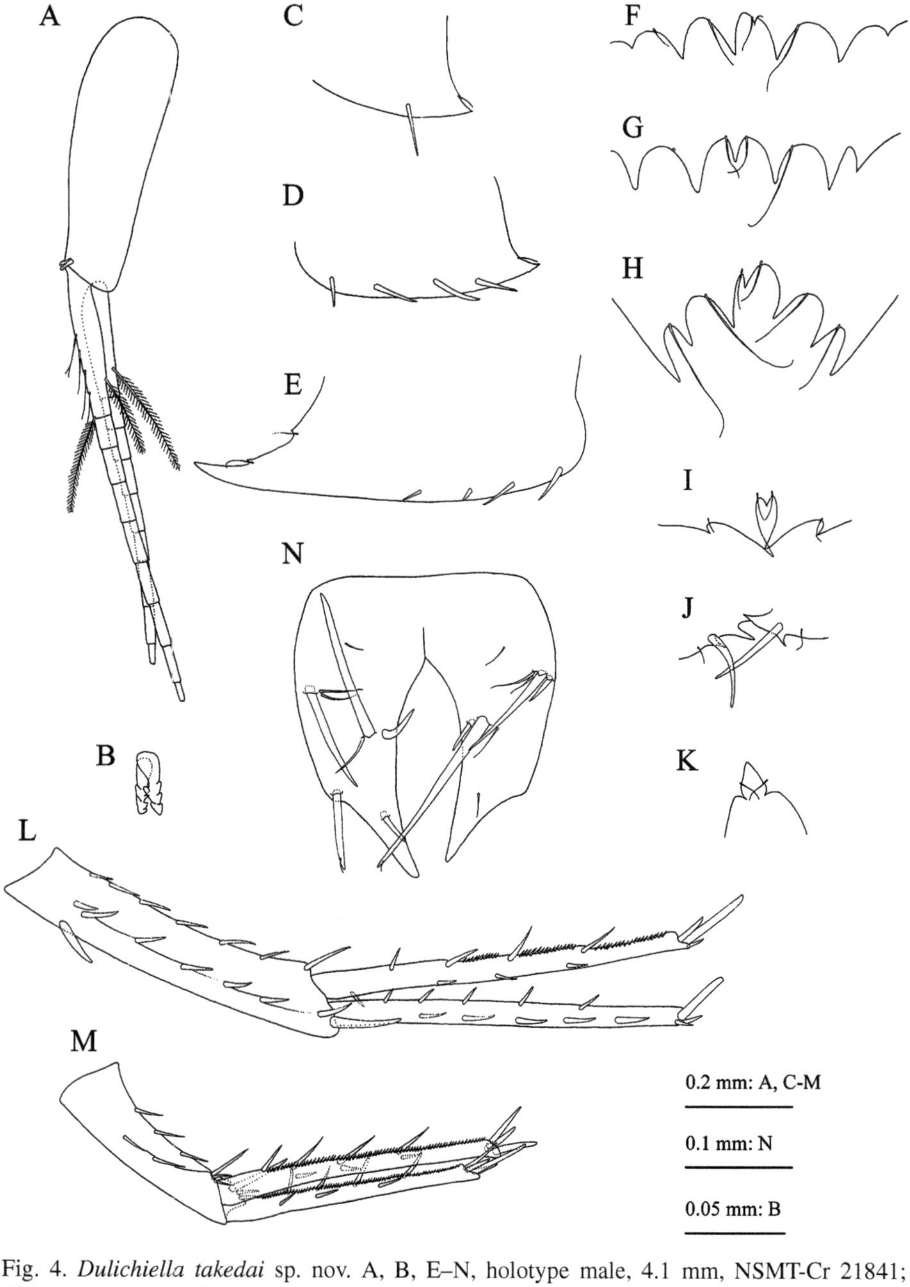

Fig. 4. *Dulichiella takedai* sp. nov. A, B, E–N, holotype male, 4.1 mm, NSMT-Cr 21841; C, D, paratype female, 5.6 mm, NSMT-Cr 21842. A, pleopod 1, medial view; B, retinacula on peduncle of pleopod 1, medial view; C–E, epimeral plates 1-3, lateral views; F–H, dorsal margins of pleonites 1-3, dorsal views; I–K, dorsal margins of urosomites 1-3, dorsal view; L, uropod 1, dorsal view; M, uropod 2, dorsal view; N, telson, dorsal view.

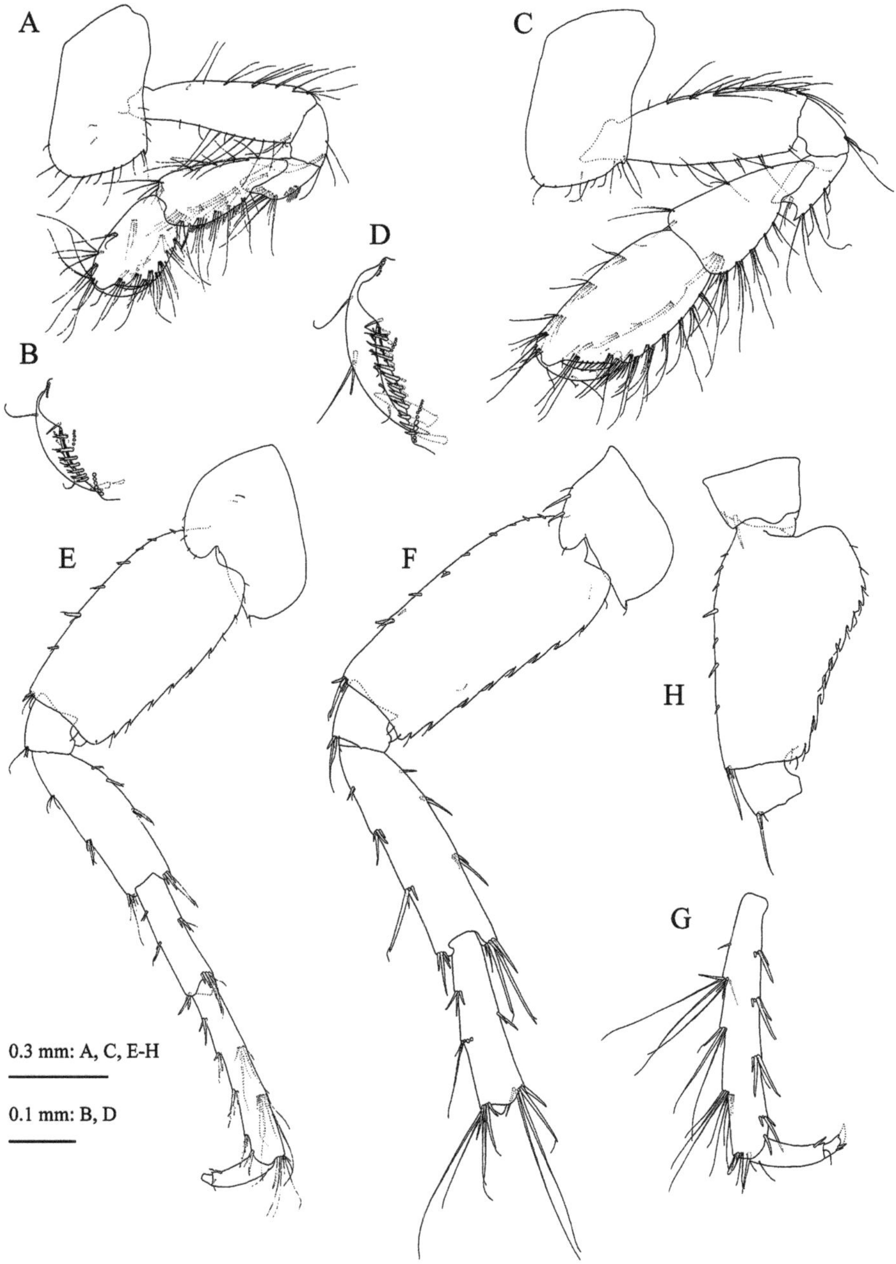

Fig. 5. *Dulichiella takedai* sp. nov., paratype female, 5.6 mm, NSMT-Cr 21842. A, gnathopod 1, lateral view; B, palmar margin and dactylus of gnathopod 1, lateral view; C, gnathopod 2, lateral view; D, palmar margin and dactylus of gnathopod 2, lateral view; E, pereopod 5, lateral view; F, coxa–carpus of pereopod 6, lateral view; G, propodus and dactylus of pereopod 6, lateral view; H, coxa–ischium of pereopod 7, lateral view.

DISCUSSION

Dulichiella takedai sp. nov. is morphologically similar to *D. appendiculata* (Say, 1818) from the USA and *D. cuvettensis* Appadoo & Myers, 2005 from Mauritius in having pleonites 1-3 without dense dorsal setae; 3 robust setae on the posterior margin of peduncular article 1 of antenna 1; 4 spines on the distolateral crown of the larger male gnathopod 2 and the dactylar ungui of pereopods 3-7 with 2 accessory spines. However, *D. takedai* differs from *D. appendiculata* and *D. cuvettensis* by the following features. It differs from *D. appendiculata* (features of which are shown in parentheses) by the slender setae on urosomite 1 shorter (vs. longer) than spines, a developed (vs. weak) fourth spine on the distolateral crown of larger male gnathopod 2 and the basis and merus of male pereopods 5-7 without (vs. with) bunches of long slender setae. It differs from *D. cuvettensis* (features of which are shown in parentheses) in a minutely serrate (vs. smooth) posteroventral margin of epimeral plate 3, mandibular palp article 3 shorter (vs. longer) than article 2, with more than eight (vs. six) slender setae, and equally (vs. unequally) developed third and fourth spines on the distolateral crown of larger male gnathopod 2.

Dulichiella takedai is also similar to *D. pacifica* Lowry and Springthorpe, 2007 from Singapore, New Guinea, New Caledonia, the Solomon Islands and Australia. *Dulichiella takedai* is distinguished from *D. pacifica* (features of which are shown in parentheses) in having two (vs. one) accessory spine(s) on the dactyli of pereopods 3-7, blunt (vs. acute) tip of the fourth spine on the distolateral crown of the propodus of larger male gnathopod 2, and a minutely serrate (vs. smooth) posteroventral margin of epimeral plate 3.

Dulichiella takedai is the second species of the genus known from Japan. This new species is easily distinguished from another Japanese species, *D. tomioka* Lowry and Springthorpe, 2007 in having 4 (vs. 2 in *D. tomioka*) spines on the distolateral crown of the propodus of the larger male gnathopod 2.

ACKNOWLEDGEMENTS

We express our sincere gratitude to Dr. Masatsune Takeda of the National Museum of Nature and Science, Tokyo, for his encouragement and cooperation to our study. This study was partly supported by grant from the Nihon Kyoiku Koumuin Kousaikai.

REFERENCES

APPADOO, C. & A. A. MYERS, 2005. Amphipods of the genera *Ceradocus*, *Dulichiella*, *Melita* and *Nuuanu* (Crustacea: Melitidae) from Mauritius, Indian Ocean. Records of the Australian Museum, **57**: 221-236.

BRADBURY, J. H. & W. D. WILLIAMS, 1999. Key to and checklist of the inland aquatic amphipods of Australia. Technical Report of the Australian Museum, **14**: 1-21.

HIRAYAMA, A., 1986. Taxonomic studies on the shallow water gammaridean Amphipoda of West Kyushu, Japan. VI. Lysianassidae (*Orchomene*), Megaluropus family group, Melitidae (*Cottesloe*, *Jerbarnia*, *Maera*, *Ceradocus*, *Eriopisella*, *Dulichiella*). Publications of the Seto Marine Biological Laboratory, **3**(1/2): 1-35.

HIRAYAMA, A. & T. KIKUCHI, 1979. The first record of *Melita appendiculata* (Say) 1818, (Crustacea: Amphipoda: Gammaridae) from Japan. Publications from the Amakusa Marine Biological Laboratory, **5**(1): 67-77.

LOWRY, J. K. & R. T. SPRINGTHORPE, 2007. A revision of the tropical/temperate amphipod genus *Dulichiella* Stout, 1912, and the description of a new Atlantic genus *Verdeia* gen. nov. (Crustacea: Amphipoda: Melitidae). Zootaxa, **1424**: 1-62.

SAY, T., 1818. An account of the Crustacea of the United States. Journal of the Academy of Natural Sciences of Philadelphia, **1**: 374-401.

STOUT, V. R., 1912. Studies in Laguna Amphipoda. Annual Report of the Laguna Marine Laboratory, **1**: 134-149.

First received 5 September 2011.
Final version accepted 25 January 2012.

www.ingramcontent.com/pod-product-compliance
Ingram Content Group UK Ltd.
Pitfield, Milton Keynes, MK11 3LW, UK
UKHW022209051026
15624UKWH00001B/70